Basics of Mutation Breeding

Basics of Mutation Breeding

Editor

Carin Fisker

scitus
academics

Basics of Mutation Breeding
Edited by **Carin Fisker**

ISBN: 978-1-68117-417-4
Library of Congress Control Number: 2016937428

SCITUS Academics LLC,
www.scitusacademics.com
Box No. 4766, 616 Corporate Way,
Suite 2, Valley Cottage,
NY 10989

Printed in United States of America on Acid Free Paper

Preface

Plant breeding is estimated to have begun 9,000 – 11,000 years ago and mutation breeding is simply a new device in the breeder's toolkit. Mutation breeding has been used since the 1930s. Mutation breeding, sometimes referred to as "variation breeding", is the process of exposing seeds to chemicals or radiation in order to generate mutants with desirable traits to be bred with other cultivars. Plants created using mutagenesis are sometimes called mutagenic plants or mutagenic seeds. Normal breeding involves cross breeding, from the parents, through a multi-generational process that may take four to five years to eliminate an unwanted genome and develop the sought-after traits before trial crops are tested, adding more time. Mutation breeding is based on selfing mutants until the induced character has a stable expression in the advanced mutant generations. Mutation breeding has many comparative advantages. It is cost effective, quick, proven and robust. In addition, mutation breeding is transferrable, ubiquitously applicable, non-hazardous and environmentally friendly.There are different kinds of mutagenic breeding such as using chemical mutagens like EMS and DMS, radiation and transposons are used to generate mutants. Mutation breeding is commonly used to produce traits in crops such as larger seeds, new colors, or sweeter fruits that either cannot be found in nature or have been lost during evolution. The book Basics of Mutation Breeding covers information on various types of mutagens and their effects, procedures for using mutagens for crop improvement, types of mutations (micro and macro) with statistical techniques to handle the mutation population. The book will be useful for both undergraduate and post graduate students of agriculture.

Table of Contents

Chapter 1

Mutation Breeding of Extracellular Polysaccharide-Producing MicroalgaCrypthecodinium cohnii by a Novel Mutagenesis with Atmospheric and Room Temperature Plasma

Bin Liu [1,2], Zheng Sun [2,3,†], Xiaonian Ma [2,†], Bo Yang [1,2], Yue Jiang [4], Dong Wei [1] and Feng Chen [2,5,*]

[1]*School of Light Industry and Food Sciences, South China University of Technology, Guangzhou 510641, China*

[2]*Institute for Food and Bioresource Engineering, College of Engineering, Peking University, Beijing 100871, China*

[3]*College of Fisheries and Life Science, Shanghai Ocean University, Shanghai 201306, China*

[4]*School of Food Science, Jiangnan University, Wuxi 214122, China*

[5]*Singapore-Peking University Research Centre for a Sustainable Low-Carbon Future, CREATE Tower 138602, Singapore*

ABSTRACT

Extracellular polysaccharides (EPS) produced by marine microalgae have the potential to be used as antioxidants, antiviral agents, immunomodulators, and anti-inflammatory agents. Although the marine microalga *Crypthecodinium cohnii* releases EPS during the process of docosahexaenoic acid (DHA) production, the yield of EPS remains relatively low. To improve the EPS production, a novel mutagenesis of *C. cohnii* was conducted by atmospheric and room temperature plasma (ARTP). Of the 12 mutants obtained, 10 mutants exhibited significantly enhanced EPS yield on biomass as compared with the wild type strain. Among them, mutant M7 was the best as it could produce an EPS volumetric yield of 1.02 g/L, EPS yield on biomass of 0.39 g/g and EPS yield on glucose of 94 mg/g, which were 33.85%, 85.35% and 57.17% higher than that of the wild type strain, respectively. Results of the present study indicated that mutagenesis of the marine microalga *C. cohnii* by ARTP was highly effective leading to the high-yield production of EPS.

Keywords: extracellular polysaccharides; atmospheric and room temperature plasma (ARTP); mutation; microalgae; *Crypthecodinium cohnii*

1. INTRODUCTION

Polysaccharides are polymeric carbohydrate molecules consisting of long chains of monosaccharide units. As a major class of biological macromolecules, they play important roles in a vast number of biochemical activities [1]. With the development of glycomics, the significant contribution of polysaccharides in human health has received increasing attention in recent years. There are a growing number of studies demonstrating the bio-modulatory effects of polysaccharides, e.g., anti-oxidative, anti-inflammatory, anti-adhesive, anti-coagulant, anti-cancer, anti-viral, and immunomodulatory properties [2,3,4]. Marine algae are known as one of the most abundant sources of polysaccharides: for example, Chrysophyta [5], *Phaeodactylum tricornutum* [6], *Nannochloropsis oculata* [7], *Porphyridium cruentum* [8,9], and *Gyrodinium impudium* [10] have been reported as good producers, and the yielded polysaccharides included alginates, carrageenans and agarose [2]. More importantly, the production of polysaccharides from marine microalgae is practically advantageous because the process is easily controlled and the harvesting does not depend on the climate or season [11].

Generally, polysaccharides from marine microalgae can be divided into structural polysaccharides, energy polysaccharides and extracellular polysaccharides based on their physiological roles [12]. From the commercialization point of view, extracellular polysaccharides (EPS) are the most promising ones because they can be diffused into the culture medium, free from cell debris, and easy to collect and purify [2,13]. In the present work, we used a novel mutagenesis tool named atmospheric and room temperature plasma (ARTP) to obtain algal strains with enhanced EPS content. The ARTP technique uses the helium radio-frequency glow discharge plasma jets to change the microbial DNA sequences, which possesses a number advantages over most traditional mutation systems [14]. In 2013 and 2014, ARTP has been successfully utilized for microbial breeding such as bacteria, fungi, and the cyanobactrium *Spirulina platensis* [14,15,16].

In this study, we applied it to the mutagenesis of eukaryotic microalgae. The strain we examined was the dinoflagellate microalga *Crypthecodinium cohnii*. *C. cohnii* can secrets EPS which have a great application potential in functional foods and pharmaceuticals. Although the EPS contentof *C. cohnii* is less than seaweeds, a type of well-known EPS producers, it is higher than most microalgae [17,18,19]. Besides, this strain is known as a prolific producer of docosahexaenoic acid (DHA), a high-value omega-3 polyunsaturated fatty acids possessing various physiological and nutritional functions [20]. This unique characteristic makes *C. cohnii* an advantageous candidate for the integrated utilization of EPS and high-value components will significantly bring down the cost of production. The aim of the present study was to investigate the possibility of using ARTP for the mutagenesis of *C. cohnii* and evaluate the EPS production potential of the selected *C. cohnii*mutants.

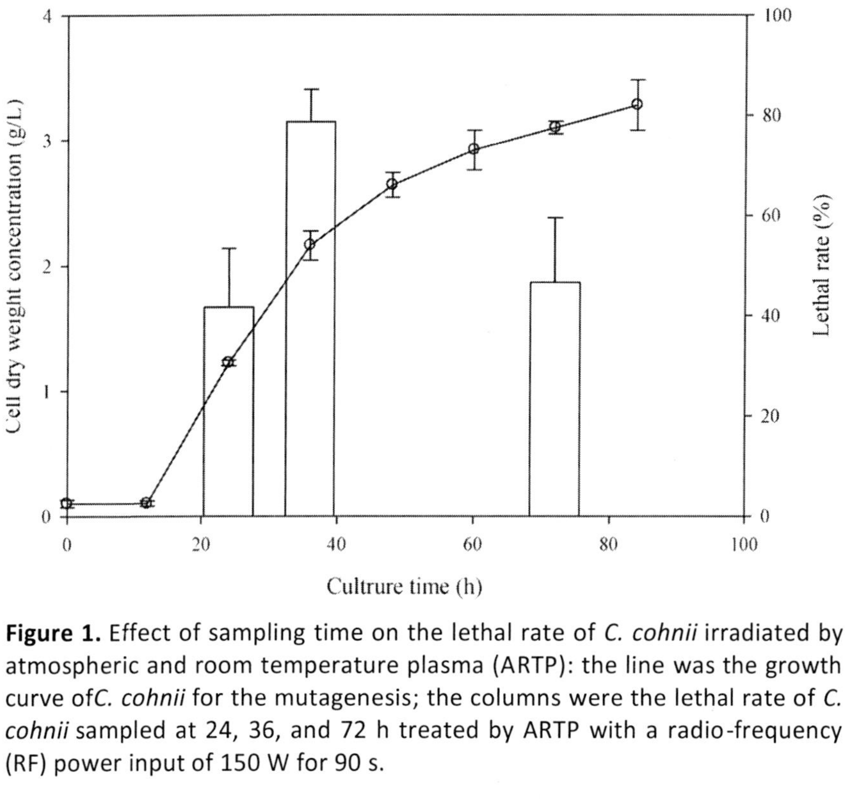

Figure 1. Effect of sampling time on the lethal rate of *C. cohnii* irradiated by atmospheric and room temperature plasma (ARTP): the line was the growth curve of*C. cohnii* for the mutagenesis; the columns were the lethal rate of *C. cohnii* sampled at 24, 36, and 72 h treated by ARTP with a radio-frequency (RF) power input of 150 W for 90 s.

2. RESULTS

2.1. Determination of the Optimal Sampling Time

To find the best sampling time, *C. cohnii* that grew after 24, 36, and 72 h were sampled to perform the ARTP treatment, which were corresponding to the beginning of log phase, the end of log phase, and the end of stationary phase, respectively.

As shown in Figure 1, the lethal rate of cells sampled at 36 h reached 78.68% after irradiating 90 s by APTP, which was much higher than those sampled at 24 and 72 h, indicating that *C. cohnii* sampled at 36 h was more sensitive to the plasma. Therefore, the plasma of ARTP system may easily lead to a lethal, cell permeability change or DNA damage of *C. cohnii* at the end of exponential phase.

2.2. Determination of the Optimal ARTP Treatment Duration

In order to obtain mutants with enhanced EPS content, the lethal rate should be taken into account. The curve of lethal rate against treatment duration is shown in Figure 2. The lethal rate was over 90% when the exposure time exceeded 100 s. There was no cell survival after treating for 120 s. Although it could meet the recommended lethal rate of 90%–99% after treating for 100 s, the cell suspension was almost dried up because of the exposure to the helium plasma flow.

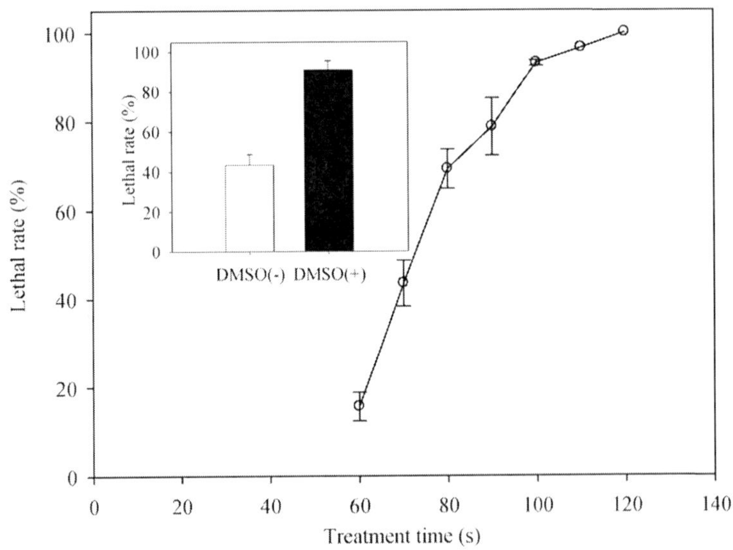

Figure 2. The relationship between treatment conditions of ARTP and lethal rate of *C. cohnii*. The curve representing lethal rate against treatment time; the white column: the lethal rate of cells diluted by By+ medium without dimethylsulfoxide (DMSO) treated for 70 s; the black column: the lethal rate of cells diluted by By+ medium with DMSO treated for 70 s; the irradiation conditions: *C. cohnii* sampled at 36 h, RF power input of 150 W, helium gas flow rate of 10 L/min, distance between the *C. cohnii* cells and the plasma source of 2 mm.

Therefore, in order to protect cells from damage caused by water evaporation and increase the cell penetrability of plasma, DMSO, a reagent to enhance the cell membrane permeability, was mixed in the suspension [21]. In the presence of DMSO, the lethal rate reached to 90.74% when treated under the same conditions for 70 s (black bar in Figure 2). In contrast, the lethal rate was only 43.49% when exposed to the plasma for 70 s without DMSO (white bar in Figure 2). Taken together,*C. cohnii* sampled at 36 h were suspended in By+ medium with 5% DMSO and used for mutation by ARTP.

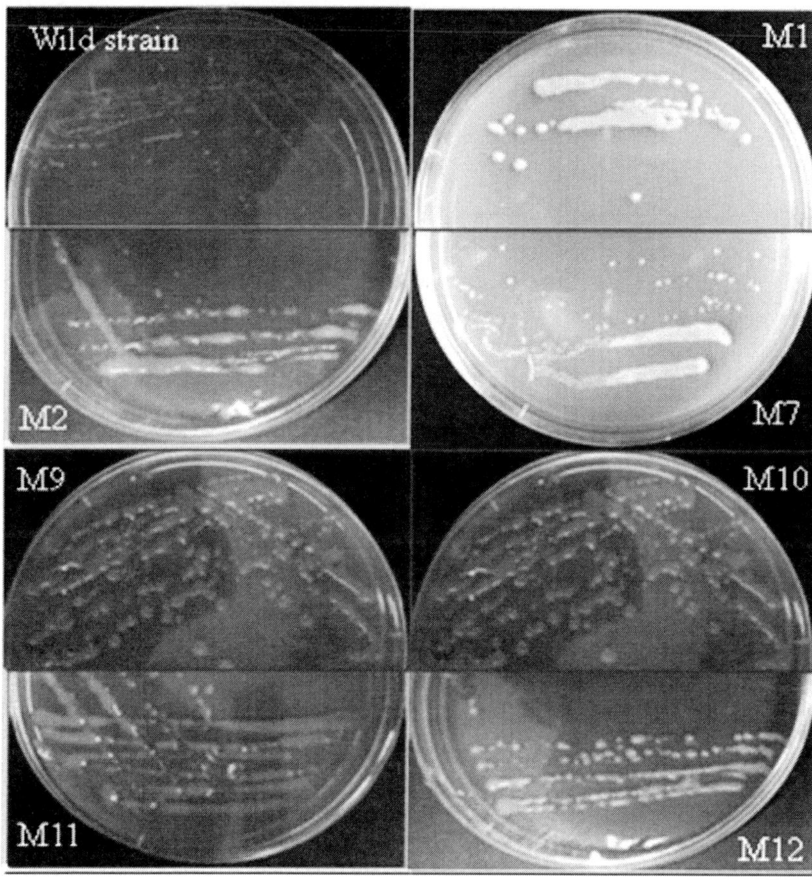

Figure 3. Colonies of wild type of *C. cohnii* and the mutants: the wild type and selected mutants of *C. cohnii* were grown on solid agar plates and incubated for 14 days.

2.3. Screening for Mutants with Enhanced EPS Content

After the treatment by ARTP, colonies of 12 mutants of *C. cohnii*, named M1 to M12, with smooth and moist surface were picked out and transferred to fresh agar plates. Compared with the wild type, the mutants formed larger colonies and most of them secreted more EPS (Figure 3).

2.4. EPS Production of the Mutants

The wild type strain and the 12 mutants of *C. cohnii* were cultured in the modified By+ medium. After 96 h, the crude EPS in the culture were precipitated by ethanol and then lyophilized. The volumetric yield of EPS by 8 mutants (M1, M2, M3, M4, M7, M8, M9 and M12) increased significantly as compared with the wild type (Figure 4A). Mutant M7 produced the highest EPS volumetric yield of 1.02 g/L with an increase of 33.85%. The EPS volumetric yield of M3 was the second highest among all mutants with an increase of 22.03%.

The EPS yield based on biomass was also calculated to evaluate the EPS production ability. Results revealed that 10 mutants produced higher EPS yield on biomass (Figure 4B). Mutant M7 and M11 produced the highest EPS yield based on biomass. Nevertheless, M11 was not suitable for EPS production because its EPS volumetric yield was very low due to its low biomass concentration (Table 1).

Considering the consumption of glucose, the EPS yield on glucose of 7 mutants was enhanced as compared with the wild type (Figure 4C). M7 produced higher EPS yield on glucose of 94.3 mg/g with an increase of 57.17%. Although M11 produced significantly higher EPS yield on glucose (*i.e.*, 107.8 mg/g, with an increase of 79.67%) than M7 ($p < 0.01$), M11 only produced 71.29% EPS volumetric yield of M7.

Taken together, among all mutants, mutant M7 was most suitable for EPS production because of its highest EPS volumetric yield. The EPS volumetric yield, EPS yield on biomass and EPS yield on glucose of M7 were improved by 33.85%, 85.35% and 57.17%, respectively (Figure 4). Besides, M1 and M4 increased significantly in terms of EPS yield on biomass, EPS volumetric yield and EPS yield on glucose.

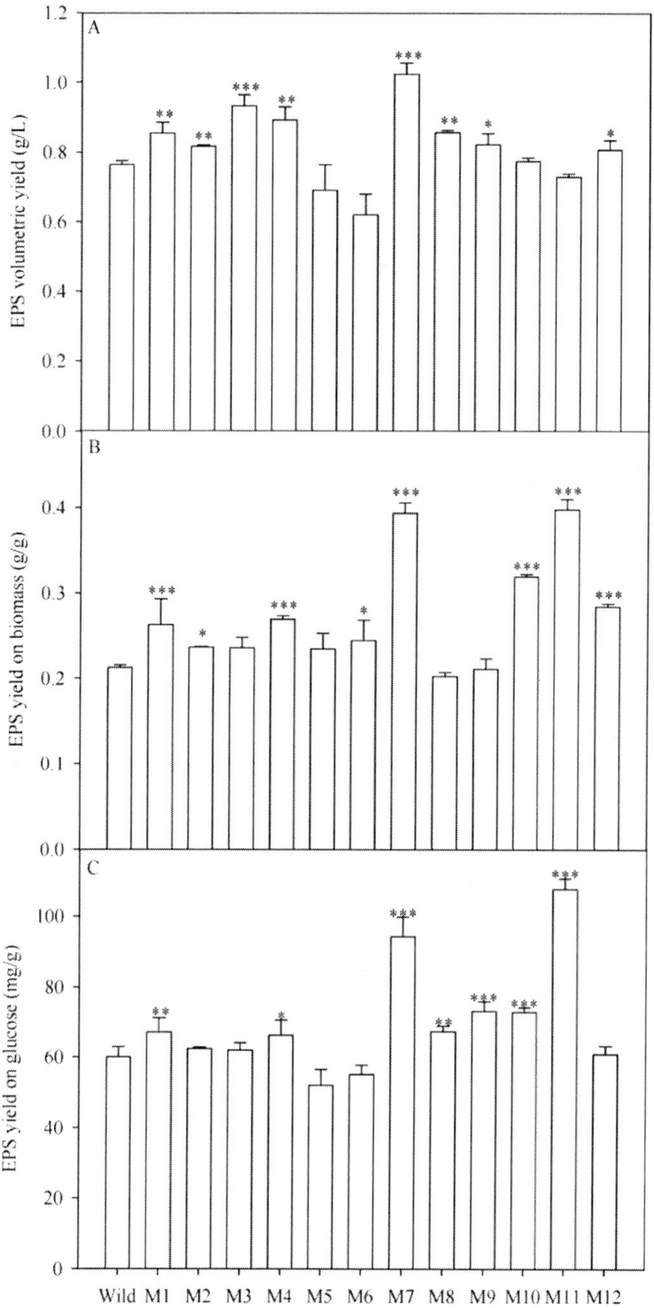

Figure 4. EPS yield of the wild type and 12 mutants of *C. cohnii*. (**A**) EPS volumetric yield; (**B**) EPS yield on biomass; (**C**) EPS yield on glucose; asterisk (* $p < 0.05$, ** $p < 0.01$, and *** $p < 0.001$) means significant differences between wild type and the mutants.

Table 1. The cell dry weight concentration and growth yield on glucose of the wild type and 12 mutants of *C. cohnii*.

Species	Cell Dry Weight Concentration (g/L)	Growth Yield on Glucose (g/g)
wild type	3.60 ± 0.09	0.28 ± 0.02
M1	3.25 ± 0.29	0.26 ± 0.02
M2	3.46 ± 0.22	0.26 ± 0.00
M3	3.96 ± 0.32 ***	0.27 ± 0.01
M4	3.31 ± 0.10	0.25 ± 0.02
M5	2.95 ± 0.11	0.22 ± 0.01
M6	2.54 ± 0.02	0.23 ± 0.01
M7	2.60 ± 0.02	0.24 ± 0.01
M8	4.24 ± 0.09 ***	0.33 ± 0.00 ***
M9	3.90 ± 0.12 **	0.35 ± 0.02 ***
M10	2.43 ± 0.05	0.23 ± 0.01
M11	1.83 ± 0.08	0.27 ± 0.02
M12	2.84 ± 0.07	0.21 ± 0.01

Asterisk (** $p < 0.01$ and *** $p < 0.001$) means that there were significant differences between the wild type and the mutants.

2.5. Monosaccharide Composition

The composition of EPS produced by the wild type and M7 was characterized (Table 2). Glucose and galactose turned out to be the predominant monosaccharides in the EPS of the wild type (*C. cohnii* ATCC 30556). Other sugars such as mannose, ribose and fucose, were also detected in the crude EPS (Table 2).

Monosaccharide composition of the EPS was also analyzed, and there was no significant difference between the wild type and M7 in monosaccharide composition (Table 2). These results indicated that mutant M7 secreted larger amounts of EPS whereas the monosaccharide composition remained unchanged as compared with the wild type.

Table 2. Monosaccharide composition (in mol %) of crude EPS from the wild type and mutant M7 of *C. cohnii*.

Species	Mannose	Ribose	Glucose	Galactose	Fucose	Unknown
Wild	6.53 ± 0.15	5.99 ± 0.51	61.73 ± 0.59	20.15 ± 0.06	3.48 ± 0.08	2.13 ± 0.05
M7	6.05 ± 0.01	4.72 ± 0.12	63.77 ± 0.12	19.61 ± 0.07	3.65 ± 0.28	2.2 ± 0.04

3. DISCUSSION

ARTP is a novel, high-efficient and environment-friendly mutagenesis tool as compared with transgenic techniques or traditional chemical and physical mutagens [14]. The plasma from ARTP contains significant amounts of active chemical species which can improve cell permeability, penetrate the cell wall/membrane and cause DNA/protein damage [22,23]. In contrast to the traditional mutagens, the plasma has higher positive genotoxic response [15]. Besides, the plasma shows different genotoxic characteristics depending on the irradiation dose [15]. In the present investigation, the ARTP was conducted for the mutagenesis of eukaryotic microalga *C. cohnii*.

For the purpose of optimizing mutation conditions, the irradiation dose of ARTP should be taken into account. A lethal rate between 90%–99% favors the generation of mutants with higher EPS contents [14]. In the present study, the RF input power was set at 150 W, higher than the power of 100 W that was normally used [14,15,16]. We found the higher RF input power provided the plasma with high energy, which allows the plasma to efficiently penetrate through the cell wall of *C. cohnii* and thereby alert the genome.

The plasma also shows different genotoxic characteristics depending on the cultivation phase of the wild type strain. In the present study, *C. cohnii* cells in log phase were found to be most sensitive to the plasma. These results were consistent with previous reports suggesting cells grown at log phase are more suitable for mutation [16]. Although *C. cohnii* ATCC 30772 produced 2.7 g/L crude EPS, the EPS yield based on biomass was only 0.15 g/g [18]. For comparison, EPS yields on biomass for the marine microalgae *Aurantiochytrium*, *Schizochytrium*, *Ulkenia*, and *Thraustochytrium* (from 0.015 to 0.119 g/g) were even lower than *C. cohnii* [17]. In contrast, the EPS yield on biomass of mutant M7 of *C. cohnii* (ATCC 30556) could reach to 0.39 g/g which was 160% higher than that of *C. cohnii* ATCC 30772.

The genetic stability of mutant M7 was also investigated. After a 9-genertation culture, M7 still maintained high EPS yield on biomass of 0.38 ± 0.01 g/g. Although further research is necessary to explain the genome mutation mechanism by ARTP mutagenesis, the target products-producing ability of mutants also remained very stable after more than ten generations [16,24].

Among enzymes involved in the biosynthesis of EPS, alpha-phosphoglucomutase (PMG) and UDP-glucose pyrophosphorylase (UGP) play key roles, whose activities are highly correlated with the EPS level [25]. Overexpression of*PGM* gene leads to high production of EPS with higher transcription levels of *UGP* gene [26]. In future, the contribution of these two enzymes towards *C. cohnii* mutant needs to be determined, which may help elucidate the underlying mechanism of ARTP for improving EPS production of *C. cohnii*.

4. EXPERIMENTAL SECTION

4.1. Algal Species and Culture Conditions

The alga *Crypthecodinium cohnii* (ATCC 30556) was obtained from the American Type Culture Collection (Rockville, MD, USA). The strain was heterotrophically maintained in By+ medium (containing 15 g sea salt, 1 g yeast extract, 1 g tryptone, and 5 g glucose per liter) at 16 °C and subcultured every month. The inocula were prepared by culturing the alga in 500-mL Erlenmeyer flasks containing 100 mL By+ medium at 25 °C for 48 h with orbital shaking at 150 rpm in the dark. The seeds of wild type and mutants were inoculated on to 100 mL fresh modified By+ medium (containing 15 g sea salt, 0.5 g yeast extract, 0.5 g tryptone, and 20 g glucose per liter) at a 5% (*v*/*v*) inoculum size for batch culture in 500-mL Erlenmeyer flasks. The pH of the modified By+ medium was adjusted to 6.5 prior to autoclaving. After inoculating, flasks were incubated at 25 °C in an orbital shaker at 150 rpm in the dark for 96 h.

4.2. Mutation by ARTP

The artificial plasma used in the present mutation was generated by the helium-based ARTP mutation system (Si Qing Yuan Biotechnology Co., Ltd., Wuxi, China). About 50,000 cells at the end of log phase were suspended in 20 μL By+ medium (with 5% dimethylsulfoxide (DMSO)) and placed on a stainless steel minidisc. The plate was then exposed to the plasma and the operating parameters of ARTP system were set as follows: radio-frequency (RF) power input of 150 W, the helium gas flow rate of 10 L/min, treating distance of 2 mm, and the treating time ranged from 60 to 120 s. The control was treated for 120 s with the same parameters except no RF power input. After being treated by the artificial plasma for various time intervals, the *C. cohnii* samples were moved into By+ medium and diluted to an appropriate concentration. Cells were transferred on to a By+ solid medium and cultured at 25 °C in the dark for two weeks. The strategies for screening high-EPS mutants were based on the moisture, smoothness, and size of the colony growing on the solid plates. The lethal rate was calculated as follows: lethal rate = $(1 - N_1/N_0) \times 100\%$, where N_0 is colony number of the control and N_1 is colony number of mutation group growing on the plates. All treatments were determined in triplicates.

4.3. Isolation of Extracellular Polysaccharides

Cultures were centrifuged at 4000× *g* for 10 min. Two volumes of absolute ethanol were added to the supernatant to precipitate the EPS. After 6 h of treatment at room temperature (25 °C), the precipitation was harvested by centrifuging at 1000× *g* for 5 min. After being washed twice by 66% (*v*/*v*) ethanol in water, the precipitated crude EPS was redissolved in ddH_2O and lyophilized [27]. The EPS production was determined by weighing the lyophilized crude EPS.

4.4. Analysis of Monosaccharide Composition

The crude EPS (0.5 mg) was hydrolyzed at 110 °C in 200 μL of 2 M trifluoroacetic acid (TFA) solution for 2 h. After cooling to room temperature, 200 μL of methanol was added to the hydrolysate which was then dried under N_2, and this procedure was repeated twice to remove TFA. The hydrolyzed sugars were dissolved in 100 μL of 0.3 M NaOH solutions and mixed with 100 μL of 0.5 M 3-methyl-1-phenyl-2-pyrazoline-5-one (PMP) in methanol to derive the monosaccharide at 70 °C for 100 min. The reaction solution was then cooled to room temperature whose pH was adjusted to 7.0 by 0.3 M HCl solution and the volume was adjusted to 1 mL by ddH_2O. Finally, the derived monosaccharide solution was filtered (with a 0.22 μm membrane) for high performance liquid chromatography (HPLC) analysis. The derived monosaccharide was measured by HPLC (2695, Waters Corporation, Milford, MA, USA) at 40 °C using a C18 column (4.6 mm × 250 mm, Waters, USA) and photo-diode array (PDA) detector. Phosphate buffer (0.1 M, pH 6.7, 83%) and acetonitrile (17%) were used as mobile phases with the flow rate at 1.0 mL/min. Five pure sugars including mannose, ribose, glucose, galactose, and fucose (purchased from Sigma-Aldrich, St. Louis, MO, USA), were used as the standards [28].

4.5. Determination of Residual Glucose Concentration and Cell Dry Weight

The cells were centrifuged at 3000× *g* for 5 min. The concentration of glucose in the supernatant was determined according to the reported methods [29]. The pellets were washed with ddH_2O twice and filtered through pre-dried filter papers (Whatman GF/C, GE Healthcare Bio-Sciences, Pittsburgh, PA, USA). Then the cells on the discs were dried to constant weight in a vacuum oven at 80 °C.

4.6. Statistical Analysis

All data were obtained from three repetitions and analyzed by using the one-way analysis of variance (ANOVA) with subsequent post hoc multiple comparison by least-significant difference (LSD) test, which were calculated by the software of IBM SPSS Statistics 20 (IBM Corporation, Armonk, NY, USA). Statistically significant values were defined as $p < 0.05$.

5. CONCLUSIONS

In conclusion, a high EPS production and stable mutant of *C. cohnii* was obtained by using the novel mutation technique of ARTP. Results of the present work supported that ARTP has the potential to enhance EPS yield of microalgae as well as other similar microorganisms. Further studies are needed to figure out the potential biological activities of EPS from the *C. cohnii* mutants.

ACKNOWLEDGMENTS

This study was partially supported by the 985 Project of Peking University, the 863 Plan of Ministry of Science and Technology of China (2012AA023107) and the National Natural Science Foundation of China (31471717).

AUTHOR CONTRIBUTIONS

Bin Liu, Zheng Sun, Xiaonian Ma and Feng Chen conceived the study, performed data analysis and wrote the manuscript; Bo Yang performed the Monosaccharide composition analysis by HPLC; Yue Jiang and Dong Wei contributed with valuable discussions. All authors read and approved the final manuscript.

CONFLICTS OF INTEREST

The authors declare no conflict of interest.

REFERENCES

1. Boddohi, S.; Kipper, M.J. Engineering nanoassemblies of polysaccharides. Adv. Mater. 2010, 22, 2998–3016.
2. Laurienzo, P. Marine polysaccharides in pharmaceutical applications: An overview. Mar. Drugs 2010, 8, 2435–2465. [PubMed]
3. Wu, J.Y.; Chen, X.; Siu, K.C. Isolation and structure characterization of an antioxidative glycopeptide from mycelial culture broth of a medicinal fungus. Int. J. Mol. Sci. 2014, 15, 17318–17332.
4. Yao, Y.; Shi, Z.X.; Ren, G.X. Antioxidant and immunoregulatory activity of polysaccharides from quinoa (Chenopodium quinoa Willd.). Int. J. Mol. Sci. 2014, 15, 19307–19318.
5. Sun, L.Q.; Wang, L.; Li, J.; Liu, H.H. Characterization and antioxidant activities of degraded polysaccharides from two marine Chrysophyta. Food Chem. 2014, 160, 1–7.
6. Guzman, S.; Gato, A.; Lamela, M.; Freire-Garabal, M.; Calleja, J.M. Anti-inflammatory and immunomodulatory activities of polysaccharide from Chlorella stigmatophora and Phaeodactylum tricomutum. Phytother. Res. 2003, 17, 665–670. [PubMed]
7. Samarakoon, K.W.; Ko, J.Y.; Shah, M.M.R.; Lee, J.H.; Kang, M.C.; O-Nam, K.; Lee, J.B.; Jeon, Y.J. In vitro studies of anti-inflammatory and anticancer activities of organic solvent extracts from cultured marine microalgae. Algae 2013,28, 111–119.
8. Gardeva, E.; Toshkova, R.; Yossifova, L.; Minkova, K.; Gigova, L. Cytotoxic and apoptogenic potential of red microalgal polysaccharides. Biotechnol. Biotechnol. Equip. 2012, 26, 3167–3172.
9. Raposo, M.F.D.; de Morais, A.; de Morais, R. Influence of sulphate on the composition and antibacterial and antiviral properties of the exopolysaccharide from Porphyridium cruentum. Life Sci. 2014, 101, 56–63.
10. Kim, M.; Yim, J.H.; Kim, S.Y.; Kim, H.S.; Lee, W.G.; Kim, S.J.; Kang, P.S.; Lee, C.K. In vitro inhibition of influenza A virus infection by marine microalga-derived sulfated polysaccharide p-KG03. Antivir. Res. 2012, 93, 253–259.

11. Raposo, M.F.D.; de Morais, R.; de Morais, A. Health applications of bioactive compounds from marine microalgae.Life Sci. 2013, 93, 479–486.

12. Pignolet, O.; Jubeau, S.; Vaca-Garcia, C.; Michaud, P. Highly valuable microalgae: Biochemical and topological aspects. J. Ind. Microbiol. Biotechnol. 2013, 40, 781–796.

13. Dewapriya, P.; Kim, S.-K. Marine microorganisms: An emerging avenue in modern nutraceuticals and functional foods. Food Res. Int. 2014, 56, 115–125.

14. Zhang, X.; Zhang, X.F.; Li, H.P.; Wang, L.Y.; Zhang, C.; Xing, X.H.; Bao, C.Y. Atmospheric and room temperature plasma (ARTP) as a new powerful mutagenesis tool. Appl. Microbiol. Biotechnol. 2014, 98, 5387–5396.

15. Fang, M.Y.; Jin, L.H.; Zhang, C.; Tan, Y.Y.; Jiang, P.X.; Ge, N.; Li, H.P.; Xing, X.H. Rapid mutation of Spirulina platensisby a new mutagenesis system of atmospheric and room temperature plasmas (ARTP) and generation of a mutant library with diverse phenotypes. PLoS ONE 2013, 8, e77046.

16. Wang, Q.; Feng, L.R.; Wei, L.; Li, H.G.; Wang, L.; Zhou, Y.; Yu, X.B. Mutation breeding of lycopene-producing strainBlakeslea trispora by a novel atmospheric and room temperature plasma (ARTP). Appl. Biochem. Biotechnol. 2014, 174, 452–460.

17. Chang, K.J.L.; Nichols, C.M.; Blackburn, S.I.; Dunstan, G.A.; Koutoulis, A.; Nichols, P.D. Comparison of thraustochytrids Aurantiochytrium sp., Schizochytrium sp., Thraustochytrium sp., and Ulkenia sp. for production of biodiesel, long-chain omega-3 oils, and exopolysaccharide. Mar. Biotechnol. 2014, 16, 396–411.

18. De Swaaf, M.E.; Grobben, G.J.; Eggink, G.; de Rijk, T.C.; van der Meer, P.; Sijtsma, L. Characterisation of extracellular polysaccharides produced by Crypthecodinium cohnii. Appl. Microbiol. Biotechnol. 2001, 57, 395–400.

19. Mohamed, S.; Hashim, S.N.; Rahman, H.A. Seaweeds: A sustainable functional food for complementary and alternative therapy. Trends Food Sci. Technol. 2012, 23, 83–96.

20. Mendes, A.; Reis, A.; Vasconcelos, R.; Guerra, P.; da Silva, T.L. Crypthecodinium cohnii with emphasis on DHA production: A review. J. Appl. Phycol. 2009, 21, 199–214.

21. Hazen, K.C. Influence of DMSO on antifungal activity during susceptibility testing in vitro. Diagn. Microbiol. Infect. Dis. 2013, 75, 60–63.

22. Laroussi, M.; Lu, X. Room-temperature atmospheric pressure plasma plume for biomedical applications. Appl. Phys. Lett. 2005, 87.

23. Leduc, M.; Guay, D.; Coulombe, S.; Leask, R.L. Effects of non-thermal plasmas on DNA and mammalian cells. Plasma Process. Polym. 2010, 7, 899–909.

24. Lu, Y.; Wang, L.Y.; Ma, K.; Li, G.; Zhang, C.; Zhao, H.X.; Lai, Q.H.; Li, H.P.; Xing, X.H. Characteristics of hydrogen production of an Enterobacter aerogenes mutant generated by a new atmospheric and room temperature plasma (ARTP). Biochem. Eng. J. 2011, 55, 17–22.

25. Degeest, B.; de Vuyst, L. Correlation of activities of the enzymes alpha-phosphoglucomutase, UDP-galactose 4-epimerase, and UDP-glucose pyrophosphorylase with exopolysaccharide biosynthesis by Streptococcus thermophilusLY03. Appl. Environ. Microbiol. 2000, 66, 3519–3527.

26. Xu, J.W.; Ji, S.L.; Li, H.J.; Zhou, J.S.; Duan, Y.Q.; Dang, L.Z.; Mo, M.H. Increased polysaccharide production and biosynthetic gene expressions in a submerged culture of Ganoderma lucidum by the overexpression of the homologous alpha-phosphoglucomutase gene. Bioprocess Biosyst. Eng. 2015, 38, 399–405.

27. Grobben, G.; van Casteren, W.; Schols, H.; Oosterveld, A.; Sala, G.; Smith, M.; Sikkema, J.; de Bont, J. Analysis of the exopolysaccharides produced by Lactobacillus delbrueckii subsp. bulgaricus NCFB 2772 grown in continuous culture on glucose and fructose. Appl. Microbiol. Biotechnol. 1997, 48, 516–521.

28. Dai, J.; Zhu, S.; Tang, J.; Wang, M.; Yin, H.P.; Chen, S.W. Analysis of monosaccharide composition in polysaccharides from D. salina by pre-column derivatization high performance liquid chromtography. J. Instrum. Anal. 2007, 26, 206–210.

29. Miller, G.L. Use of dinitrosalicylic acid reagent for determination of reducing sugar. Anal. Chem. 1959, 31, 426–428.

Chapter 2

A Substantial Fraction of Barley (*Hordeum vulgare* L.) *Low Phytic Acid* Mutations Have Little or No Effect on Yield across Diverse Production Environments

Victor Raboy [1,†,*], Kevin Peterson [1,†,‡], Chad Jackson [2,†], Juliet M. Marshall [2], Gongshe Hu [1,†], Hirofumi Saneoka [3,†] and Phil Bregitzer [1]

[1]*USDA-ARS Small Grains and Potato Research Unit, 1600 South 2700 West, Aberdeen, ID 83210, USA*

[2]*Idaho Falls Research and Extension Center, University of Idaho, 1776 Science Center Drive, Suite 205, Idaho Falls, ID 83402, USA*

[3]*Graduate School of Biosphere Science, Hiroshima University, 1-4-4 Kagamiyama, Higashi-Hiroshima 739-8528, Japan*

ABSTRACT

The potential benefits of the *low phytic acid* (*lpa*) seed trait for human and animal nutrition, and for phosphorus management in non-ruminant animal production, are well documented. However, in many cases the *lpa*trait is associated with impaired seed or plant performance, resulting in reduced yield. This has given rise to the perception that the *lpa* trait is tightly correlated with reduced yield in diverse crop species. Here we report a powerful test of this correlation. We measured grain yield in lines homozygous for each of six barley (*Hordeum vulgare* L.) *lpa*mutations that greatly differ in their seed phytic acid levels.

Performance comparisons were between sibling wild-type and mutant lines obtained following backcrossing, and across two years in five Idaho (USA) locations that greatly differ in crop yield potential. We found that one *lpa* mutation (*Hvlpa*1-1) had no detectable effect on yield and a second (*Hvlpa*4-1) resulted in yield losses of only 3.5%, across all locations. When comparing yields in three relatively non-stressful production environments, at least three *lpa* mutations (*Hvlpa*1-1, *Hvlpa*3-1, and *Hvlpa*4-1) typically had yields similar to or within 5% of the wild-type sibling isoline. Therefore in the case of barley, *lpa* mutations can be readily identified that when simply incorporated into a cultivar result in adequately performing lines, even with no additional breeding for performance within the *lpa* line. In conclusion, while some barley *lpa* mutations do impact field performance, a substantial fraction appears to have little or no effect on yield.

Keywords: barley; *Hordeum vulgare*; low phytic acid; grain yield

1. INTRODUCTION

The potential benefits of the *low phytic acid* (*lpa*) crop seed phenotype, in the contexts of human and animal nutrition and in the context of management of phosphorus (P) in animal agricultural production, are well documented [1]. However, there is a commonly encountered view that this seed trait is conditionally associated with reduced seed performance and reduced seed yield. This view often reflects sound science since in many cases the *lpa* trait is associated with impaired seed or plant performance [2]. However, many studies reporting on field performance of *lpa* genotypes present relatively limited data using lines obtained following limited or no breeding, which are often evaluated in a growth chamber or greenhouse bench, or one or a few field locations or replications over years. This limited experimental approach can also lead to the opposite but equally problematic and often not well-supported conclusion that a given *lpa* mutation has little or no impact on yield [3]. Further, comparisons are often made between lines that greatly differ in genetic background other than the *lpa* loci. Even comparisons made between a wild-type (for seed phytic acid) cultivar and an *lpa* mutation isolated following mutagenesis of that cultivar, or following most transgenic approaches [4], are less than optimal. Many genetic differences that can impact field performance, other than the *lpa* alleles, may exist between these types of lines. The strongest test of the relationship between the *lpa* trait and any other trait, whether it be yield or nutritional quality, is comparisons between sibling near-isogenic lines, obtained following sufficient backcrossing, and following selection for field performance in fixed genotypes that are either homozygous wild-type or homozygous mutant. Such sibling near-isogenic lines are genetically similar at most loci other than the *lpa* loci, so any differences observed in field performance or nutritional quality can validly be attributed to the different *lpa* locus alleles.

Here we report the results of what we believe is, to date, the most robust field evaluation of *lpa* variants of any crop. Six independently isolated barley (*Hordeum vulgare* L.) *lpa* mutations, representing lesions mapped to four different chromosomes therefore representing a minimum of four loci, were backcrossed to the cultivar (*cv.* Harrington) in which they were initially induced and isolated. For each mutation a set of near-isogenic sibling lines were then established, each set consisting of one mutant and one wild-type line. Field trials were then conducted over two years at five locations with widely divergent production conditions, providing a robust test of performance across environments.

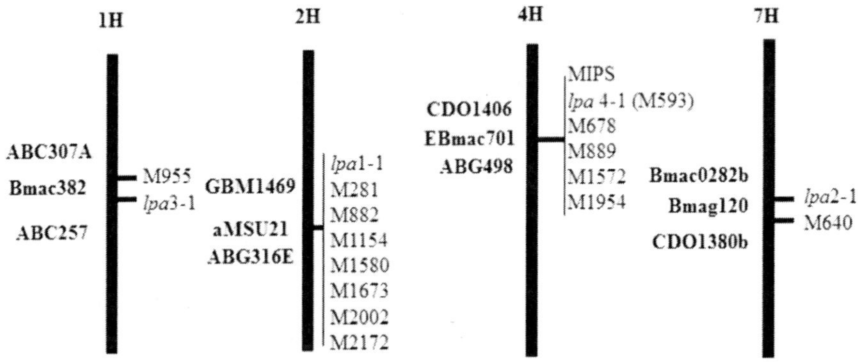

Figure 1. Chromosomal position of 17 barley *low phytic acid* mutations and the barley genome's single *myo*-Inositol-3-P_1 synthase (MIPS) gene. Map positions are approximate and only meant to be illustrative of linkage groups. As indicated, all mutations fell into four linkage groups, but allelic or non-allelic relationships within each linkage group are not yet known. The molecular markers used in mapping are on the left of each chromosome. The *lpa* mutations comprising each linkage group are on the right of each chromosome.

2. RESULTS

2.1. Barley Low Phytic Acid Mutations Selected for Isoline Development

A screen of sodium azide-induced mutations in barley for the "high seed inorganic P" phenotype associated with *lpa*mutants had identified 23 putative mutants, of which 20 were selected for initial seed P phenotype evaluation (Supplementary Methods and Supplementary Table 1), and initial visual evaluation of plant performance. These 20 were grown in a nursery in 2004 in Aberdeen, Idaho, and following grain harvest, assayed for whole grain P fractions. Of these all but one displayed the low grain phytic acid trait (Supplementary Table 1). The one exception was *Hv*-MAZ423, which appears to be a high grain total P mutant in which all grain P fractions, including inorganic P, are elevated, as compared with wild-type. This putative high grain total P mutant is the subject of separate, ongoing study. Of the remaining 19 mutants, a total of 17 have been chromosomally mapped to date (Figure 1; Supplementary Tables 2 and 3).

Previous studies focused on four mutations that appeared to represent the range of seed P phenotypes observed, and that mapped to three linkage groups: *Hvlpa*1-1 (chromosome 2H), *Hvlpa*2-1 (chromosome 7H), *Hvlpa*3-1 (chromosome 1H), and*Hvlpa*-M955 (chromosome 1H). Based on initial observation of adequate plant performance, and to add mutations mapping to new chromosomal sites, for the present study we selected two additional mutations: *Hvlpa*-M593 and *Hvlpa*-M640. Initial quantitative analyses (Supplementary Table 1) indicated that *Hvlpa*-M593 conditioned a ~53% reduction in seed phytic acid, and that *Hvlpa*-M640 conditioned a ~47% reduction, but in the case of *Hvlpa*-M640, the effect appeared to be endosperm-specific.

Segregation analyses conducted during mapping indicated that these additional two mutants displayed an inheritance pattern of recessive alleles at single gene (Supplementary Table 2). Linkage analyses (Supplementary Table 3) revealed that*Hvlpa*-M593 mapped to chromosome 4H, defining a new barley *lpa* locus. Therefore hereafter, *Hvlpa*-M593 will be referred to as *Hvlpa*4-1. *Hvlpa*-M640 mapped to chromosome 7H, linked to *lpa*2-1. However, the allelic relationship if any between *lpa*2-1 and *lpa*M640 on chromosome 7H, or between *lpa*3-1 and *lpa*-M955 on chromosome 1H, is not known at present and is the subject of ongoing study. Thus these six mutations represent a total of four linkage groups and a minimum of four loci.

2.2. Confirmation of Isoline Seed Phytic Acid Phenotype

The seed inorganic P test was used for routine genotyping (homozygous wild-type *versus* homozygous mutant) of *lpa*isolines used in this study (Supplementary Figure 1 and Experimental Section). However, to provide a more in-depth confirmation of the seed P and phytic acid P phenotype of grain produced by these lines when grown at the locations used here, a quantitative analysis was conducted with grain samples from test plots at two 2011 locations, Aberdeen and Soda Springs (Table 1). These two locations were selected for this analysis because they represent the extremes in yield potential over all locations in this study (see below). This analysis confirmed that when grown in these specific locations, the predicted *lpa* phenotype of each mutant line (based on previous analyses such as given in Supplementary Table 1) was observed. Apparent reductions in grain phytic acid P, as compared with levels observed in grain produced by sibling wild-type lines, ranged from a low of ~22% in *Hvlpa*2-1, to the moderate reductions of ~35% in *Hvlpa*4-1, ~40% in *Hvlpa*-M640 and ~48% in *Hvlpa*1-1, to the more substantial reductions of ~57% in *Hvlpa*3-1 and ~80% in *Hvlpa*-M955. These are referred to as "apparent reductions" since the method used here probably overestimates phytic acid by 5% to 10%, an inflation resulting from reading components of the "extraction matrix" as phytic acid P when using this method. The best illustration of this is the data for *Hvlpa*-M955. Previous analyses of the grain phytic acid levels in this mutant using more accurate methods consistently indicated that the reduction in phytic acid P in this mutant, as compared with wild-type, is >90%. This analysis also confirmed that with one exception, *Hvlpa*1-1, the *lpa* mutations did not greatly impact grain total P: the reductions in grain phytic acid P were largely matched by molar-equivalent (in terms of P) increases in grain inorganic P. In the case of*Hvlpa*1-1, this analysis also confirmed that homozygosity for the mutation causes a reduction in whole-grain total P of 15% to 20% (Table 1).

2.3. Yield Evaluations

To provide the most valid statistical approach to test for differences between sibling near-isogenic lines, the Analysis of Variance of the grain yield data presented in Table 2 was conducted using only the data for the six isoline pairs. The greatest variation observed was due to the main effects of Year and Location and the interaction of Year × Location. The large main effect of Year is clearly seen in the greater overall productivity in 2011 as compared with 2010; overall mean grain yield in 2011 (5.47 ± 0.28 $Mg \cdot ha^{-1}$) was ~20% greater than that observed in 2010 (4.52 ± 0.22 $Mg \cdot ha^{-1}$) (Table 3 and Table 4). The large main effect of Location is also clear in the data, and critical to the main objective of this study. Clear and large differences in overall mean grain yield per location, a measure of location yield potential, were observed in both years and the location ranking were similar in both years (from highest overall yield to lowest): Aberdeen > Idaho Falls > Rupert > Ashton > Soda Springs. The interaction effect of Year × Location was largely due to the very large difference in overall yield at one location, Ashton, where the yield in 2010 (2.43 ± 0.21 $Mg \cdot ha^{-1}$) was only 40% of that observed in 2011 (5.96 ± 0.32 $Mg \cdot ha^{-1}$). The large differences in productivity observed across the two years and five locations provides a powerful test of the impact on yield of an *lpa* mutation, the primary objective of this study.

Table 1. Seed total phosphorus (Total P), phytic acid P and inorganic P (P_i) in pairs of sibling barley near-isogenic and cultivar controls grown in Aberdeen and Soda Springs, ID, in 2011.

Isoline Pair	Genotype	Idaho Field Location					
		Aberdeen			Soda Springs		
		Total P	Phytic Acid P	Inorganic P	Total P	Phytic Acid P	Inorganic P
		mg·g^{-1}					
1	WT	2.96	2.86	0.29	3.10	2.62	0.26
1	*Hvlpa1-1*	2.52	1.51	1.05	2.44	1.35	0.83
2	WT	2.86	2.47	0.24	3.09	2.75	0.25
2	*Hvlpa3-1*	2.76	1.10	1.45	2.88	1.15	1.66
3	WT	2.78	2.32	0.22	3.12	2.52	0.24
3	*Hvlpa*-M955	2.91	0.53	2.12	3.33	0.51	2.21
4	WT	2.75	2.38	0.23	2.93	2.54	0.25
4	*Hvlpa2-1*	2.93	1.83	0.67	3.18	1.99	0.90
5	WT	3.10	2.49	0.39	2.98	2.59	0.27
5	*Hvlpa4-1*	2.91	1.53	1.18	3.31	1.76	1.34
6	WT	2.86	2.44	0.26	3.15	2.69	0.26
6	*Hvlpa*-M640	2.98	1.61	0.80	3.36	1.51	1.32
Cultivar Controls							
Harrington		2.73	2.37	0.26	3.07	2.59	0.30
Baronesse		2.89	2.41	0.26	3.12	2.65	0.28
Conrad		3.01	2.52	0.19	3.49	3.10	0.31
Std. Dev. of Mean		0.12	0.14	0.07	0.15	0.14	0.05
LSD $p = 0.05$		0.34	0.41	0.20	0.43	0.39	0.15

Table 2. Analysis of variance of yield of twelve isolines grown in five locations in Idaho in two years.

Source of Variation	*df*	Mean Square	*F* Value	Prob. > F
Model	132	10.20	28.82	<0.0001
Year	1	97.24	274.70	<0.0001
Location	4	224.79	635.05	<0.0001
Year × Loc	4	61.43	173.54	<0.0001
Rep (Loc)	15	0.65	1.85	0.03
Isoline	11	3.70	10.45	<0.0001
Year × Line	11	0.22	0.62	0.82
Loc × Line	44	0.73	2.06	0.0002
Year × Loc × Line	42	0.47	1.31	0.11
Residual Error	330	0.35		

Table 3. Yield of barley *lpa* near-isogenic lines and three cultivar controls when grown in five locations in Idaho, 2010. Data represent means of four replicates at each location.

Isoline Pair	Genotype	Idaho Location					Line Mean
		Aberdeen	Ashton	Idaho Falls	Rupert	Soda Springs	
		Grain Yield (Mg·ha^{-1})					
1	WT	6.04	2.52	6.15	6.20	2.13	4.61
1	*Hvlpa*1-1	6.61	2.78	6.57	5.47	2.00	4.69
2	WT	5.83	2.37	5.84	6.20	2.22	4.49
2	*Hvlpa*3-1	6.17	2.03	5.71	5.15	1.88	4.19
3	WT	5.72	2.59	5.98	5.45	2.80	4.60
3	*Hvlpa*-M955	5.27	1.88	5.71	4.74	1.47	3.81
4	WT	6.48	2.34	6.15	5.89	2.41	4.91
4	*Hvlpa*2-1	6.51	2.59	5.98	5.23	1.81	3.86
5	WT	6.14	2.32	6.45	5.71	2.52	4.63
5	*Hvlpa*4-1	6.06	2.30	6.74	5.45	1.96	4.50
6	WT	6.59	2.30	6.77	5.89	2.10	4.73
6	*Hvlpa*-M640	6.30	2.40	6.45	4.44	1.61	3.58
		Cultivar Controls					
Harrington		6.38	2.42	5.54	5.54	2.22	4.42
Baronesse		7.17	3.57	6.81	7.45	2.59	5.52
Conrad		7.06	2.37	6.08	6.91	2.05	4.89
Location Mean ± SE		6.29 ± 0.34	2.43 ± 0.21	6.19 ± 0.29	5.78 ± 0.22	2.10 ± 0.19	4.52 ± 0.22
LSD $p = 0.05$		0.98	0.59	0.83	0.63	0.55	0.64

Table 4. Yield of barley *lpa* near-isogenic lines and three cultivar controls when grown in five locations in Idaho, 2011. Data represent means of four replicates at each location.

Isoline Pair	Genotype	Idaho Location					Line Mean
		Aberdeen	Ashton	Idaho Falls	Rupert	Soda Springs	
		Grain Yield (Mg·ha^{-1})					
1	WT	6.51	5.79	6.10	5.22	3.05	5.33
1	*Hvlpa*1-1	7.35	6.69	5.57	5.57	3.25	5.68
2	WT	6.80	6.15	4.93	4.98	3.55	5.37
2	*Hvlpa*3-1	6.98	5.32	4.96	5.01	2.78	5.01
3	WT	6.40	6.34	5.57	4.84	3.59	5.35
3	*Hvlpa*-M955	5.07	4.83	4.86	4.61	2.47	4.37
4	WT	7.03	6.32	5.52	5.10	3.42	5.48
4	*Hvlpa*2-1	7.51	4.57	6.27	5.03	2.78	5.23
5	WT	7.80	6.37	5.96	5.25	3.44	5.76
5	*Hvlpa*4-1	6.95	5.84	5.27	5.88	2.99	5.51
6	WT	7.29	6.01	5.40	5.48	3.20	5.48
6	*Hvlpa*-M640	6.98	5.03	5.93	4.95	2.47	5.21
		Cultivar Controls					
Harrington		6.96	6.27	5.69	5.30	3.17	5.44
Baronesse		8.61	7.13	6.86	6.08	4.05	6.55
Conrad		8.00	6.71	6.64	6.74	3.03	6.22
Location Mean ± SE		7.08 ± 0.37	5.96 ± 0.32	5.69 ± 0.29	5.34 ± 0.36	3.16 ± 0.21	5.47 ± 0.28
LSD $p = 0.05$		1.05	0.93	0.83	1.02	0.59	0.79

A simple, statistically-valid approach to evaluating the impact of these barley *lpa* mutations on yield, and one that also reflects a good genetics design, is to conduct pair-wise "*T*-tests", using the "LSD $p = 0.05$" value, to compare yields of sibling backcross lines that are either homozygous wild-type or homozygous mutant. Using this approach, if the difference in yield between wild-type and mutant sibling isolines is greater than the "LSD $p = 0.05$" value, one can conclude that the yield difference is statistically significant at the $p = 0.05$ level and is attributable to the allelic variant at the given *lpa* locus.

While variation due to isoline was of a lesser magnitude as compared with the main effects of Year and Location, it still was highly statistically significant. It is first important to note that no statistically-significant differences in yield were observed between the six wild-type isolines obtained after backcrossing and the original cultivar Harrington included as a control (Table 3 and Table 4). Yet differences are clear between the six wild-type isolines and Harrington on the one hand, and the cultivars Baronesse and Conrad on the other. This supports the experimental model that any observed differences in yield between a given wild-type and sibling *lpa* line are most likely attributable to the *lpa* allelic difference and not due to background genomic effects [5].

Of importance to this study are the two interactions of Year × Line and Location × Line. The Year × Line interaction was not statistically significant, indicating that isoline performances ranked similarly in each year. In contrast, the statistically significant Location × Line interaction indicates that isoline performance was differentially impacted by environment. This can be readily observed in the yield data when viewed over both years (Table 3 and Table 4): while the ranking of lines by yield might not have greatly differed between years, the magnitude of these differences in line yield varied. As a result, the large Year main effect adds power to the evaluation of the magnitude of differences in line yields between locations.

Inspection of the yield data meaned over locations within each year indicates that one line, *Hvlpa*1-1, displayed grain yield equal to or greater than that of its sibling wild-type isoline in both years (Table 3 and Table 4). *Hvlpa*1-1 yielded as much or more than its sibling wild-type line in seven out of ten location/years, only yielding less in statistically-significant terms in one location/year (Rupert in 2010). In fact the results obtained for the Rupert 2010 location/year are anomalous. Despite being a presumed "non-stressful" location, all *lpa* isolines except for *lpa*4-1 yielded less than their wild-type siblings in that location/year, including the *lpa*1-1 isoline. We have no explanation for this anomalous result other than that it might be due to an unknown or unrecognized stress (or stresses) impacting the *lpa* lines more than the wild-type lines, or that it might indicate an example of unpredictability for *lpa* line yield, itself potentially important.

Despite this caveat, *Hvlpa*1-1 was similarly productive, as compared with its sibling wild-type isoline, in nearly all locations/years, including both stressful (Soda Springs in both years and Ashton in 2010) and non-stressful (Aberdeen, Idaho Falls, Rupert/2011) environments. A second mutation, *Hvlpa*4-1, yielded within 4% of its sibling wild-type isoline across all locations and years. Further, when yield comparisons are made between *lpa* and sibling wild-type lines when grown in the three least stressful, most productive environments (Aberdeen, Idaho Falls and Rupert), for three of the six *lpa*lines (*Hvlpa*1-1, *Hvlpa*3-1 and *Hvlpa*4-1), yields were consistently within 5% of the given sibling wild-type line.

Where the impact of the *lpa* trait on grain yield was most pronounced was in the most stressful, least productive environments, which in this experiment was Soda Springs, a "Dry-Land" production site where no irrigation was provided. Statistically-significant impacts on yield were observed for eight of the ten possible isoline/year pair-wise comparisons between mutant and wild-type at Soda Springs. The exceptions were *Hvlpa*1-1 in both years, and for *Hvlpa*3-1 in 2010 and*Hvlpa*4-1 in 2011.

In broad terms there does appear to be a correlation between the magnitude of grain phytic acid reduction conditioned by a given mutation and impact on yield. The consistently poorest performer was *lpa*-M955, which as a homozygote conditions a reduction in grain phytic acid that is >80%, and results in a ~20% reduction in grain yield, as compared with its sibling wild-type line. In the most stressful, poorest yield-potential location, Soda Springs, homozygosity for *lpa*-M955 results in a 30% to 50% reduction in yield. However, the correlation between the extent of grain phytic acid reduction conditioned by a mutant and its yield breaks down when one considers mutations, which condition more moderate reductions in grain phytic acid. The three best performers, *Hvlpa*1-1, *Hvlpa*3-1 and *Hvlpa*4-1 condition reductions in grain phytic acid ranging from ~40% to 60% (depending on environment), as compared with sibling wild-type lines, whereas the two mutations with intermediate performance, *Hvlpa*2-1 and *Hvlpa*-M640, condition reductions in the range of ~30%. Also, the best performer,*Hvlpa*1-1, is known to impact grain phytic acid accumulation in a tissue-specific manner, perturbing accumulation only in the endosperm. Therefore while the magnitude of grain phytic acid reduction conditioned by a mutation may be an important determinant of yield impact, gene function and tissue-specificity also play an important role.

3. DISCUSSION

A previous study evaluated field performance of less advanced versions of four of these pairs of barley *lpa* near-isogenic lines, and was conducted at four sites in Idaho (USA) that include three used in the present study

[5]. *Hvlpa*1-1, *Hvlpa*2-1 or*Hvlpa*3-1, conditioning seed phytic acid reductions of 30% to 60%, had yields similar to their wild-type sibling lines when grown in two relatively non-stressful production environments that utilized irrigation. *Hv*-M955, with greater than 90% reduction in seed phytic acid, yielded 13% less than its wild-type isoline in these environments. In the two non-irrigated, more stressful production environments included in the initial study, only *Hvlpa*1-1 displayed yields statistically similar to wild-type, but its yield was nevertheless 11% less than that recorded for its wild-type sibling. Furthermore, *Hvlpa*2-1 and*Hvlpa*3-1 each had 25% reduction in seed yield and *Hv*-M955 had nearly 35% reduction in seed yield. Thus in this initial study three out of four barley *lpa* isolines performed well in non-stressful, "high-yield potential" environments, but yield in more stressful environments was clearly reduced, and a trend for a yield reduction was observed even for the best performing *Hvlpa* line. This initial study was conducted over two years at two sites but only in one year at two additional sites.

The good field performance of *Hvlpa*1-1 lines observed in initial studies led to the breeding and registration of the first *lpa*cultivar of any crop species [6]. *Hvlpa*1-1 probably encodes a putative member of the sulfate transporter gene family [7]. Perturbation of this gene in barley also results in a second desirable crop seed phenotype, reduced seed total P [8]. Therefore this gene might prove to be an excellent target for engineering the ideal seed P phenotype, low-phytic acid and low total P, in other crops species.

In the present study performance comparisons were made using more advanced versions of the initial four sets of near-isogenic lines, obtained following one additional backcross, and an additional two sets of near-isogenic lines developed using *Hvlpa*-M640 and *Hvlpa*4-1. Furthermore, all six sets of isolines were evaluated across two years in five Idaho (USA) locations that greatly differ in crop yield potential. Confirming the results of the initial study [5], *Hvlpa*1-1 performed well in all locations, and had no detectable negative impact on yield.

A second mutation, *Hvlpa*4-1, resulted in yield losses of only 3.5%. Furthermore, when comparing yields in three relatively non-stressful production environments, at least three *lpa* mutations (*Hvlpa*1-1, *Hvlpa*3-1, and *Hvlpa*4-1) typically had yields similar to or within 5% of the wild-type sibling isoline. Thus while some *lpa* mutations do greatly or consistently impact field performance, a substantial fraction of barley *lpa* mutations appear to have little or no effect on yield. Therefore in the case of barley, *lpa* mutations can be readily identified that when simply incorporated into a cultivar result in adequately performing lines, at least in many production environments, even with no additional breeding for performance within the *lpa* line. Nevertheless, *lpa* line yield in more stressful environments remains problematic, as does consistency of performance for any given line in any environment. Therefore additional breeding via recurrent selection for field performance would still be needed to produce lines that are consistently high-yielding across diverse production environments (see below).

*Hvlpa*1-1 differs from the other three genotypes evaluated in the initial study [5] in that its recessive alleles result in a localized, tissue-specific block in phytic acid accumulation. Here we report that *Hvlpa*M640 is phenotypically similar to barley *lpa*1-1 in that it also appears to be endosperm-specific. However, the *Hvlpa*-M640 isoline did not perform as consistently well as did the *Hvlpa*1-1 isoline. This may indicate that it is not solely the endosperm-specificity of *Hvlpa*1-1 that results in its consistent performance, but rather that there are gene/function-specific differences impacting yield. Nevertheless, in contrast to the results of the present study yield trials indicate that a low-phytate barley cultivar produced using *Hv*-M640 has yields comparable with conventional cultivars in a wide variety of production environments [9]. Thus conventional genetics and breeding approaches can achieve at least one of the advantages provided by genetic engineering; genotypes in which blocks in the phytic acid pathways are restricted to specific tissues of the seed, such as the embryo or aleurone layer, thus minimizing the negative impact on whole plant performance and yield.

The only statistically valid experimental approach to attributing a quantitative difference in field performance, or in any trait, to a specific allelic variant of an *lpa* gene, is via comparisons using wild-type and mutant sibling near-isogenic lines. This is true regardless of the method used in generating the novel allelic variants that confer the *lpa* trait; whether the *lpa*trait is conferred by recessive alleles obtained following chemical or radiation mutagenesis, or obtained using any method that involves transformation or "genome modification", methods that require tissue culture, regeneration or the use of selectable markers such as those conferring herbicide tolerance. A good example of this is "genome modification" [4], a powerful and highly valuable method that targets specific endogenous gene sequences and thus in theory avoids non-linked genomic change such as that resulting from chemical mutagenesis, or from non-homologous insertion/mutation resulting from earlier genetic modification methods. Genome modification has been used to target one of the maize (*Zea mays* L.) genome's *IPK1* genes, which encodes an inositol-1,3,4,5,6-pentakisphosphate 2-kinase [4]. However, since tissue culture and regeneration can greatly impact genome expression and epigenetics in offspring [10], as can expression of reporter genes that confer herbicide resistance [11], assessing field performance using *lpa* progeny obtained via this method still first requires sufficient backcrossing, certainly a minimum of four to six backcrosses to selected recurrent parents, followed by isolation and evaluation of near-isogenic lines.

Even the use of near-isogenic lines has a potentially serious limitation. If there was no selection for performance during or after the incorporation of an allele from a divergent donor line into an elite line or cultivar, the resulting *lpa* or wild-type sibling lines might differ in yield due to inheritance of undesirable allelic variants at loci not subject to selection, or via "linkage-drag" of undesirable loci linked to the *lpa* locus in the donor parent [12]. This is particularly important because elite field performance reflects inheritance of favorable alleles, both genetic and epigenetic, at multiple loci throughout the genome. Thus simply incorporating an *lpa* allele via backcrossing, regardless of its source or method of induction, without some form of accompanying selection for performance, might result in lines with reduced performance that is not linked to the *lpa* trait or the *lpa* allele. Clearly, recurrent selection for field performance, within an *lpa* genotype over several generations, is required to obtain a line with favorable alleles genome-wide, both genetic and epigenetic, that retains the desired *lpa* phenotype and that also display elite performance. Despite this caveat, the use of near-isogenic lines, even without additional breeding, remains the most statistically appropriate approach to attributing a difference in yield or field performance to inheritance of an allelic difference at an *lpa* locus.

Few studies of field performance of *lpa* types have in fact made comparisons between truly near-isogenic sibling lines that, that to the best approximation available with current science, essentially differ only in the allelic variants of a single *lpa*locus. The first study reporting evaluation of the impact on crop performance or yield of an *lpa* type was Ertl *et al.* [12]. In this study, the maize *lpa*1-1 allele (*Zmlpa*1-1), originally isolated in a "Flint corn-like" genetic background referred to as "Early ACR" [13], was backrossed to the BC_3 (four crosses to the recurrent parent inbred line), into a sufficient number of diverse elite maize "Dent corn" inbred lines to produce a total of 14 pairs of near-isognenic hybrids, each pair consisting of a wild-type and *Zmlpa*1-1 version. The performance evaluation of these pairs of maize hybrids consisted of studies conducted in one year, at three locations with two replications, in central Iowa, USA. In six of the 14 near-isogenic pairs no statistically-significant effect of the *Zmlpa*1-1 allele on yield was observed. In eight of the 14 near-isogenic pairs a yield loss due to the*Zmlpa*1-1 was observed, and overall the yield loss was 6%, as compared with wild-type. Thus this first report of performance evaluation of low-phytate versions of a crop reported results similar to those reported here. However, while ground-breaking, this first study had inherent limitations. The production environments and practices were similar with no clearly stressful or low-productivity environments, and replication was minimal. More importantly the Flint and Dent corn genetic backgrounds used greatly differ, and only four crosses to the recurrent parent (the Dent corn inbred lines) were completed. Therefore the resulting pairs of lines can only be considered "early generation" near-isogenic lines, with residual genomic contribution of the very divergent donor parent of approximately 6.25%. In the present study the *lpa* mutations were isolated in the same genetic background that they were subsequently backcrossed to, cv. Harrington. Thus in this case even "early-generation" backcross lines, such as the *Hvlpa*4-1 and *Hvlpa*-M640 lines, are more genetically similar to their wild-type sibling lines than those evaluated in the Ertl *et al.* study [12].

A third series of studies that evaluated field performance of *lpa* types using near-isogenic lines addressed issues of germination and field emergence in *lpa* soybean [*Glycine max* (L.) Merr.] types [14,15,16]. Germination and emergence is reduced in many soybean *lpa* genotypes, and the effect is highly pronounced when seed is produced in tropical *versus*temperate environments, a phenomenon referred to as the "seed source effect" [14]. When low-phytate soybean lines were produced via incorporation of recessive alleles of the soybean MIPS-encoding gene (the LR33 mutations) [17], or the *pha1*and *pha2* alleles first isolated as the M153 mutant [18], and seed was produced in a tropical environment, subsequent field emergence was greatly reduced when compared to that of seeds of the same genotype when tested seed was produced in a more moderate season in Puerto Rico, or in a temperate region [14,16]. While the extreme effects on germination and emergence that maturation in a tropical environment produces has not been overcome via subsequent breeding, evaluation of BC_4-derived near-isogenic lines indicated that selection could identify lines with improved performance [16].

The studies with soybean and barley *lpa* genotypes illustrate the potential of classical genetics and breeding approaches for overcoming agronomic problems associated with the *lpa* trait. Selection within lines segregating for a given *lpa* allele can identify lines with enhanced agronomic performance [15]. Identification of genes that, if perturbed, have little impact on agronomic performance, such as barley *lpa*1 and M640 [6,9], or soybean *lpa*-ZC-2 [19], can lead to low-phytate cultivars with superior agronomic performance.

4. EXPERIMENTAL SECTION

4.1. Isoline Pair Development and Cultivar Controls

Each of six *lpa* mutations were crossed and subsequently backcrossed to varying degrees to the cultivar Harrington in which they were originally isolated. BC_xF_2 progenies were obtained following self-pollination of a given BC_xF_1, and sibling homozygous wild-type and homozygous mutant plants were identified. These were then used for increases to provide seed for the present studies. Seed used in the 2010 yield trial were: BC_5F_4 (*Hvlpa*1-1 and its wild-type sibling line); BC_4F_4 (*Hvlpa*2-1, *Hvlpa*3-1, and *Hvlpa*-M955 and their wild-type sibling lines); F_4 (*Hvlpa*4-1 and *Hvlpa*-M640). Seed used in the 2011 yield trial were advanced further generations as follows: BC_5F_5 (*Hvlpa*1-1 and its wild-type sibling line); BC4F5 (Hvlpa2-1, Hvlpa3-1, and Hvlpa-M955 and their wild-type sibling lines); BC1F3 (Hvlpa4-1 and *Hvlpa*-M640). Since each mutation was first isolated in the genetic background represented by the cultivar Harrington, then crossed or backcrossed to the same cultivar, following which sibling plants either homozygous wild-type or homozygous mutant were increased, each pair of sibling lines can be considered near-isogenic. For confirming the *lpa* genotype and phenotype of a given plant or line a microtitre plate assay for seed inorganic P was used (Supplemental Figure 1): homozygous wild-type lines typically have relatively low levels of inorganic P (<0.5 mg·g^{-1}), whereas homozygous *lpa* lines typically have greatly elevated seed inorganic P (>1.0 mg·g^{-1} in most cases). This "high seed inorganic P" phenotype provides a simple assay visual inspection of which is sufficient to genotype lines and test for homogeneity (Supplemental Figure 1).

In addition to the six pairs of near-isogenic lines, three barley cultivars were included in the field trials as controls. These were: (1) Harrington, a "2-rowed malting barley" that represents the genetic background in which these mutations were isolated; (2) Baronesse, a "two-rowed feed barley" that is observed to be one of the highest yielding barleys when grown in the Pacific Northwest region; and (3) Conrad, a "2-rowed malting barley" that has displayed higher yields than Harrington when produced in the Pacific Northwest region.

4.2. Test Sites

The six pairs of isolines, and three cultivar controls (see below) were grown in field trials in two consecutive years (2010 and 2011) at five locations in Idaho that represent diverse environments (soil types, elevation, growing season, temperature and water availability) and diverse production practices (irrigation *versus* non-irritation, planting density). These location differences result in clearly diverse yield capacities. These test sites were: (1) Aberdeen Research and Extension Center, Coordinates 42°57'47.23" N, 112°49'9.31" W, elevation 1341 M, soil type Declo fine sandy loam with sprinkler irrigation; (2) Ashton, Coordinates 44°04'53.96" N, 111°19'8.16" W, elevation 1603 M, soil type #92 rin silt loam, sprinkler irrigation; (3) Idaho Falls, Coordinates 43°28'33.15" N, 112°07'14.93" W, elevation 1430 M, soil type #22 Pancheri silt loam, sprinkler irrigated; (4) Rupert, Coordinates 42°43'14.95" N, 113°30'36.13" W, elevation 1297 M, soil type #36 Sluka silt loam, sprinkler irrigation; and (5) Soda Springs, Coordinates 42°42'13.10" N, 111°36'30.80" W, elevation 1826 M, non-irrigated "Dry-land".

Prevailing agronomic practices were used at all sites. Seed for planting was not pre-treated with fungicide. Planting and harvest dates varied between locations reflecting prevailing local conditions in each year. Planting density was 90 kilos of seed·ha^{-1} at irrigated sites, and 68 kilos of seed·ha^{-1} at non-irrigated sites. Row spacing was 18 cm for all sites. Harvested plots at all locations except for Aberdeen were 1.5 M by 3.0 M. At Aberdeen plots were 1.5 M by 2.8 M. All entries were replicated 4 times at each location in a randomized complete block (RCB) design. Nitrogen fertilizer in irrigated locations was managed according to the following methodology: yield goals were set for each class at each location using historical yield data. These yield goals were used to calculate optimal fertility amounts according to the following methods: nitrogen needed to produce 1.7 times the yield goal. Details of fertilizer application and meteorological data (temperature and rainfall) are available upon request.

4.3. Data Collection

Agronomic data collected includes grain yield (Mg·ha^{-1}), test weight (g 30 L^{-1}), plant height, "heading date" as the date when 50% of heads are fully emerged, and lodging as a visually determined percentage of each plot that is not standing straight at harvest. Additional agronomic data not reported here are available upon request.

Seed P fractions were assayed as follows. Samples of mature seeds were ground with a coffee grinder. Total P was determined following wet-ashing of aliquots of flour (150 mg) and colorimetric assay of digest P [20]. For determination of seed phytic acid P and inorganic P, aliquots of flour (50 mg) were extracted in 1.0 mL of extraction media (0.2 M HCL:10% Na_2SO_4) in 1.5 mL Eppendorf tubes overnight at 4 °C with shaking. Extracts were centrifuged (4500 RCF, 20 min). Supernatant phytic acid P was determined using a modification of the method as described [21] and supernatant inorganic P was determined colorimetrically [20].

4.4. Statistical Analyses

Statistical analyses were conducted using the General Linear Models procedures developed by the SAS Institute (Raleigh, NC, USA).

5. CONCLUSIONS

The results presented here demonstrate that in the case of barley, *lpa* mutations can readily be identified that even with no additional breeding have little or no effect on yield. This conclusion is supported by a very stringent test; a multi-year, multi-location yield trial that compared yields between six sets of sibling near-isogenic lines. This represents the most thorough field trial of *lpa* crop variants published to date. Thus the perception that the *lpa* trait is obligately correlated with significantly reduced yield should be moderated; many *lpa* mutations do greatly impact plant and seed performance and thus have a significant impact on yield, but some have little or no effect on yield and can be used to breed crops with substantially enhanced nutritional quality. However it is also possible that the relative lack of impact on yield observed here simply reflects a difference between barley and other cereal and legume crop species: that the relatively small impact on yield of barley *lpa* mutations is somehow a unique characteristic of barley.

ACKNOWLEDGMENTS

The authors would like to thank Amanda Harvey for her excellent laboratory and field assistance needed to complete this work.

AUTHOR CONTRIBUTIONS

V.R. designed and supervised the research. K.P. and H.S. conducted the initial set of seed phosphorus fraction analyses. G.H. conducted all chromosomal mapping work. C.J. supervised and conducted the field trials. J.M. and P.B. also provided supervision for line development and field evaluation.

CONFLICTS OF INTEREST

The authors declare no conflict of interest.

REFERENCES

1. Cichy, K.; Raboy, V. Evaluation and development of low-phytate crops. In Modification of Seed Composition to Promote Health and Nutrition; Krishnan, H., Ed.; American Society of Agronomy: Madison, WI, USA, 2009; pp. 177–200.

2. Raboy, V. Approaches and challenges to engineering seed phytate and total phosphorus. Plant Sci. 2009, 177, 281–296.

3. Shi, J.; Wang, H.; Schellin, K.; Li, B.; Faller, M.; Stoop, J.M.; Meeley, R.B.; Ertl, D.S.; Ranch, J.P.; Glassman, K. Embryo-specific silencing of a transporter reduces phytic acid content of maize and soybean seeds. Nat. Biotechnol. 2007, 25, 930–937.

4. Shukla, V.K.; Donyon, Y.; Miller, J.C.; DeKelver, R.C.; Moehle, E.A.; Worden, S.E.; Mitchell, J.C.; Arnold, N.L.; Gopolan, S.; Meng, X.; et al. Precise genome modification in the crop species Zea mays using zinc-finger nucleases.Nature 2009, 459, 437–443.

5. Bregitzer, P.; Raboy, V. Effects of four independent Low-Phytate mutations on barley agronomic performance. Crop Sci. 2006, 46, 1318–1322.

6. Bregitzer, P.; Raboy, V.; Obert, D.E.; Windes, J.M.; Whitmore, J.C. Registration of “Herald” Barley. Crop Sci. 2007, 47, 441–442.

7. Ye, H.; Zhang, X.-Q.; Broughton, S.; Westcott, S.; Wu, D.; Lance, R.; Li, C. A nonsense mutation in a putative sulphate transporter gene results in low phytic acid in barley. Funct. Integr. Genomics 2011, 11, 103–110.

8. Ockenden, I.; Dorsch, J.A.; Reid, M.M.; Lin, L.; Grant, L.K.; Raboy, V.; Lott, J.N.A. Characterization of the storage of phosphorus, inositol phosphate and cations in grain tissues of four barley (Hordeum vulgare L.) low phytic acidgenotypes. Plant Sci. 2004, 167, 1131–1142.

9. Bregitzer, P.; Raboy, V.; Obert, D.E. Registration of “Clearwater” low-phytate hulless spring barley. J. Plant Regist.2008, 2, 1–4.

10. Stelpflug, S.C.; Eichten, S.R.; Hermanson, P.J.; Springer, N.M.; Kaeppler, S.M. Consistent and heritable alterations of DNA methylation are induced by tissue culture in maize. Genetics 2014, 198, 209–218.

11. Brown, R.H.; Raboy, V.; Bregitzer, P. Unintended consequences: High phosphinothricin acetyltransferase activity related to reduced fitness in barley. In Vitro Cell. Dev. Biol. 2013, 49, 240–247.

12. Ertl, D.; Young, K.A.; Raboy, V. Plant genetic approaches to phosphorus management in agricultural production. J. Environ. Qual. 1998, 27, 299–304.

13. Raboy, V.; Gerbasi, P.F.; Young, K.A.; Stoneberg, S.D.; Pickett, S.G.; Bauman, A.T.; Murthy, P.P.N.; Sheridan, W.F.; Ertl, D.S. Origin and seed

phenotype of maize low phytic acid 1-1 and low phytic acid 2-1. Plant Physiol. 2000, 124, 355–368.

14. Meis, S.J.; Fehr, W.R.; Schnebly, S.R. Seed source effect on field emergence of soybean lines with reduced phytate and raffinose saccharides. Crop Sci. 2003, 43, 1336–1339.

15. Oltmans, S.E.; Fehr, W.R.; Welke, G.A.; Raboy, V.; Peterson, K.L. Agronomic and seed traits of soybean lines with low-phytate phosphorus. Crop Sci. 2005, 45, 593–598.

16. Anderson, B.P.; Fehr, W.R. Seed source affects field emergence of low-phytate soybean lines. Crop Sci. 2008, 48, 929–932.

17. Hitz, W.; Carlson, T.J.; Kerr, P.S.; Sebastian, S.A. Biochemical and molecular characterization of a mutation that confers a decreased raffinosaccharide and phytic acid phenotype on soybean seeds. Plant Physiol. 2002, 128, 650–660.

18. Wilcox, J.R.; Premachandra, G.S.; Young, K.A.; Raboy, V. Isolation of high seed inorganic P, low-phytate soybean mutants. Crop Sci. 2000, 40, 1601–1605.

19. Yuan, F.J.; Zhao, H.J.; Ren, X.L.; Zhu, S.L.; Fu, X.J.; Shu, Q.Y. Generation and characterization of two novel low phytate mutations in soybean (Glycine max L. Merr.). Theor. Appl. Genet. 2007, 115, 945–957.

20. Dorsch, A.J.; Cook, A.; Young, A.K.; Anderson, N.J.; Bauman, T.A.; Volkmann, J.C.; Murthy, P.N.P.; Raboy, V. Seed phosphorus and inositol phosphate phenotype of barley low phytic acid genotypes. Phytochemistry 2003, 62, 691–706.

21. Haug, W.; Lantzsch, H.-J. Sensitive method for the rapid determination of phytate in cereals and cereal products. J. Sci. Food Agric. 1983, 34, 1423–1426.

Chapter 3

Understanding Crop Genetic Diversity Under Modern Plant Breeding

Yong-Bi Fu*

Plant Gene Resources of Canada, AAFC Saskatoon Research Centre, 107 Science Place, Saskatoon, SK S7N0X2 Canada

INTRODUCTION

Abstract

Electronic supplementary material

Plant breeding since the early 1900s has made a profound impact on food production and will continue to play a vital role in the world food security (Borlaug 1983; Tester and Langridge 2010). However, it has also introduced crop uniformity across the farm fields, which is genetically vulnerable to biotic and abiotic stresses (Day 1973; Duvick 1984; Vellve 1993; Tripp 1996; Keneni et al. 2012). Such risks have been well documented with the occurrence of epidemics such as the Irish potato blight in the 1840s and the U.S.A. corn blight in the 1970s (National Academy of Sciences 1972; Ullstrup 1972). The threat of the extremely virulent new race of stem rust Ug99 from East Africa to genetically uniform wheat is currently evident (Borlaug 2007; Babiker et al. 2015). Thus, it is important, although challenging, to compromise between maximizing crop yield under a given set of conditions and minimizing the risk of crop failure when conditions change and to develop effective strategies for sustainable agriculture (Hallauer 1985; Shukla and Mattoo 2013). Such a compromise requires a better understanding of the impacts of modern plant breeding on crop genetic diversity (Duvick et al. 2004; Fu 2006).

Efforts have been made over the last three decades to assess crop genetic diversity using molecular marker technologies. These assessments have generated considerable knowledge about the extent and nature of genetic diversity present in conserved and/or actively utilized germplasm of various crops (Rauf et al. 2010). These assessments not only facilitate our efforts in germplasm conservation, but also provide guidance for better germplasm utilization for genetic improvement. However, some assessments have also revealed some temporal patterns of crop genetic diversity (e.g., see van de Wouw et al. 2010a, b) that are largely inconsistent with our perception that modern plant breeding reduces crop genetic diversity (Fu et al. 2003; Gepts 2006). For example, a metaanalysis of 44 published diversity assessments indicated that a gradual narrowing of the genetic base of the varieties released by breeders could not be observed (van de Wouw et al. 2010b). One would expect that the intensive selection in modern plant breeding programs within a narrow range of plant germplasm with limited allele introgressions over time (Hallauer 1985; Allard 1999) would have reduced genetic diversity. Also it is evident that newly released crop varieties are phenotypically more uniform than before, implying a genetic diversity reduction (e.g., Duvick 1984; Bowman et al. 2003). Such a discrepancy suggests that we may still be far away from understanding genetic diversity of crops developed under modern plant breeding.

In this review, we attempt to explain this discrepancy by examining empirical assessments of crop genetic diversity and theoretical investigations on genetic diversity changes under artificial selection. Specifically, we hope to address the following questions: (1) Why diversity assessments have often not revealed diversity reduction from modern plant breeding? (2) Does plant breeding truly reduce crop genetic diversity? (3) How much is known theoretically about genetic diversity changes under artificial selection? and (4) What research is needed to fill the knowledge gap in this area of study? We have organized the review to address these questions with our arguments and thoughts.

Empirical assessments of crop genetic diversity

Over the last three decades, concerns have been expressed about crop uniformity (Duvick 1984; Vellve1993; Swanson 1996; Tripp 1996) and there have been an increased number of crop genetic diversity assessments (Donini et al. 2000; Reeves et al. 2004; Fu 2006; Rauf et al. 2010). Earlier assessments based on phenotypic (Rodgers et al. 1983; Ortiz et al. 2003) and pedigree data (Cox et al. 1985; van Beuningen and Busch 1997) had demonstrated that the substantial progress achieved in improving yield and other traits resulted in a reduction in the genetic diversity of improved gene pools (Cox et al. 1986; Smith et al. 2004). Advances in molecular markers such as random amplified polymorphic DNAs (RAPDs), amplified fragment length polymorphisms (AFLPs), and simple sequence repeats (SSRs) have made crop diversity assessments more attainable and informative than before. These molecular assessments, although rarely using genome-wide SNP markers, have generated a lot of knowledge about the nature and extent of genetic diversity present in various crops. Specific reviews on these assessments have been made with respect to marker application (Mondini et al. 2009) and crop genetic diversity (Reeves et al. 2004; Fu 2006; van de Wouw et al. 2010a; Rauf et al. 2010). The highlights of their findings are summarized in the following.

Fu (2006) reviewed 23 articles with the applications of RAPDs, AFLPs, and SSRs that were published from 2000 to 2005 in eight major journals associated with plant breeding. These articles revealed different impacts of plant breeding on improved gene pools, not only narrowing or widening their genetic base, but also resulting in genetic shifts. Overall, the genome-wide reduction of crop genetic diversity accompanying genetic improvement over time was minor, but allelic reduction at individual chromosomal segments was substantial. This review was not exhaustive, but focused more on the understanding of the impact of plant breeding on the genome.

van de Wouw et al. (2010a) reviewed about 110 publications associated with crop genetic diversity and agricultural modernization and concluded that different views exist on the concept of crop genetic erosion. Genetic erosion of cultivated diversity occurs in two stages: the initial replacement of landraces by modern cultivars; and further trends in diversity as a consequence of modern breeding practices. Genetic erosion may also occur at three levels of integration: crop, variety, and allele. They further argued that there is a reduction in diversity due to the replacement of landraces by modern cultivars, but no further reduction after this replacement has been completed. To support their argument, they performed a metaanalysis of 44 published papers and showed that a gradual narrowing of the genetic base of the varieties released by breeders was not observed (van de Wouw et al. 2010b). Specifically, a significant diversity reduction of 6 % occurred before the 1960s, and after the 1960s, increased diversity was found from plant breeding.

Rauf et al. (2010) reviewed about 230 publications associated with plant breeding and genetic diversity to understand the diversity impacts of different plant methods such as introduction, selection, and hybridization. In general, this review showed that losses of genetic diversity occurred but followed spatial and sometimes temporal trends. More losses of genetic diversity were found in released cultivars, followed by wild germplasm and landraces. Different plant breeding methods showed different impacts on plant genetic diversity. Plant introduction increased genetic diversity. Selection enhanced genetic differentiation at the expense of genetic diversity. Intraspecific hybridization lowered genetic diversity.

It is clear that no consensus has been reached on the overall impact of modern plant breeding on crop genetic diversity. The temporal patterns of crop genetic diversity are largely inconsistent with our perception that modern plant breeding reduces crop genetic diversity (Gepts 2006) and are also incompatible with the fact that newly released crop varieties become phenotypically more uniform (Duvick 1984; Bowman et al. 2003). Such a discrepancy suggests that we may still be far away from understanding crop genetic diversity under modern plant breeding. An attempt was made here to explain this discrepancy as outlined in Fig. 1 and discussed below.

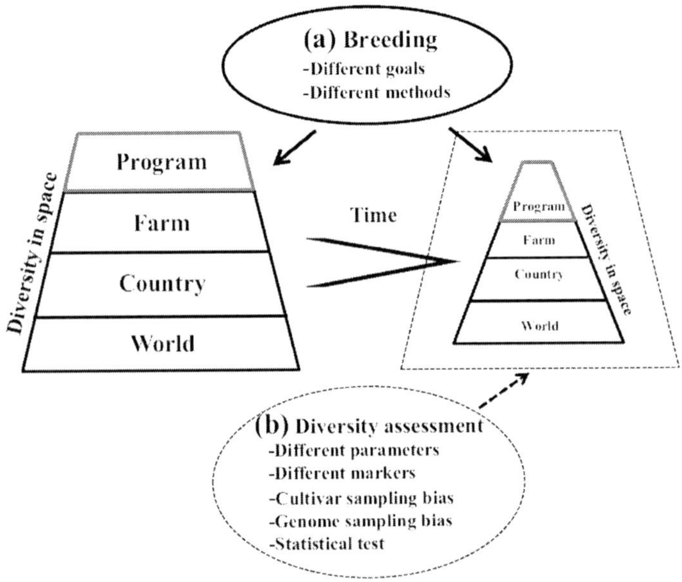

Figure 1. Illustration of the spatial and temporal changes (*solid line* and some highlight in *red*) in crop genetic diversity generated by modern plant breeding with variable goals and methods (**a**) and how they are obscured (*broken line*) under diversity assessments **...**

Nature of modern plant breeding

Plant breeding since 1900s has had a profound impact on food production through developing and deploying new cultivars on a worldwide basis (Borlaug 1983). New cultivars have been developed through applications of many effective breeding methods, ranging from introduction, phenotypic selection on natural variants, selection with controlled mating, to marker-assisted selection for desirable genes (Allard 1999). The core of all plant breeding can be characterized as follows: (1) the conscious introduction of genetic diversity into breeding populations by intercrossing selected plants with outstanding characters that complement one another and; (2) the selection of superior plants with genes for desired traits until higher levels of improved adaptation, genetic uniformity, and agronomic stability are reached

(Breseghello 2013). The choice for use of a breeding methodology is determined mainly by the mode of crop reproduction (selfing or crossing) and the breeding objectives to be achieved.

Crop breeding has been largely aimed at the improvement of yield, adaptation, resistance to biotic and abiotic stresses, and end-use quality. However, breeding objectives have changed over the years beyond yield improvement. New cultivars need to be developed with the capacity to achieve high yields in reduced chemical-input systems and with the genetic diversity needed to maintain yield stability under fluctuating climatic conditions (Heinemann et al. 2014). Many novel traits improved for sustainable agriculture include improved weed suppression ability, enhancement of nutritional value, and optimization of plant interactions with microbial communities in the soil, among others (e.g., see Brummer et al. 2011). To meet these challenges, conventional plant breeding has evolved by adopting approaches from different scientific disciplines, allowing breeders to increase their efficiency and exploit genetic resources more thoroughly. Among these new approaches are haploid generation (Kasha and Kao 1970); the use of sterility systems and transgenic technology (Salick 1995); apomixis (Spillane et al. 2004); and molecular marker-assisted breeding (Moose and Mumm 2008).

Modern plant breeding has been evolving from conventional breeding to molecular breeding for various breeding goals and diverse breeding methods have been applied over time (Gepts and Hancock 2006). As a consequence, selective pressure within breeding populations differs at various breeding stages for different breeding programs, so genetic diversity present in released cultivars of a crop may vary (Rauf et al. 2010). More heterogeneity is expected in varietal genetic diversity between selfing and outcrossing crops.

Variation in genetic diversity measures

To our knowledge, genetic diversity is a term not well defined. It is broadly referred to as any variation in the nucleotides, genes, chromosomes, or genomes of a species at a level of individual, population, species, or region for a given time. Such a broad definition certainly invites misunderstanding and misinterpretation. Accordingly, its measurements are not unique (Mondini et al. 2009). Genetic diversity within a population is commonly measured by (1) allelic polymorphism; (2) heterozygosity; and (3) theta parameter ($\vartheta = 4Ne\mu$ for diploid genes, where *Ne* is effective population size and μ is per generation mutation rate). Genetic variation among populations, reflected in the differences in allele and genotype frequencies, is frequently measured using several different metrics. They are (1) *F*st and analogs; (2) genetic distance such as Nei's D; and (3) sequence divergence (e.g., see Hedrick 2011). There has been considerable discussion on the proper uses of these diversity measures (e.g., see Jost 2008; Whitlock 2011).

Crop genetic diversity has traditionally been analyzed using morphological traits, particularly those agro-morphological traits of interest to users. To minimize the impact of environmental factors in the analysis, biochemical techniques such as isozyme and protein electrophoresis (Hunter and Merkert 1957) were later employed. Since 1990, various molecular techniques such as RAPD, AFLP, and SSR, have been used to measure genetic variation (Mondini et al. 2009). These molecular markers not only avoid the influence of environment, but also provide better sampling of the plant genome, thus increasing the resolution of measurements of genetic variation. Currently, there are more than 30 types of molecular markers available for assessing genetic diversity (Mondini et al. 2009). These markers have been widely applied to measure genetic diversity in crop plants and have played an important role in the characterization of crop genetic variation. However, genome-wide SNP markers with better sampling of plant genomes have not fully been applied to assess crop genetic diversity (Hyten et al. 2006).

Based on the broad definition of genetic diversity, the use of different diversity parameters, and the application of different molecular markers, it is not surprising that there is considerable heterogeneity in reported diversity assessments (Fu 2006; Aremu 2011). It is difficult to interpret and generalize the findings from estimation of different diversity parameters using different markers, even on a crop species (Rauf et al.2010). Specifically, not all of the genetic diversity measures applied have been equally sensitive in detecting diversity changes from plant breeding practices, and different diversity measures may have different levels of accuracy and precision (Mohammadi and Prasanna 2003; Fu et al. 2005). Not all of the molecular markers applied have been equally informative for diversity assessments, as illustrated in oat using AFLP and SSR markers (Fu et al. 2003, 2004). Thus, discrepancies can be expected, even for the same assessment using different diversity parameters.

Technical considerations in diversity assessment

An analysis of the published diversity assessments since 2000 shows that a majority of published assessments were not aimed specifically at the assessment of the diversity impacts of individual breeding programs or schemes on released cultivars. For example, many assessments were made to evaluate spatial and temporal patterns of genetic diversity present among cultivars released from different breeding programs, regions, or countries (e.g., see Roussel et al. 2005; Orabi et al. 2014). These assessments can inform us in general about the nature and extent of existing crop genetic diversity for exploration of conservation strategies and for germplasm utilization in plant breeding, but cannot provide us with much evidence that modern plant breeding reduces crop genetic diversity. The informative assessments would be those that evaluate the genetic diversity changes in all the released cultivars over all the breeding periods from individual breeding programs (e.g., see Fu et al. 2003). Alternatively, the diversity comparison between on-farm landraces present before a breeding program and on-farm cultivars released over time from the breeding program would also generate findings useful for evidence of genetic diversity reduction (e.g., see Russell et al. 2000). However, there are not many such assessments available to make firm inferences due to many practical and/or unknown reasons as shown in Fig. 1 and discussed below (also see Fu 2006; Rauf et al. 2010).

Many reported temporal diversity analyses are technically not ideal to address the diversity impacts of plant breeding on released cultivars. Technical biases could further mask the reported patterns of crop genetic diversity. Several issues have been identified (Fu 2006; Aremu 2011) and they include bias of sampling cultivars from a specific breeding program, arbitrary grouping of cultivars to represent specific breeding periods, mingling of gene pools from different breeding programs, applications of different markers that are variably informative, use of different genetic diversity measures, and inadequate statistical tests of significance. For example, bias may occur in sampling cultivars. Some older, important cultivars may have been lost; some newly developed cultivars may not be accessible; and selection may favor dominant, but genetically related, cultivars (Koebner et al. 2003; Le Clerc et al. 2005). Another example is the arbitrary grouping of assayed cultivars to represent specific breeding periods or inadequate separation of the change from landraces to cultivars from modern plant breeding (Reif et al. 2005).

The lack of statistical tests for significance in many reported assessments, even with useful test methods available (e.g., see Fu et al. 2003; Fu 2010), is another notable issue, adding little confidence to the interpretation of the diversity changes, and complicating generalization of published findings. Allelic counts for various groups are sensitive to the cultivar number for each group (Fu 2010), but some SSR studies did not correct the bias of unbalanced group sizes (e.g., see Russell et al. 2000; Duvick et al. 2004), thus weakening the argument for the allelic reductions found (Lu and Bernardo 2001; Fu et al. 2003).

Informative assessments of diversity changes

Fortunately, informative assessments of diversity changes are available. Here we highlight a few cases to illustrate the importance of assessing specific, large-scale, long-term breeding programs for understanding diversity reduction from plant breeding.

A notable and commendable effort is the assessment of the diversity changes under recurrent selection (RS) schemes over many generations (Brown and Allard 1971), as RS has been widely used for maize improvement, while simultaneously maintaining genetic variability for continued selection since 1939 (Hallauer and Miranda 1988). Several investigations using molecular markers have yielded a clear picture on the diversity impacts of RS practiced over many generations (e.g., see Labate et al. 1999; Pinto et al. 2003; Hinze et al. 2005; Solomon et al. 2010; Romay et al. 2012). For example, Labate et al. (1999) investigated temporal changes in RFLP alleles over 12 generations in two reciprocally selected maize populations and reported about 10 % of the original RFLP alleles fixed and about 40 % of the total heterozygosity lost between generations 0 and 12. Similarly, Solomon et al. (2010) assessed SSR variation over 11 generations of RS in tropical maize breeding populations and found 33 % alleles lost and an 18 % reduction of within-population variance from generation 0.

Duvick et al. (2004) made a considerable effort to understand the diversity changes in the long-term commercial plant breeding program of Pioneer Hi-Bred International by using 969 SSR alleles to quantify the diversity changes in the Era hybrids and open-pollinated cultivars that were sequentially released in the west-central U.S. Corn Belt from the 1930s to 2000s. They found that the number of alleles fluctuated from decade to decade; that about 40–50 % of the 969 alleles were present in any one decade; and that there is a weak trend toward lower numbers per decade, starting in the 1980s. Based on the SSR polymorphism data, they also revealed a trend toward reduction in the average number of alleles per locus and a clear divergence between the allele profiles of the inbreds created by pedigree breeding in the Stiff Stalk and Non-Stiff Stalk heterotic groups.

The Illinois long-term selection experiment for grain protein and oil concentrations in maize has undergone 114 generations of recurrent selection since 1896, making it the longest running continuous genetics experiment in higher plants (Moose et al. 2004). Mikkilineni and Rocheford (2004) conducted a RFLP analysis of 200 plants selected from generations 65 and 91 of the Illinois High and Low Protein (IHP and ILP, respectively) strains and from generations 69 and 91 of the Reverse High and Low Protein (RHP and RLP, respectively) strains. Based on 35 RFLP probes, they found the percentage of RFLP probes with a variant fixed ranged from 22.9 to 51.4 % over 26 generations of selection for IHP; 25.7–42 % for ILP, and from 14.3 to 17.1 % over 22 generations of selection for RHP, and 14.3–22.9 % for RLP. Measuring the loss of heterozygosity relative to the original heterozygosity in generations 65 or 69 revealed that the IHP strain at generation 91 had lost 36 % of the original heterozygosity, followed by ILP 23 %, RLP 10 %, and RHP 3 %.

We conducted a series of genetic diversity analyses from 1999 to 2009 focusing on the Canadian gene pools of flax, oat, wheat, soybean, potato, and canola that were established over the last century and summarized these case studies in a book chapter of genetic erosion and biodiversity (Fu and Dong 2015). The Canadian crop gene pools displayed variable patterns and degrees of genetic diversity decline over the past 100 years breeding efforts. For example, we performed a SSR-based diversity analysis of 75 Canadian hard red spring wheat cultivars released from 1845 to 2004 by several Canadian wheat breeding programs with similar breeding goals and methods (Fu and Somers 2009). We found that (1) significant allelic reduction started as early as the 1930s; (2) 38 % of 2010 SSR alleles detected in the 20 cultivars released before 1910 were retained, 18 % are new, and 44 % were lost in the 20 cultivars released after 1990; (3) allelic reduction occurred in every part of the wheat genome and a majority of the reduced alleles resided in only a few early cultivars; (4) a significant genetic shift was also observed in the gene pool in response to the long-term breeding pressure; and (5) these allelic changes were associated with long-term wheat trait improvements (Fu and Somers 2011).

Theoretical investigations on diversity changes under artificial selection

Modern breeding programs of plants and animals have heavily relied on the theories of quantitative genetics and artificial selection in finite populations as guides to improve quantitative traits of interest (Falconer and Mackay 1996; Hill 2010). These theories allow us to understand the genetic basis of a quantitative trait, the prediction of selection responses, the limit of artificial selection, and the maintenance of genetic variation at the selected loci (Lynch and Walsh 1998). Discussing all of these theories is beyond the scope of this review, except the relevant and influencing theory of selection limit (Robertson 1960). This theory states that artificial selection in a small population, such as a plant breeding population, is expected to increase the frequency of favorable alleles, along with the chance fixation of other less desirable and selectively neutral alleles. Consequently, considerable efforts have been made to investigate the effects of selection and genetic drift in finite populations from theoretical reasoning, computer simulation, and empirical evaluation (Jones et al.1968; Hill and Caballero 1992; Wray and Goddard 1994; Walsh 2004; Hill 2014).

However, our literature search revealed several interesting observations. First, previous research into selection and drift in finite populations focused more on selection response and limit, and less on companying diversity change (Hill 2014), so diversity dynamics in small populations are poorly understood. Second, some theoretical queries investigated diversity changes at selected or relevant loci through genetic variance, and did not consider the diversity impacts on genetic background through linkage and recombination (Hill and Robertson 1966; Felsenstein 1974; Charlesworth et al. 1993), so a little is known about the genome-wide diversity changes under artificial selection. Third, little attention has been paid to theoretical queries on the long-term diversity impact from specific breeding methods (Fu et al. 1998), as plant breeding programs often employ a mix of different artificial selection procedures over breeding periods to reach breeding goals.

Heterozygosity changes in a small population

If modern plant breeding can be characterized as the directional selection that humans have performed in a small population to improve a trait of interest, the early theoretical research on small populations, particularly for those by Wright-Fisher, is highly relevant. Related population genetic theories for small populations (Hedrick 2011) predict that small population size can lead to the loss of neutral genetic variation and fixation of mildly deleterious alleles, thereby reducing population fitness. The predicted heterozygosity (H_t) due to genetic drift alone is

$$H_t = 2q_0(1 - q_0)\left(1 - \frac{1}{2N}\right)^t, \quad (1)$$

where q_0 is the initial allele frequency, N is the population size, and t is the generation (Hedrick 2011). A numerical illustration is given in Fig. 2a, where the heterozygosity for a neutral allele of frequency (q = 0.1 or 0.4) is predicted to reduce over generations in a population of size (N = 20 or 50). These predictions were empirically confirmed with some *Drosophila* experiments (e.g., see Buri 1956).

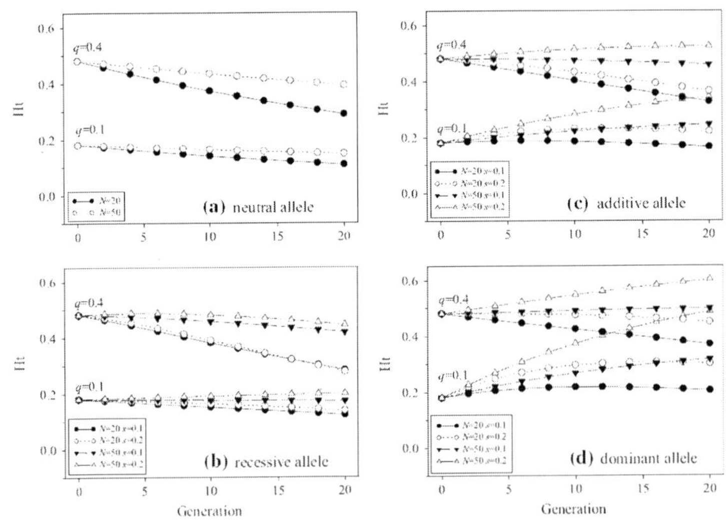

Figure 2. Predicted heterozygosity (H_t) over generations for an allele of various characteristics (**a** neutral alleles; **b** recessive allele; **c** additive allele; and **d** dominant allele) under genetic drift and/or selection in a finite population of size N. The prediction ...

The single-locus theory of selection in a finite population was formulated by Kimura in 1957 using a diffusion model to predict the fixation of a favorable allele and was extended by Robertson (1960) to predict the limit (or selection potential) to directional selection. The effects of genetic drift and selection have been theoretically investigated on the selection responses and limits, but less on the maintenance of genetic variance (Arunachalam 1974). Silvela (1980) applied conditional probabilities and moment-generating matrices to derive the expected single-locus heterozygosity (H_t) over generations of artificial selection with a finite population (see the Eq. (5) of Silvela 1980). Replacing with the directional selection fitness model (1 + *s*, 1 + *sh*, 1 for AA, Aa, aa genotypes), we can formulate his derived heterozygosity as below:

$$H_t = 2q_0(1-q_0)A^t\left\{1 + \left[t + \frac{4N-3}{5N-3}\left(\frac{1-B^tC^t}{1-BC} - t\right) - \frac{8N-6}{2N-1}\frac{1-B^tC^t}{1-BC}q_0(1-q_0)\right]s\left(h-\frac{1}{2}\right) + \frac{1-B^t}{1-B}(1-2q_0)\frac{s}{2}\right\}, 2$$

where $A = (1 - \frac{1}{2N})$, $B = (1 - \frac{2}{2N})$, $C = (1 - \frac{3}{2N})$, *s* is the selection coefficient, and *h* is the level of dominance.

To understand the changes of expected heterozygosity over generations of directional selection in a finite population, Eq. (2) was specified with some sample parameter values and illustrated in Fig. 2b-d. It is clear that the heterozygosity for a favorable allele of moderate frequency ($q = 0.4$) in a smaller population ($N = 20$) will be reduced over generations, but could be increased in a larger population ($N = 50$). However, the expected heterozygosity for a favorable allele of low frequency ($q = 0.1$) in a small population can be maintained or increased over generations, depending on the type of alleles (or the level of dominance *h*). Both the population size and selection coefficient work in the same direction to enhance the heterozygosity changes (either decrease or increase) over generations. Generally, heterozygosity for a neutral allele in a small population will be reduced over generations, but those for a selective allele can be increased or decreased, depending on the initial frequency and type of alleles.

However, a little is known about the heterozygosity changes in a multi-locus system with linkage, epistasis, and inbreeding (Robertson 1970; Arunachalam 1974). Heterozygosity in a multi-locus system is known to be less sensitive to rare alleles and thus is a less accurate measure of genetic diversity (Crow and Kimura 1970). There are no direct theoretical models available to predict the allelic richness in a finite population under directional selection over generations (Caballero and García-Dorado 2013). Some theoretical studies, mainly using computer simulation (e.g., Hedrick 1970), have been conducted to predict the maintenance of genetic variation in quantitative traits under artificial selection (e.g., Hill and Rasbash 1986; Zhang et al. 2002), but we are still far from understanding the dynamics of genetic variance under artificial selection (Brotherstone and Goddard 2005; Hill 2014).

Heterozygosity changes in a simulated breeding population

A breeding population may be more complex than the single-locus or multiple-loci model under directional selection in a small population as discussed above, because selection, genetic drift, non-random mating, migration, and mutation can all contribute and interact to alteration of genetic diversity (Falconer and Mackay1996). Although studies on recurrent selection in maize breeding populations have advanced our understanding of diversity changes as mentioned above, insufficient attention has been paid to theoretical investigations of the diversity impact of specific breeding schemes. Much less is known on how a breeding scheme affects the genetic background, along with the selected loci (Allard 1988).

To confirm our theoretical expectation on diversity reduction under plant breeding for this review, a computer simulation of a small breeding population developed for the improvement of a quantitative trait of interest was carried out and diversity changes over generations were assessed. In the simulation, five breeding schemes were applied to a breeding population over 20 generations (Fig. 3a). The breeding schemes, Self and Half-sib, used a pure mating type over generations (i.e., continuous selfing and half-sib mating, respectively) and the others (Self + Half-sib, Half-sib + Self, and Half-sib + Self + Half-sib) applied a mix of both selfing and half-sib mating, alternating over generations. These simplified breeding schemes, although knowingly deviating from existing plant breeding programs, were arbitrarily chosen mainly for theoretical confirmation (Fu et al. 1998).

We considered each individual had n unlinked selective loci each with two alleles a and A. At each locus, we assigned three genotypes aa, Aa, and AA to relative genetic values of 1, $1 + hs$, and $1 + s$, respectively, where s is the selective disadvantage of AA and h is the level of dominance. The genetic value of a progeny $i$with the trait was given in the multiplicative fitness model by the expression:

$$w_i = (1 + hs)^y (1 + s)^z, \tag{3}$$

where y and z are the numbers of loci with Aa and AA in the progeny, respectively. We also assumed that selection took place in the diploid stage of the life cycle, that no mutation at these loci occurred during the period of breeding, and that selection and dominance parameters were the same at all loci. Moreover, environmental variation and its interactions with genotypes were not considered.

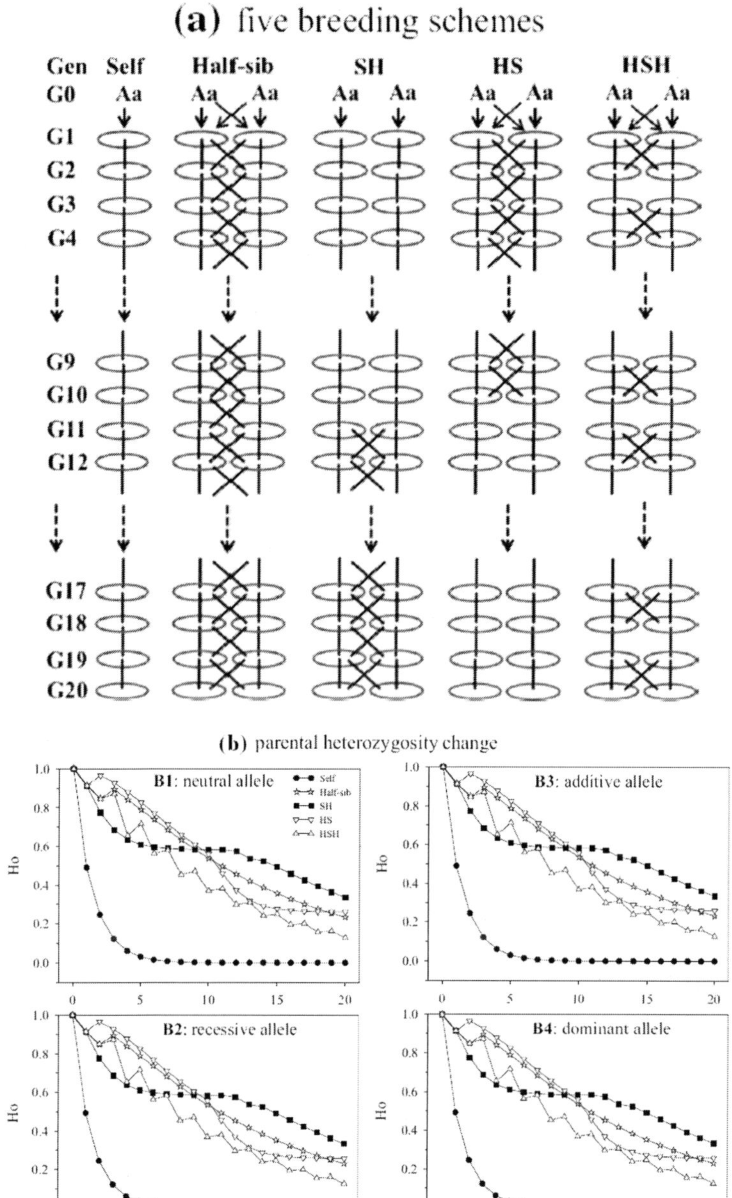

Figure 3. Breeding schemes (**a**) and parental heterozygosity (H_o) changes (**b**) over 20 generations in simulated breeding programs to improve a quantitative trait of interest. The simulation considered five breeding schemes (Self = selfing; Half-sib = half-sib; ...

For the breeding scheme Self, the simulation started with a parental generation of two unrelated individuals for each replicate. For this pair of parents, one of the four initial alleles at each locus was randomly designated as having a selective disadvantage as a homozygote; the other three alleles had no disadvantage and were selectively equivalent. From each mating, progeny of given size were generated. For each progeny, the genotype was determined, locus by locus, by randomly choosing one of two gametes (with equal probability) from each of two parents to form a zygote. For each genotype, the numbers of homozygotes and heterozygotes of alleles were counted over loci, from which the relative genetic value of a progeny for the trait of interest was determined using Eq. (3), in combination with various sets of genetic parameter values. The progeny with the largest genetic value for each full-sib family was selected as the parent for the next generation and was self-fertilized to produce the G1 progeny. This process was followed for 20 generations and 100 replicates were run for each combination of genetic parameters and progeny sizes. In this study, 5000 loci were simulated and only the first 20 loci contributed multiplicatively to the trait of interest. Seven genetic models were examined: neutral ($s = 0, h = 0$), weakly recessive ($s = 0.1, h = 0$), strongly recessive ($s = 0.2, h = 0$), weakly additive ($s = 0.1, h = 0.5$), strongly additive ($s = 0.2, h = 0.5$), weakly dominant ($s = 0.1, h = 1$), and strongly dominant ($s = 0.2, h = 1$). Two progeny sizes (20, 50) were used. Thus, there were 14 combinations for each breeding scheme.

For the other breeding schemes, the same procedures as in Self were applied, but they differed in the use of two parental individuals heterozygous for all the loci to form two full-sib families and the selection of the two progeny with the largest genetic values for the trait as the parents of the next generation from each full-sib family, as described in the breeding schemes above. The simulations were done with an R script (R Development Core Team 2014) that was written specifically for this investigation and is available from the author upon request.

The simulated results on progeny size 50 are shown in Table S1 for parental genetic value improvements and Table S2 for heterozygosity changes. Several diversity patterns are clear. First, as expected, artificial selection improved the trait with increased genetic value (Table S1), while decreased the genetic diversity with reduced heterozygosity (Table S2). For example, the parental heterozygosity approached zero after seven successive selfings (Fig. 3b; Table S2). Second, comparisons among the five breeding schemes showed half-sib mating was the most effective scheme to achieve higher genetic gain with lower reduction of heterozygosity (Fig. 3b). Third, the genetic fitness models examined were associated with the trait improvement, but not directly with the patterns of heterozygosity reduction (Fig. 3b). Fourth, the patterns of trait improvement and diversity reduction were essentially the same for both progeny sizes of 20 and 50 (Fig. 3b). These findings help to confirm our expectation that plant breeding can theoretically reduce genetic diversity. Also, our simulation, although preliminary, appears to have utility in understanding gain potential and diversity dynamics under a breeding scheme.

Future investigations into genetic diversity dynamics

Concerns about crop uniformity and the possible risk of crop failure (Ullstrup 1972; Day 1973) have continued in the past decade, particularly with the emerging threat of wheat stem rust Ug99 from East Africa (Borlaug 2007; Babiker et al. 2015). More questions have been raised concerning sustainability and innovation in staple crop production based on the narrowing genetic base (Kahane et al. 2013; Heinemann et al. 2014). Sustainable agriculture requires a compromise between maximizing crop yield and minimizing the risk of crop failure; thus, there is a need for a better understanding of the plant breeding impacts on crop genetic diversity. Here we propose several lines of research that would fill the gaps present in our knowledge of crop genetic diversity occurring under plant breeding.

First, a standardization of crop genetic diversity assessments should be sought to avoid technical biases identified in this review and applied for more informative assessment of genetic diversity dynamics (FAO2013). It is technically feasible to develop and establish standard assessments, given the recent advances in next generation sequencing. Genotyping-by-sequencing methods are available to allow the acquisition of thousands of genome-wide SNPs in non-model plant species for diversity assessment (e.g., see Peterson et al.2014). Heterozygosity should be measured and reported on SNP data. Also, computer simulations can be performed using different diversity parameters, genome coverages, and sampling strategies to evaluate the accuracy and precision of a diversity assessment.

Second, continuous assessments on crop genetic diversity using advanced genomic techniques are needed to facilitate the effective monitoring of newly released cultivars and on-farm crop diversity. This could be done with an assay of cultivars released over time from specific breeding programs and quantification of diversity changes. More emphasis should be given to the assessments of crop genetic diversity at the breeding program and farm levels, and less on country or world scale. More can be learned from the temporal diversity analysis of long-term public and commercial breeding programs such as the Illinois long-term maize selection experiment and Pioneer Hi-Bred International maize improvement program. Empirical assessments of genomic response to artificial selection like those in animal breeding (Flori et al. 2009; Kim et al. 2013) would provide more insight into breeding impacts on the plant genome. The major challenge is to determine whether the alleles eliminated from plant breeding are of any adaptive value, are genetically associated with any traits of future importance, or are linked to any selected genes at nearby loci (Fu and Somers 2009).

Third, theoretical research into trait improvement and diversity change under different breeding schemes should be pursued, particularly by computer simulation (Caballero and García-Dorado 2013). The preliminary simulation described in this review can be expanded to consider complex breeding schemes that are compatible with existing breeding programs. Genotype-by-environment interaction can be modeled to reflect the complex nature of a quantitative trait expressed in target environments (Marigorta and Gibson2014). Focus should be placed on the development of breeding strategies that allow for continuous development of superior genotypes with minimal loss of genetic variation (Hallauer and Miranda 1988). These studies will not only enhance our understanding of crop genetic diversity dynamics under artificial selection, but also may generate guidance for plant breeders on how to compromise between maximizing crop yield and minimizing the risk of crop failure.

CONCLUDING REMARKS

Our review has highlighted the fact that crop genetic diversity under modern plant breeding has been improperly and insufficiently investigated, either from an empirical aspect or theoretical background, and thus is poorly understood. Many reported assessments of crop genetic diversity were not aimed at the assessment of diversity impacts from specific breeding programs and some were experimentally inadequate and had technical biases from the sampling of cultivars and genomes. Little attention has been paid to theoretical investigations on crop genetic diversity changes under plant breeding. A computer simulation of five simplified breeding schemes showed substantial breeding effects on the retention of heterozygosity over generations. It is clear that more efforts are needed to investigate crop genetic diversity in space and time under plant breeding for sustainable crop production.

AUTHOR CONTRIBUTION STATEMENT

YBF conceived of the research, performed the computer simulation and wrote the paper.

ACKNOWLEDGMENTS

I would like to thank Bruce Coulman, Bill Biligetu, Santosh Kumar, and Paul Gepts for their helpful comments on the earlier version of this manuscript. The research is financially supported by an A-Base Research Project of Agriculture and Agri-Food Canada.

COMPLIANCE WITH ETHICAL STANDARDS

Conflict of interest

The authors declare that they have no conflict of interest.

REFERENCES

1. Allard RW. Genetic changes associated with the evolution of adaptedness in cultivated plants and their wild progenitors. J Hered. 1988;79:225–238.
2. Allard RW. Principles of plant breeding. 2. New York: Wiley; 1999.
3. Aremu CO. Genetic diversity: a review for need and measurements for intraspecies crop improvement.J Microbiol Biotech Res. 2011;1:80–85.
4. Arunachalam V. A study of inbreeding in some models of artificial selection. J Genet. 1974;61:87–107.
5. Babiker EM, Gordon TC, Chao S, Newcomb M, Rouse MN, Jin Y, Wanyera R, Acevedo M, Brown-Guedira G, Williamson S, Bonman JM. Mapping resistance to the Ug99 race group of the stem rust pathogen in a spring wheat landrace. Theor Appl Genet. 2015;128:605–612.
6. Borlaug NE. Contributions of conventional plant breeding to food production. Science.1983;219:689–693.
7. Borlaug N. Sixty-two years of fighting hunger: personal recollections. Euphytica. 2007;157:287–297.
8. Bowman DT, May OL, Creech JB. Genetic uniformity of the U.S. upland cotton crop since the introduction of transgenic cottons. Crop Sci. 2003;43:515–518.
9. Breseghello F. Traditional and modern plant breeding methods with examples in rice (*Oryza sativa* L.)J Agric Food Chem. 2013;61:8277–8286.
10. Brotherstone S, Goddard M. Artificial selection and maintenance of genetic variance in the global dairy cow population. Phil Trans R Soc B. 2005;360:1479–1488.
11. Brown AHD, Allard RW. Effect of reciprocal recurrent selection for yield on isozyme polymorphisms in maize (*Zea mays* L.) Crop Sci. 1971;11:888–893.
12. Brummer EC, Barber WT, Collier SM, Cox TS, Johnson R, Murray SC, Olsen RT, Pratt RC, Thro AM. Plant breeding for harmony between agriculture and the environment. Front Ecol Environ.2011;9:561–568
13. Buri P. Gene frequency in small populations of mutant Drosophila. Evolution. 1956;10:367–402.
14. Caballero A, García-Dorado A. Allelic diversity and its implications for the rate of adaptation.Genetics. 2013;195:1373–1384.

15. Charlesworth B, Morgan MT, Charlesworth D. The effect of deleterious mutations on neutral molecular variation. Genetics. 1993;134:1289–1303.
16. Cox TS, Kiang YT, Gorman MB, Rodgers DM. Relationship between coefficient of parentage and genetic similarity indices in soybean. Crop Sci. 1985;25:529–532.
17. Cox TS, Murphy JP, Rodgers DM. Changes in genetic diversity in the red and winter wheat regions of the United States. Proc Natl Acad Sci (USA) 1986;83:5583–5586.
18. Crow JF, Kimura M. An introduction to population genetics theory. New York: Harper and Row; 1970.
19. Day PR. Genetic variability of crops. Annu Rev Phytopathol. 1973;11:293–312.
20. Donini P, Law JR, Koebner RMD, Reeves JC. Temporal trends in the diversity of UK wheat. Theor Appl Genet. 2000;100:912–917.
21. Duvick DN. Genetic diversity in major farm crops on the farm and in reserve. Econ Bot.1984;38:161–178.
22. Duvick DN, Smith JSC, Cooper M, et al. Changes in performance, parentage, and genetic diversity of successful corn hybrids, 1930 to 2000. In: Smith CW, et al., editors. Corn: origin, history, technology and production. Hoboken: Wiley; 2004. pp. 65–97.
23. Falconer DS, Mackay TFC. An introduction to quantitative genetics. Essex: Pearson Prentice Hall; 1996.
24. FAO (2013) Draft genebank standards for plant genetic resources for food and agriculture.http://www.fao.org/docrep/meeting/027/mf804e.pdf. Accessed March 23 2015
25. Felsenstein J. The evolutionary advantage of recombination. Genetics. 1974;78:737–756
26. Flori L, Fritz S, Jaffrézic F, Boussaha M, Gut I, et al. The genome response to artificial selection: a case study in dairy cattle. PLoS One. 2009;4:e6595.
27. Fu YB. Impact of plant breeding on genetic diversity of agricultural crops: searching for molecular evidence. Plant Genet Resour. 2006;4:71–78.
28. Fu YB. FPTest: a SAS routine for testing allelic counts. Mol Ecol Resour. 2010;10:389–392.
29. Fu YB, Dong Y. Genetic erosion under modern plant breeding: case studies in Canadian crop gene pools. In: Ahuja MR, Jain SM, editors. Genetic erosion and biodiversity. Cham: Springer International Publishing AG; 2015.
30. Fu YB, Somers DJ. Genome-wide reduction of genetic diversity in wheat breeding. Crop Sci.2009;49:161–168.

31. Fu YB, Somers DJ. Allelic changes in bread wheat cultivars were associated with long-term wheat trait improvements. Euphytica. 2011;179:209–225
32. Fu YB, Namkoong G, Carlson JE. Comparison of breeding strategies for purging inbreeding depression via simulation. Conserv Biol. 1998;12:856–864.
33. Fu YB, Peterson GW, Scoles G, Rossnagel B, Schoen DJ, Richards KW. Allelic diversity changes in 96 Canadian oat cultivars released from 1886 to 2001. Crop Sci. 2003;43:1989–1995.
34. Fu YB, Kibite S, Richards KW. Amplified fragment length polymorphism analysis of 96 Canadian oat cultivars released between 1886 and 2001. Can J Plant Sci. 2004;84:23–30.
35. Fu YB, Peterson GW, Ricahrds KW, Somers D, DePauw RM, Clarke JM. Allelic reduction and genetic shift in the Canadian hard red spring wheat germplasm released from 1845 to 2004. Theor Appl Genet. 2005;110:1505–1516.
36. Gepts P. Plant genetic resources conservation and utilization: the accomplishments and future of a societal insurance policy. Crop Sci. 2006;46:2278–2292.
37. Gepts P, Hancock J. The future of plant breeding. Crop Sci. 2006;46:1630–1634.
38. Hallauer AR. Compendium on recurrent selection methods and their application. Crit Rev Plant Sci.1985;3:1–33.
39. Hallauer AR, Miranda JB. Quantitative genetics in maize breeding. 2. Ames: Iowa State University Press; 1988.
40. Hedrick PW. Selection in finite populations. II. The selection limit and rate of response for a Monte Carlo simulation model. Genetics. 1970;65:175–186.
41. Hedrick PW. Genetics of populations. 4. Sudbury: Jones and Bartlett; 2011.
42. Heinemann JA, Massaro M, Coray DS, Agapito-Tenfen SZ, Wen JD. Sustainability and innovation in staple crop production in the US Midwest. Int J Agric Sustain. 2014;12:71–88.
43. Hill WG. Understanding and using quantitative genetic variation. Phil Trans R Soc B. 2010;365:73–85.
44. Hill WG. Applications of population genetics to animal breeding, from Wright, Fisher and Lush to genomic prediction. Genetics. 2014;196:1–16. doi: 10.1534/genetics.112.147850.
45. Hill WG, Caballero A. Artificial selection experiments. Ann Rev Ecol Syst. 1992;23:287–310.
46. Hill WG, Rasbash J. Models of long term artificial selection in finite population. Genet Res.1986;48:41–50.
47. Hill WG, Robertson A. The effect of linkage on limits to artificial selection. Genet Res. 1966;8:269–294.

48. Hinze LL, Kresovich S, Nason JD, Lamkey KR. Population genetic diversity in a maize reciprocal recurrent selection program. Crop Sci. 2005;45:2435–2442. doi: 10.2135/cropsci2004.0662.
49. Hunter RL, Merkert CL. Histochemical demonstration of enzymes separated by zone electrophoresis in starch gels. Science. 1957;125:1294–1295.
50. Hyten DL, Song Q, Zhu Y, Choi IY, Nelson RL, Costa JM, Specht JE, Shoemaker RC, Cregan PB. Impacts of genetic bottlenecks on soybean genome diversity. Proc Natl Acad Sci USA.2006;103:16666–16671..
51. Jones LP, Frankham R, Barker JSF. The effects of population size and selection intensity in selection for a quantitative character in *Drosophila melanogaster*. II. Long-term response to selection. Genet Res. 1968;12:249–266.
52. Jost L. Gst and its relatives do not measure differentiation. Mol Ecol. 2008;17:4015–4026.
53. Kahane R, Hodgkin T, Jaenicke H, Hoogendoorn C, Hermann M, Keatinge JDH, d'Arros Hughes J, Padulosi S, Looney N. Agrobiodiversity for food security, health and income. Agron Sustain Dev.2013;33:671–693.
54. Kasha KJ, Kao KN. High frequency haploid production in barley (*Hordeum vulgare* L.) Nature.1970;225:874–876.
55. Keneni G, Bekele E, Imtiaz M, Dagne K. Genetic vulnerability of modern crop cultivars: causes, mechanism and remedies. Int J Plant Res. 2012;2:69–79.
56. Kim E-S, Cole JB, Huson H, Wiggans GR, Van Tassell CP, et al. Effect of artificial selection on runs of homozygosity in U.S. Holstein cattle. PLoS One. 2013;8:e80813.
57. Kimura M. Some problems of stochastic processes in genetics. Ann Math Stat. 1957;28:882–890.
58. Koebner RMD, Donini P, Reeves JC, Cooke RJ, Law JR. Temporal flux in the morphological and molecular diversity of UK barley. Theor Appl Genet. 2003;106:550–558.
59. Labate JA, Lamkey KR, Lee M, Woodman ML. Temporal changes in allele frequencies in two reciprocally selected maize populations. Theor Appl Genet. 1999;99:1166–1178.
60. Le Clerc V, Bazante F, Baril C, Guiard J, Zhang D. Assessing temporal changes in genetic diversity of maize varieties using microsatellite markers. Theor Appl Genet. 2005;110:294–302.
61. Lu H, Bernardo R. Molecular marker diversity among current and historical maize inbreds. Theor Appl Genet. 2001;103:613–617.
62. Lynch M, Walsh JB. Genetics and analysis of quantitative traits. Sunderland: Sinauer Assocs., Inc.; 1998.

63. Marigorta UM, Gibson G. A simulation study of gene-by-environment interactions in GWAS implies ample hidden effects. Front Genet. 2014;5:225.
64. Mikkilineni V, Rocheford TR. RFLP variant frequency differences among Illinois long-term selection protein strains. Plant Breed Rev. 2004;24:111–131.
65. Mohammadi SA, Prasanna BM. Analyses of genetic diversity in crop plants-salient statistical tools and considerations. Crop Sci. 2003;43:1235–1248.
66. Mondini L, Noorani A, Pagnotta MA. Assessing plant genetic diversity by molecular tools. Diversity.2009;1:19–35.
67. Moose SP, Mumm RH. Molecular plant breeding as the foundation for 21st century crop improvement. Plant Physiol. 2008;147:969–977.
68. Moose SP, Dudley JW, Rocheford TR. Maize selection passes the century mark: a unique resource for 21st century genomics. Trends Plant Sci. 2004;9:358–364.
69. National Academy of Sciences . Genetic vulnerability of major crops. Washington: National Academy of Sciences; 1972.
70. Orabi J, Jahoor A, Backes G. Changes in allelic frequency over time in European bread wheat (*Triticum aestivum* L.) varieties revealed using DArT and SSR markers. Euphytica. 2014;197:447–462.
71. Ortiz R, Lund B, Anderson S-B. Breeding gain and changes in morphotype of Nordic spring wheat (1901-1993) under contrasting environments. Genet Resour Crop Evol. 2003;50:455–459.
72. Peterson GW, Dong Y, Horbach C, Fu YB. Genotyping-by-sequencing for plant genetic diversity analysis: a lab guide for SNP genotyping. Diversity. 2014;6:665–680
73. Pinto LR, Vieria MLC, de Souza Jr CL, de Souza AP. Reciprocal recurrent selection effects on the genetic structure of tropical maize populations assessed at microsatellite loci. Genet Mol Biol.2003;26:355–364.
74. R Development Core Team . R: a language and environment for statistical computing. Vienna: R Foundation for Statistical Computing; 2014.
75. Rauf S, da Silva JT, Khan AA, Naveed A. Consequences of plant breeding on genetic diversity. Int J Plant Breed. 2010;41:1–21.
76. Reeves JC, Chiapparino E, Donini P, Ganal M, Guiard J, Hamrit S et al (2004) Changes over time in genetic diversity of four European crops—a report from the Gediflux Framework 5 project. In: Vollmann J, Grausgruber H and Ruckenbauer P (eds) Genetic Variation for Plant Breeding, Proceedings of the 17th EUCARPIA General Congress, 8–11 September 2004, Tulln, Austria. Autria: BOKU-University of Natural Resources and Applied Life Sciences, Vienna, pp 3–8
77. Reif JC, Zhang P, Dreisigacker S, Warburton ML, van Ginkel M, Hoisington D, Bohn M, Melchinger AE. Wheat genetic diversity trends

during domestication and breeding. Theor Appl Genet.2005;110:859–864.
78. Robertson A. A theory of limits in artificial selection. Proc R Soc Lond B. 1960;153:234–249.
79. Robertson A. A theory of limits in artificial selection with many linked loci. In: Kojima K, editor.Mathematical topics in population genetics. Berlin: Springer; 1970. pp. 246–268.
80. Rodgers DM, Murphy JP, Frey KJ. Impact of plant breeding on the grain yield and genetic diversity of spring oats. Crop Sci. 1983;23:737–740.
81. Romay MC, Butron A, Ordas A, Revilla P, Ordas B. Effect of recurrent selection on the genetic structure of two broad-based Spanish maize populations. Crop Sci. 2012;52:1493–1502.
82. Roussel V, Leisova L, Exbrayat F, Stehno Z, Balfourier F. SSR allelic diversity changes in 480 European bread wheat varieties released from 1840 to 2000. Theor Appl Genet. 2005;111:162–170.
83. Russell JR, Ellis RP, Thomas WTB, Waugh R, Provan J, Booth A, Fuller J, Lawrence P, Young G, Powell W. A retrospective analysis of spring barley germplasm development from 'foundation genotypes' to currently successful cultivars. Mol Breed. 2000;6:553–568.
84. Salick J. Toward an integration of evolutionary ecology and economic botany—personal perspectives on plant/people interactions. Ann Missouri Bot Garden. 1995;82:25–33.
85. Shukla V, Mattoo AK. Developing robust crop plants for sustaining growth and yield under adverse climatic changes. In: Tuteja N, Gill SS, editors. Climate change and plant abiotic stress tolerance.Weinheim: Wiley-VCH Verlag GmbH & Co. KGaA; 2013.
86. Silvela L. Genetic changes with generations of artificial selection. Genetics. 1980;95:769–782.
87. Smith JSC, Duvick DN, Smith OS, Cooper M, Feng L. Changes in pedigree backgrounds of Pioneer brand maize hybrids widely grown from 1930 to 1999. Crop Sci. 2004;44:1935–1946.
88. Solomon F, Martin I, Zeppa A, Keating V (2010) Effect of recurrent selection on genetic diversity patterns in tropical maize breeding populations. In: Proceeding of the 1st Australian Summer Grains Conference, Gold Coast, Australia, June 21–24, 2010
89. Spillane C, Curtis MD, Grossniklaus U. Apomixis technology development—virgin births in farmers' fields? Nat Biotech. 2004;22:687–691.
90. Swanson T. Global values of biological diversity: the public interest in the conservation of plant genetic resources for agriculture. Plant Genet Resour Newsl. 1996;105:1–7.
91. Tester M, Langridge P. Breeding technologies to increase crop production in a changing world.Science. 2010;327:818–822.

92. Tripp R. Biodiversity and modern crop varieties: sharpening the debate. Agri Human Values.1996;13:48–63.

93. Ullstrup AJ. The impacts of the southern corn leaf blight epidemics of 1970-1971. Ann Rev Phytopath. 1972;10:37–50.

94. van Beuningen LT, Busch RH. Genetic diversity among North American spring wheat cultivars: I. Analysis of the coefficient of parentage matrix. Crop Sci. 1997;37:570–579.

95. van de Wouw M, Kik C, van Hintum T, van Treuren R, Visser B. Genetic erosion in crops: concept, research results and challenges. Plant Genet Resour. 2010;8:1–15.

96. van de Wouw M, van Hintum T, Kik C, van Treuren R, Visser B. Genetic diversity trends in twentieth century crop cultivars: a meta analysis. Theor Appl Genet. 2010;120:1241–1252.

97. Vellve R. The decline of diversity in European agriculture. The Ecologist. 1993;23:64–69.

98. Walsh B. Population- and quantitative-genetic models of selection limits. Plant Breed Rev.2004;24:177–225.

99. Whitlock M. G'st and D do not replace Fst. Mol Ecol. 2011;20:1083–1091.

100. Wray NR, Goddard ME. Increasing long-term response to selection. Genet Sel Evol. 1994;26:431–451.]

101. Zhang X-S, Wang J, Hill WG. Pleiotropic model of maintenance of quantitative genetic variation at mutation–selection balance. Genetics. 2002;161:419–433.

Chapter 4

Marker-Assisted Breeding of Indonesia Local Rice Variety Siputeh for Semi-Dwarf Phonetype, Good Grain Quality and Disease Resistance to Bacterial Blight

Yanchang Luo1,3, Sabaruddin Zakaria2, Bakhtiar Basyah2, Tingchen Ma3, Zefu Li3, Jianbo Yang3 and Zhongchao Yin1,4*

[1]*Temasek Life Sciences Laboratory, 1 Research Link, National University of Singapore, Singapore 117604, Republic of Singapore.*

[2]*Department of Agrotechnology, Agriculture Faculty, Syiah Kuala University,*

Darussalam-Banda, Aceh 23111, Indonesia.

[3]*Rice Research institute, Anhui Academy of Agricultrural Sciences, Hefei 230031, China.*

[4]*Department of Biological Sciences, 14 Science Drive, National University of Singapore, Singapore 117543, Republic of Singapore.*

ABSTRACT

Background

Rice is one of the most important staple food crops in Asia. Since the first green revolution beginning in 1960s, high-yield semidwarf modern rice varieties have been widely planted; however, traditional rice varieties with tall plant type are still grown in many countries due to their good grain quality and adaptation to local climate and environment. Siputeh, a local rice variety mainly planted in Java and Sumatra islands of Indonesia, produces long grain rice with good cooking and eating quality. However, the variety has low yield with tall plant type and long growth duration and is highly susceptible to biotic and abiotic stress.

Results

Siputeh as the recurrent female was crossed with the donor line WH421, an elite paternal line of hybrid rice containing the *sd1*, *Wx*^{b}, *Xa4* and*Xa21* genes, followed by backcrossing and self-pollination. TS4, a BC3F4 line derived from the breeding program, was obtained through marker-assisted selection for the *sd1*, *Wx*^{b}, *Xa4* and *Xa21* loci. TS4 has semi-dwarf phenotype and short growth duration. TS4 conferred disease resistance to multiple *Xanthomonas oryzae* pv. *oryzae* (*Xoo*) strains collected from different countries around the world. TS4 achieved higher grain yield than Siputeh in two field trials conducted in Banda Aceh, Indonesia and Lingshui, China, respectively. Finally, TS4 has better grain quality than Siputeh in terms of degree of chalkiness and amylose content.

Conclusion

An improved rice line, designed as TS4, has been developed to contain semi-dwarf gene *sd1*, low amylase content gene *Wx*^{b} and bacterial light resistance genes *Xa4* and *Xa21* through marker-assisted selection. TS4 has semi-dwarf phenotype with reduced growth duration, produces high yield with good grain quality and provides broad-spectrum resistance to *Xoo* strains. The development of TS4 enriches the diversity of local rice varieties with high yield potential and good grain quality.

Keywords: Rice Marker-assisted selection Semi-dwarf Amylose content Bacterial blight

BACKGROUND

Rice (*Oryza sativa* L.) is a staple food crop in most of the Asia countries. The introduction of semi-dwarf rice led to record yield increases throughout Asia in the 1960s. The recessive semi-dwarfing gene, *sd-1*, is one of the most important genes deployed in modern rice breeding. Rice varieties with the *sd1* gene have a shortened culm with improved lodging resistance and a greater harvest index, allowing for the increased use of nitrogen fertilizers (Jennings [1964]). The *sd1* gene has been cloned from rice and its dominant wild-type allele encodes a gibberellin (GA)-20 oxidase-2 (GA20ox-2) (Ashikari et al. [2002]; Sasaki et al. [2002]; Spielmeyer et al. [2002]). The deletion or substitution mutation of the *Os20ox2* gene results in loss of function of the enzyme, which ultimately leads to a deficiency in plant growth hormones gibberellins that control plant height (Ashikari et al. [2002]; Sasaki et al. [2002]; Spielmeyer et al. [2002]). The different alleles of the *sd1* gene have remained the predominant semi-dwarf genes in modern rice cultivars (Asano et al. [2007]).

There are two different types of starch, amylose and amylopectin, in rice endosperm and the amylose content affects rice eating and cooking quality (Cruz and Khush [2000]). With content ranging from 5% to 15%, the low-amylose rice is characterized by fluffy texture, glossy appearance of the cooked rice, soft texture of cooled rice, and excellent expansibility for food processing (Zhu et al. [2003]). Genetic studies have shown that the rice *Wx* gene determines the amylose content in the endosperm. *Wx* encodes granule-bound starch synthase I (GBSSI), a key enzyme for amylose biosynthesis in the endosperm, and *wx* mutant endosperm contains almost exclusively amylopectin (Hori et al. [2007]; Sano [1984]; Wang et al. [1995]). In addition to controlling amylose content, *Wx* also affects gel consistency and gelatinization temperature of rice starch (Su et al. [2011]). Wx^a and Wx^b are two predominantly distributed functional *Wx* alleles in rice (Sano [1984]; Wang et al. [1995]). Wx^a is widely distributed in *indica* rice with high amylose content while Wx^b is mainly found in *japonica* rice with low or intermediate amylose content (Sano [1984]; Wang et al. [1995]). A single nucleotide substitution (G-to-T) at the splice donor site of the first intron in Wx^b potentially alters the splicing site and decrease the splicing efficiency of the first intron of *Wx*, resulting in the low level of both mature transcript of Wx^b and amylose content (Cai et al. [1998]; Larkin and Park [1999]; Tian et al. [2009]; Wang et al. [1995]). Based on the polymorphism at the splicing site of the first intron, a co-dominant CAPS (cleaved amplified polymorphic sequence) molecular marker PCR-*Acc* I was developed for selection of *Wx* alleles in rice breeding (Cai et al. [2002]).

Bacterial blight of rice, caused by *Xanthomonas oryzae* pv. *oryzae* (*Xoo*), is one of the most important bacterial diseases prevalent throughout the world (Gnanamanickam et al. [1999]). Rice yield losses caused by bacterial blight can reduce yield by 20-30% and by up to 50% in some areas of Asia (Mew et al. [1993]). The utilization of host disease resistance (*R*) genes is the most economic and efficient method for controlling the disease. *Xa4* is a dominant *R* gene that provides durable resistance to bacterial blight. Cultivars with *Xa4* conferred resistance to almost all Chinese patho-types of *Xoo* except for patho-type C5 (Zhang [2009]). *Xa21* is another dominant bacterial blight gene, which was originally discovered in wild rice specie *Oryza longistaminata* (Ikeda et al. [1990]). IRBB21, an *Xa21* line in IR24 genetic background, showed resistance to all the known races of *Xoo* collected from India and the Philippines (Ikeda et al. [1990]). Both *Xa4* and *Xa21* genes are widely exploited *R* genes in Asian rice breeding programs for bacterial blight resistance (Datta et al. [2002]; Huang et al. [1997]; Luo et al. [2012]; Luo and Yin [2013]; Singh et al. [2001]; Suh et al. [2013]; Zhang et al. [2006]).

Although modern rice varieties have been planted worldwide, elite local rice varieties are still popular in many rice growing countries due to their good grain quality and adaptation to the local climate and environment. Cultivar Siputeh is a local rice variety still planted in Java and Sumatra islands of Indonesia. Siputeh produces long grain rice with good cooking and eating quality. However, the variety has low yield with tall plant type and long growth duration and is highly susceptible to biotic and abiotic stress (Suhartini [1991]). Here we report the genetic improvement of Siputeh for semi-dwarf phenotype, good grain quality and disease resistance to bacterial blight through marker-assisted breeding.

RESULTS

Breeding of TS4

Marker-assisted breeding was conducted between cultivar Siputeh as the recurrent female and cultivar WH421 as the donor line for the *sd1*, Wx^b, *Xa4* and *Xa21* genes (Figure 1). The genotypes at the *sd1*, *Wx*, *Xa4* and *Xa21* loci in each plant at different generations were determined with molecular markers as described in Table 1. In summary, 10 F1 plants, 3 BC1F1 plants, 2 BC2F1 plants and 3 BC3F1 were identified from F1, BC1F1, BC2F1 and BC3F1 generations, respectively (Figure 1). All these plants carried heterozygous alleles at the *sd1*, Wx^b, *Xa4* and*Xa21* loci (Figure 1). One of the 3 BC3F1 plants (BC3F1 #6) was selected for self-pollination to generate a BC3F2 population, which consisted of 74 BC3F2 individuals. Three BC3F2 plants were identified from the BC3F2 population and they all carried homozygous alleles at the *sd1*and *Xa21* loci and heterozygous alleles at the Wx^b and *Xa4* loci (Figure 1). Plant BC3F2 #38 was selected to produce a BC3F3 population, which consisted of 146 BC3F3 individuals. Nine plants (BC3F3 #1, 3, 4, 6–11) were identified from the BC3F3 population and they all carried homozygous alleles at the *sd1*, Wx^b, *Xa4* and *Xa21* loci (Figures 1 and 2). Plant BC3F3 #1 was selected to produce a BC3F4 population. The 48 individual plants in the BC3F4 population showed uniform morphological phenotype and similar growth duration. Plant BC3F4 #30 was designated as TS4 and selected for disease evaluation for resistance to bacterial blight and field trials (Figure 1).

Table 1. Molecular markers used in this study

Marker	Gene-of-interest	DNA sequence	Type of marker	Reference
sd1STS	*sd1*	F: 5'ACAAATACCCCACCCTCCTGCC3' R: 5'TAGTAGCCTCGCTCCACGCCC3'	STS, co dominant	(Luo and Yin 2013)
21	*Xa21*	F: 5'ATAGCAACTGATTGCTTGG3' R:5' GATCGGTATAACAGCAAAAC3'	STS, co dominant	(Chen et al. 2000)
RM224[a]	*Xa4*	F:5' ATCGATCGATCTTCACGAGG3' R:5' TGCTATAAAAGGCATTCGGG3'	SSR, co dominant	(Sun et al. 2003)
PCR *AccI*[b]	*Wx*	F: 5'GCTTCACTTCTCTGCTTGTG3' R: 5'ATGATTTAACGAGAGTTGAA3'	CAPS, co dominant	(Cai et al. 2002)

F, forward primer; R, reverse primer; STS, sequence-tagged site; SSR, simple sequence repeat; CAPS, Cleaved amplified polymorphic sequence.

[a]The genetic distance between RM224 and the *Xa4* locus is 1.1 cM.

[b]CAPS marker at the *Wx* locus by digesting PCR product with *AccI* to show polymorphism.

Siputeh(*Sd1Sd1*, Wx^aWx^a, *xa4xa4*, *xa21xa21*) x WH421(*sd1sd1*, Wx^bWx^b, *Xa4Xa4*, *Xa21Xa21*)
↓
Siputeh(*Sd1Sd1*, Wx^aWx^a, *xa4xa4*, *xa21xa21*) x F1(*Sd1sd1*, Wx^aWx^b, *Xa4xa4*, *Xa21xa21*)
↓
Siputeh(*Sd1Sd1*, Wx^aWx^a, *xa4xa4*, *xa21xa21*) x BC1F1(*Sd1sd1*, Wx^aWx^b, *Xa4xa4*, *Xa21xa21*), 3/25
↓
Siputeh(*Sd1Sd1*, Wx^aWx^a, *xa4xa4*, *xa21xa21*) x BC2F1(*Sd1sd1*, Wx^aWx^b, *Xa4xa4*, *Xa21xa21*), 2/18
↓
BC3F1(*Sd1sd1*, Wx^aWx^b, *Xa4xa4*, *Xa21xa21*), 3/35
↓ Selfing
BC3F2(*sd1sd1*, Wx^aWx^b, *Xa4xa4*, *Xa21Xa21*), 3/74
↓ Selfing
BC3F3(*sd1sd1*, Wx^bWx^b, *Xa4Xa4*, *Xa21Xa21*), 9/146
↓ Selfing
BC3F4(*sd1sd1*, Wx^bWx^b, *Xa4Xa4*, *Xa21Xa21*), 1/48
↓
BC3F4 #30(*sd1sd1*, Wx^bWx^b, *Xa4Xa4*, *Xa21Xa21*)
(TS4)

Figure 1. Marker-assisted breeding of TS4. The number of positive plants over the number of total plants screened for presence of molecular makers are indicated in the generation undergone marker-assisted selection. *Sd1* and *sd1*, the dominant and recessive alleles of the*Sd1* gene; Wx^a and Wx^b, the alleles *a* and *b* of the rice *Wx* gene; *Xa4* and *xa4*, dominant and recessive alleles of the *Xa4* gene, respectively; *Xa21* and *xa21*, dominant and recessive alleles of the *Xa21* gene, respectively.

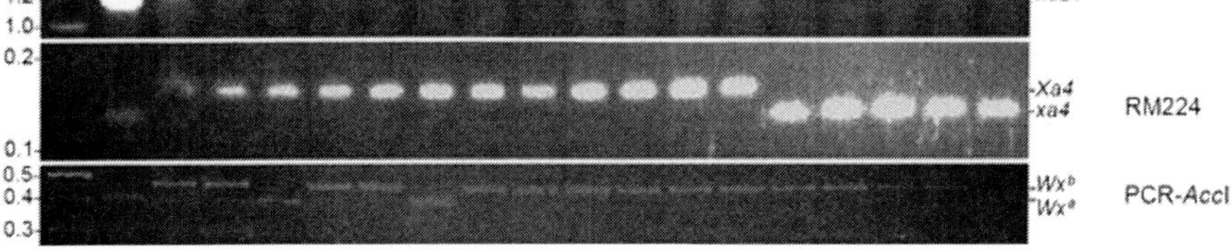

Figure 2. Detection of molecular markers at the *sd1* , *Xa4, Xa21* and Wx^b loci in the B3F3 individuals. The BC3F3 individuals were derived from the cross between Siputeh (recurrent female) and WH421 followed by backcrossing for 3 generations. The amplified PCR products were separated on a 1.5% w/v agarose gel for Marker sd1STS, a 2.0% agarose gel for Marker 21, and a 3.5% agarose gel for Marker PCR-*Acc* I and RM224.BC3F3 plants #1, 3, 4, 6–11 carry homozygous alleles at the *sd1*, *Xa4*, *Xa21* and Wx^b loci.

Disease evaluation of TS4 for bacterial blight resistance

Twenty-eight *Xoo* strains were used to inoculate WH421, Siputeh and TS4. Siputeh was susceptible to all *Xoo* strains tested while WH421 conferred resistance or moderate resistance to 22 strains and was susceptible or moderate susceptible to 1947, GD1358, A3857, JW89011, K202 and PXO99 (Table 2). TS4 provided resistance or moderate resistance to 25 of the 28 *Xoo* strains tested and was only susceptible to A3857, JW89011 and K202 (Table 2). Compared to WH421, TS4 also provided enhanced resistance with shorter disease lesions to 1947, GD1358, A3857, JW89011, K202 and PXO99 (Table 2). It should be noted that neither WH421 nor TS4 could provide complete resistant to PXO99 (Table 2). The result was not consistent with the previous finding that the *Xa21* gene in either IR24 or TP309 genetic background conferred complete resistance to PXO99 (Gu et al. [2004]; Ikeda et al. [1990]; Song et al. [1995]; Wang et al. [1996]).

Table 2 Disease evaluation of WH421, Siputeh and TS4 for resistance to ***Xoo*** strains

		Lesion length (cm) and disease phenotype[a]		
Stains	**Origin**	**WH421**	**Siputeh**	**TS4**
1947	Africa	14.8 ± 7.4 (S)	24.1 ± 6.9 (S)	4.4 ± 4.5 (MR)
Aust-2031	Australia	2.8 ± 1.3 (R)	25.5 ± 8.4 (S)	1.3 ± 0.5 (R)
Aust-R3	Australia	1.8 ± 1.1 (R)	39.5 ± 8.8 (S)	1.0 ± 0.5 (R)
GD1358	China	7.0 ± 1.7 (MS)	23.9 ± 4.7 (S)	4.8 ± 3.6 (MR)
HB17	China	0.5 ± 0.3 (R)	31.2 ± 3.8 (S)	0.9 ± 0.6 (R)
HB21	China	1.7 ± 0.9 (R)	39.6 ± 6.7 (S)	1.3 ± 0.4 (R)
HLJ72	China	1.5 ± 0.6 (R)	16.2 ± 3.3 (S)	0.7 ± 0.4 (R)
JS49-6	China	0.9 ± 0.5 (R)	13.6 ± 4.9 (S)	1.3 ± 0.6 (R)
LN57	China	0.7 ± 0.4 (R)	30.7 ± 8.5 (S)	1.1 ± 0.8 (R)
NX42	China	1.3 ± 0.8 (R)	29.7 ± 5.6 (S)	1.5 ± 1.0 (R)
ZHE173	China	0.2 ± 0.1 (R)	34.2 ± 7.8 (S)	1.1 ± 0.8 (R)
CIAT1185	Columbia	3.5 ± 2.0 (MR)	10.0 ± 3.5 (S)	2.3 ± 0.7 (R)
A3842	India	4.3 ± 1.2 (MR)	37.9 ± 13.2 (S)	3.0 ± 1.1 (R)
A3857	India	18.8 ± 4.9 (S)	38.3 ± 7.4 (S)	10.5 ± 2.4 (S)
IXO56	Indonesia	6.0 ± 2.2 (MR)	30.3 ± 6.7 (S)	3.7 ± 2.0 (MR)

H75373	Japan	0.6 ± 0.6 (R)	25.5 ± 4.9 (S)	0.4 ± 0.1 (R)
T7174	Japan	0.6 ± 0.6 (R)	29.9 ± 6.7 (S)	0.6 ± 0.5 (R)
JW89011	Korea	16.1 ± 8.9 (S)	35.9 ± 7.0 (S)	9.2 ± 4.3 (S)
NXO260	Nepal	3.8 ± 6.1 (MR)	22.1 ± 2.6 (S)	1.1 ± 1.0 (R)
K202	Korea	15.1 ± 3.7 (S)	44.9 ± 10.7 (S)	17.0 ± 4.3 (S)
PXO71	Philippines	0.4 ± 0.3 (R)	37.8 ± 14.6 (S)	1.1 ± 0.7 (R)
PXO79	Philippines	0.7 ± 0.5 (R)	20.4 ± 7.5 (S)	2.8 ± 1.5 (R)
PXO86	Philippines	1.9 ± 0.8 (R)	28.7 ± 9.6 (S)	2.8 ± 1.2 (R)
PXO99	Philippines	11.8 ± 2.8 (S)	25.0 ± 7.1 (S)	5.9 ± 3.0 (MR)
PXO112	Philippines	0.6 ± 0.9 (R)	10.8 ± 3.3 (S)	1.1 ± 2.3 (R)
PXO113	Philippines	0.6 ± 0.4 (R)	27.8 ± 7.8 (S)	0.9 ± 0.6 (R)
Thai R7	Thailand	1.3 ± 0.5 (R)	13.2 ± 6.0 (S)	0.9 ± 0.7 (R)
Thai R2	Thailand	2.8 ± 1.3 (R)	35.4 ± 5.6 (S)	2.2 ± 1.4 (R)

[a]R, resistant (Lesion length ≤3.0 cm); MR, moderately resistant (3.0 cm < Lesion length ≤6.0 cm); MS, moderately susceptible (6.0 cm < Lesion length ≤9.0 cm); S, susceptible (Lesion length >9.0 cm).

Field evaluation of TS4 and Siputeh

The agronomic traits of TS4 and Siputeh were evaluated in two field trials and different growth seasons conducted in Banda Aceh, Indonesia and Lingshui, China, respectively. Due to presence of the *sd1* gene, TS4 (123.7 ± 2.1 cm) had shorter plant height than Siputeh (208.3 ± 3.9 cm) when they were grown in Banda Aceh (Table 3). The difference in plant height between Siputeh (214.6 ± 4.4 cm) and TS4 (106.7 ± 1.1 cm) became even greater when they were grown in Lingshui under short-day condition during the winter season (Figure 3). TS4 (117 days) had shorter growth duration than Siputeh (160 days) when they were planted in Banda Aceh, Indonesia (Table 3). However, both TS4 (162 days) and Siputeh (182 days) had long growth duration when they were planted in Lingshui under short-day condition during the winter (Table 3). TS4 produced more productive panicles per plant with higher seed-setting rate but fewer grain number per panicle than Siputeh grown in both field trials (Table 3). The greater number of productive panicles per plant and higher seed-setting rate also collectively contributed to the higher grain yield of TS4 than that of Siputeh grown in both field trails (Table 3). TS4 (27.6 g) had similar 1000-grain weight to Siputeh (27.4 g) when they were grown in Banda Aceh (Table 3). However, the 1000-grain weight of TS4 (34.1 g) were slightly heavier than that of Siputeh (31.7 g) when they were grown in Lingshui (Table 3). In summary, TS4 showed better agronomic traits with shorter plant height and growth duration and higher yield than Siputeh.

Table 3. Agronomic traits of Siputeh and TS4 plants grown in field trials

	Field trial 1[a]			Field trial 2[b]		
Agronomic trait	Siputeh	TS4	t-test[c]	Siputeh	TS4	t-test
Growth duration (days)	160	117		182	162	-
Plant height (cm)	208.3 ± 3.9	123.7 ± 2.1**	33.081	214.6 ± 4.4	106.7 ± 1.1**	41.120
Productive panicles per plant	6.5 ± 0.8	9.0 ± 0.4**	4.841	4.8 ± 0.2	5.7 ± 0.5*	3.298
Grain number per panicle	137.3 ± 3.4	129.7 ± 2.9*	2.946	172.4 ± 11.5	139.0 ± 16.3*	2.902
Seed setting rate (%)	70.1 ± 2.8	80.6 ± 2.6**	4.760	70.5 ± 1.2	82.8 ± 2.0**	9.097
1000-grain weight (g)	27.4 ± 0.7	27.6 ± 0.4	0.430	31.7 ± 0.6	34.1 ± 0.5**	8.978
Yield (t/ha)	4.3 ± 0.3	6.5 ± 0.1**	12.050	6.6 ± 0.7	8.0 ± 0.2*	3.538

[a]Field trial 1 was conducted in Banda Aceh, Indonesia, in the wet season of 2011/2012 (November 2011 to April 2012).
[b]Field trial 2 was conducted in Lingshui, China, in the winter season of 2013/2014 (November 2013 to May 2013).
[c]t-test is the comparision between Siputeh and TS4. $T_{0.05, 4}$ = 2.776, $T_{0.01, 4}$ = 4.604. "**" and "*" stand for significance difference at 0.01 and 0.05 probability levels, respectively.

Siputeh TS4

Figure 3. Plant height and maturation time of Siputeh and TS4 in the field. The semi-dwarf TS4 plants matured earlier than Siputeh with tall plant type. Picture was taken in the field in Lingshui, China, at 162 days after sowing.

Comparison of grain quality between TS4 and Siputeh

The grain quality of TS4 and Siputeh was measured using rice grains harvested from the two field trials. Both TS4 and Siputeh produce long grain rice with similar grain length and length-to-width ratio (Table 4). TS4 also has significantly lower degree of chalkiness than Siputeh in both field trials (Table 4). TS4 has lower amylose content (15.3-17.8%) than Siputeh (26.3-

28.9%), due to the presence of the Wx^b gene in TS4 (Table 4). For the rice grains harvested in field trial in Banda Aceh, TS4 has soft gel consistency while Siputeh had medium gel consistency (Table 4). Both TS4 and Siputeh had intermediate gelatinization temperature even though TS4 had slightly lower akali spreading value than Siputeh (Table 4). The results collectively showed that TS4 has better grain quality than Siputeh in terms of degree of chalkiness and amylose content.

Table 4. Grain quality of Siputeh and TS4

	Field trial 1[a]			Field trail 2[b]		
Traits	Siputeh	TS4	*t*-test[c]	Siputeh	TS4	*t*-test
Grain length (mm)[d]	7.4 ± 0.1 (Long)	7.6 ± 0.1 (Long)	2.449	7.2 ± 0.1 (Long)	7.3 ± 0.1 (Long)	1.225
Length-to-width ratio[e]	3.2 ± 0.1 (Slender)	3.1 ± 0.1 (Slender)	1.225	3.4 ± 0.1 (Slender)	3.2 ± 0.1 (Slender)	2.449
Degree of chalkiness (%)[f]	23.0 ± 2.0 (9)	12.0 ± 1.0** (5)	18.249	16.3 ± 0.6 (5)	9.3 ± 0.3** (1)	18.074
Amylose content (%)[g]	26.3 ± 0.1 (High)	15.3 ± 0.2** (Low)	67.361	28.9 ± 0.2 (High)	17.8 ± 0.2** (Low)	67.973
Gel consistency (mm)[h]	51.0 ± 1.6 (Medium)	90.0 ± 2.7** (Soft)	23.735	68.8 ± 5.0 (Soft)	94.5 ± 0.9** (Soft)	8.762
Alkali spreading value and gelatinization temperature[i]	5.0 ± 0.1 (Intermediate)	4.5 ± 0.1** (Intermediate)	8.660	5.0 ± 0.0 (Intermediate)	4.0 ± 0.1** (Intermediate)	6.124

[a]Field trial 1 was conducted in Banda Aceh, Indonesia, in the wet season of 2011/2012 (November 2011 to April 2012).
[b]Field trial 2 was conducted in Lingshui, China, in the winter season of 2013/2014 (November 2013 to May 2013).
[c]*t*-test is the comparision between Siputeh and TS4. $T_{0.05,4} = 2.776$, $T_{0.01,4} = 4.604$. "**" and "*" stand for significance difference at 0.01 and 0.05 probability levels, respectively.
[d]Category of grain length: Very long, grain length > 7.5 mm; Long, 6.6 mm < grain length ≤ 7.5 mm; Medium, 5.5 mm < grain length ≤ 6.6 mm; Short, grain length ≤ 5.5 mm.
[e]Grain shape based on length-to-width ratio: Slender, length-to-width ratio > 3.0; Medium, 2.0 < length-to-width ratio ≤ 3.0; Bold, length-to-width ratio ≤ 2.0.
[f]Scale for degree of chalkiness: 0, degree of chalkiness = 0; 1, 0 < degree of chalkiness ≤ 10%; 5, 10% < degree of chalkiness ≤ 20%, 9, degree of chalkiness > 20%.
[g]Classification of amylose content: Waxy, amylose content ≤ 2%; Very low, 2% < amylose content ≤ 9%; Low, 9% < amylose content ≤ 20%; Intermediate, 20% < amylose content ≤ 25%; High, amylose content > 25%.
[h]Classification of gel consistency: Soft, gel consistency > 60 mm; Medium, 40 mm < gel consistency ≤ 60 mm; Hard, gel consistency ≤ 40 mm.
[i]Grade of gelatinization temperature estimated by alkali spreading value: High (74.5°C ≤ gelatinization temperature < 80°C), 1 ≤ alkali spreading value < 2.5; Intermediate high (74°C ≤ gelatinization temperature < 74.5°C), 2.5 ≤ alkali spreading value < 3.5; Intermediate (70°C ≤ gelatinization temperature < 74°C), 3.5 ≤ alkali spreading value < 5.5; Low (gelatinization temperature < 70°C), 5.5 ≤ alkali spreading value ≤ 7.

DISCUSSION

Marker-assisted selection is a molecular breeding process whereby a molecular marker based on DNA variation is used for indirect selection of an interest agronomic trait. Marker-assisted selection is not affected by environments and can be conducted at any plant growth stage. It is especially useful for selection of recessive genes and biochemical traits in heterozygous plants. In this study, the *sd1* gene is a partially recessive gene that controls semi-dwarf phenotype. It is difficult or impractical to select the *sd1* gene in the backcrossing progeny according to plant height, especially under greenhouse condition. Similarly, the amylose content in rice grains can only be directly detected after harvesting, however, we need to know the genotype for crossing purpose before flowering. As for the pyramiding of *Xa4* and *Xa21* for bacterial blight resistance, both *R* genes have different resistance specificities and their resistance spectrums are over-lapped. It would be impractical to select the two *R* genes in a single line by conducting disease evaluation and phenotypic selection. To overcome these difficulties, we chose marker-assisted selection approach, which enable us to perform precisely and effectively multiple gene selection and/or pyramiding in limited breeding time.

Compared with Siputeh, the significant change in TS4 is the semi-dwarf plant type, which is mainly resulted from the introduction of the *sd1*gene. The semidwarf plant type of rice indiredtly contributes to lodging resistance in the field. It also increases grain-straw ratio, which ultimately results in higher yield. In the present study, TS4 produced higher yields than Siputeh in both field trials conducted in different location (Table 3). The results are consistent with the results of previously studies on the *sd1* gene introduced in other varieties (Hedden [2003]; Luo and Yin [2013]; Spielmeyer et al. [2002]). TS4 also showed significant reduction in growth duration than Siputeh. The growth duration of TS4 was about four months when it was grown in Banda Aceh, Indonesia (Table 3), which was similar to most of the typical modern rice varieties grown in tropical region. Short-growth duration makes it easier for farmers to plan growth season, synchronize sowing and harvest time with other varieties and manage diseases and insects in the field.

Like the *Xa4* and *Xa21* donor line WH421, TS4 provided broad-spectrum resistance to multiple *Xoo* strains tested in this study. It was noted that WH421 were susceptible or moderate susceptible to 1947, PXO99 and GD1358 while TS4 provided moderate resistance to the three *Xoo*strains (Table 2). It seems that either *Xa4* or *Xa21* or both *R* genes performed better for resistance to *Xoo* in TS4 than that in WH421. Previously, it was found that the *Xa21* gene in IRBB21 conferred complete resistance to PXO99 (Gu et al. [2004]; Ikeda et al. [1990]; Song et al. [1995]), however, in this study, neither WH421 nor TS4 could confer full resistance to the *Xoo* strain. It has been reported that the *R* gene-mediated resistance to *Xoo* in rice may be influenced by rice genetic backgrounds and developmental stages (Cao et al. [2007]). For instance, the *Xa21*-containing rice line Minghui was still susceptible to PXO99 (Zhang et al. [2006]).

The introduction of the *sd1* gene to Siputeh would significantly reduce its plant height and growth duration, which might affect rice grain quality. To overcome this potential shortcoming, we also introduced the Wx^b gene into TS4 through marker-assisted selection. Indeed, the Wx^b gene significantly reduced the amylose content in TS4. Amylose content strongly affects the cooking and eating quality of rice. Rice with amylose content at 25-30% usually tends to be firm and dry after cooking, rice with amylose content at 20-25% tends to be softer and stickier and rice with amylose content at <20% is generally quite soft and sticky. TS4 has amylose content at 15.3-17.8% and shows soft gel consistency. The results indicate that TS4 is generally softer and stickier than Siputeh, which has amylose content at 26.3-28.9% and shows medium to soft gel consistency.

CONCLUSION

An improved rice line, designed as TS4, has been developed to contain semi-dwarf gene *sd1*, low amylase content gene *Wx*^{b} and bacterial light resistance genes *Xa4* and *Xa21* through marker-assisted selection. TS4 has semi-dwarf phenotype with reduced growth duration, produces high yield with good grain quality and provides broad-spectrum resistance to *Xoo* strains. The development of TS4 enriches the diversity of local rice varieties with high yield potential and good grain quality.

METHODS

Plant materials and growth condition

Rice variety Siputeh is a local cultivar collected in Aceh province, Indonesia. Rice variety WH421 carrying *sd1*, *Wx*^{b}, *Xa4* and *Xa21* genes is a paternal line of hybrid rice in Mianhui 725 backgroud (Luo et al. [2012]). Rice plants were grown in the greenhouse at 32°C for 12.5 h (light) and 25°C for 11.5 h (dark).

PCR-based molecular markers

The molecular markers used in this study include the codominant STS (Sequence-Tagged Sites) marker sd1STS for the *sd1* gene (Luo and Yin [2013]), the co-dominant microsatellite DNA marker RM224 for the *Xa4* gene (Sun et al. [2003]), the co-dominant STS marker 21 for the*Xa21* gene (Chen et al. [2000]) and the co-dominant CAPS (Cleaved Amplified Polymorphic Sequences) marker PCR-*Acc* I for the *Wx*^{a} and*Wx*^{b} genes (Cai et al. [2002]). Oligo primers for PCR amplification of the molecular markers are listed in Table 1.

PCR amplification

PCR was performed on a PTC-100 programmable thermal controller (MJ Research). The PCR reaction mixture of 20 µl consisted of 1–100 ng of rice genomic DNA, 0.2 mM each of dNTPs, 0.2 µM of each primer, 2 µl of 10 × PCR buffer [500 mM KCl, 100 mM Tris–HCl (pH 8.3), 20 mM $MgCl_2$] and 0.5 unit of Taq polymerase. For PCR amplification of markers sd1STS, 4 ul of 5 × Q-solution (QIAGEN) was added to the PCR reaction mixture. Template DNA was initially denatured at 94°C for 2 min followed by 35 cycles of PCR amplification with the following parameters: a 30 s of denaturation at 94°C, a 40 s of primer annealing at 55°C for markers 21, RM224 and PCR-*Acc* I, 65°C for maker sd1STS, and 1 min of primer extension at 72°C for markers RM224 and PCR-*Acc* I, and 1.5 min for markers 21 and sd1STS. Finally, the reaction mixture was maintained at 72°C for 5 min of primer extension before completion. The PCR product of the marker PCR-*Acc* I was digested by *Acc* I for 4 h. The amplified products were electrophoretically resolved on a 1.5% agarose gel for marker sd1STS, a 2.0% agarose gel for marker 21, and a 3.5% agarose gel for marker PCR-*Acc* I and RM224 in 1 × TAE buffer.

Bacterial inoculation and disease scoring

Xoo strains were grown on PSA medium (10 g/l peptone, 10 g/l sucrose, 1 g/lglutamic acid, 16 g/l bacto-agar, and pH7.0) for about 60 hours at 28°C. Bacterial cells were suspended in sterile water and diluted to an opticla density (OD) at 600 nm of 0.5. Plants were inoculated by the leaf-clipping method (Kauffman et al. [1973]). Lesion length (L.L.) was measured at 14 days after inoculation. The disease symptoms were scored as resistant (R, L.L. ≤ 3.0 cm), moderately resistant (MR, 3.0 cm < L.L. ≤ 6.0 cm), moderately susceptible (MS, 6.0 cm < L.L. ≤ 9.0 cm) or susceptible (S, L.L. > 9 cm).

Field trial design and collection of important agronomic traits

Field trials were conducted in Banda Aceh, Indonesia in the season of 2011/2012 (November 2011 - March 2012) and in Lingshui, China in the winter season of 2013/2014 (November 2013 - May 2014), respectively. Rice lines were tested in replicated plots in each field trial. The plot size was 4.2 m × 3.8 m and the plant spacing was 20 cm × 20 cm in Banda Aceh, while the plot size was 4 m × 4 m and the plant spacing was 16.7 cm × 16.7 cm in Lingshui. Thirty plants (3 × 10 plants/per plot) were randomly selected for each tested line and scored for important agronomic traits, including growth duration, plant height, productive panicle per plant, number of grains per panicle, seed setting rate, weight of 1000 grains and grain yield per hectare. Statistical analysis was performed using a two-tailed *t*- test for independent samples.

Evaluation of grain quality

Rice grain quality parameters, including grain length, the ratio of length to width (L/W), chalkiness degree, amylose content (AC), gel consistency (GC) and alkali spreading value (ASV) were measured by the methods described previously (Cruz and Khush [2000]).

AUTHORS' INFORMATION

YL is the Senior Research Officer of Temasek Life Science Laboratory (TLL) and the Professor of Anhui Rice Research Institute (ARRI), Anhui Academy of

Agricultural Sciences (AAAS), with over 25 years of experiences on rice breeding; SZ is the Professor of Agriculture Faculty, Syiah Kuala University (SKU); BB is the Associate Professor of Agriculture Faculty, SKU; TM is the Associate Professor of ARRI, AAAS; ZL is the Director and Professor of ARRI, AAAS; JY is the Professor of ARRI and the President of AAAS; ZY is the Senior Principal Investigator of TLL.

Acknowledgements

The authors wish to thank Mr Y. Zhou for technical assistance in field trials. This research is supported by a grant from Temasek Foundation to Z. Yin.

Competing interests

The authors declare that they have no competing interests.

Authors' contribution

YL and ZY designed the experiments. YL conducted the experiments with the help from SZ, BB, TM, ZL and JY. YL and ZY wrote the manuscript. All authors read and approved the final manuscript.

REFERENCES

1. Asano K, Takashi T, Miura K, Qian Q, Kitano H, Matsuoka M, Ashikari M: Genetic and molecular analysis of utility of *sd1* alleles in rice breeding. *Breed Sci* 2007, 57: 53–58. 10.1270/jsbbs.57.53

2. Ashikari M, Sasaki A, Ueguchi-Tanaka M, Itoh H, Nishimura A, Datta S, Ishiyama K, Saito T, Kobayashi M, Khush GS: Loss-of-function of a rice gibberellin biosynthetic gene, GA20 oxidase (GA20ox-2), led to the rice'green revolution'. *Breed Sci* 2002, 52: 143–150. 10.1270/jsbbs.52.143

3. Cai XL, Wang ZY, Xing YY, Zhang JL, Hong MM: Aberrant splicing of intron 1 leads to the heterogeneous 5′ UTR and decreased expression of waxy gene in rice cultivars of intermediate amylose content. *Plant J* 1998, 14: 459–465.

4. Cai XL, Liu QQ, Tang SZ, Cu MH, Wang ZY: Development of molecular marker for screening the rice cultivars with intermediate amylose content in *Oryza sativa* subsp. *indica*. *J Plant Physiol and Mol Biol* 2002, 28: 137–144.

5. Cao YL, Ding XH, Cai M, Zhao J, Lin YJ, Li XH, Xu CG, Wang SP: Expression pattern of a rice disease resistance gene *Xa3/Xa26* is differentially regulated by the genetic backgrounds and developmental stages that influence its function. *Genetics* 2007, 177: 523–533.

6. Chen S, Lin X, Xu C, Zhang Q: Improvement of bacterial blight resistance of 'Minghui 63', an elite restorer line of hybrid rice, by molecular marker-assisted selection. *Crop Sci* 2000, 40: 239–244. 10.2135/cropsci2000.401239x

7. Cruz ND, Khush G: Rice Grain Quality Evaluation Procedures. In *Aromatic Rices*. Edited by: Singh RK, Singh US, Khush GS. Oxford and IBH Pulishing Co. Pvt. Ltd, New Dehli, Calcutta; 2000:24–36.

8. Datta K, Baisakh N, Thet KM, Tu J, Datta SK: Pyramiding transgenes for multiple resistance in rice against bacterial blight, yellow stem borer and sheath blight. *Theor Appl Genet* 2002, 106: 1–8.

9. Gnanamanickam SS, Priyadarisini VB, Narayanan NN, Vasudevan P, Kavitha S: An overview of bacterial blight disease of rice and strategies for its management. *Curr Sci India* 1999, 77: 1435–1444.

10. Gu K, Tian D, Yang F, Wu L, Sreekala C, Wang D, Wang GL, Yin Z: High-resolution genetic mapping of *Xa27* (t), a new bacterial blight resistance gene in rice, *Oryza sativa* L. *Theor Appl Genet* 2004, 108: 800–807. 10.1007/s00122-003-1491-x

11. Hedden P: The genes of the green revolution. *Trends Genet* 2003, 19: 5–9. doi:S0168–9525(02)00009–4 doi:S0168-9525(02)00009-4 10.1016/S0168-9525(02)00009-4

12. Hori Y, Fujimoto R, Sato Y, Nishio T: A novel wx mutation caused by insertion of a retrotransposon-like sequence in a glutinous cultivar of rice (*Oryza sativa*). *Theor Appl Genet* 2007, 115: 217–224. 10.1007/s00122-007-0557-6

13. Huang N, Angeles ER, Domingo J, Magpantay G, Singh S, Zhang G, Kumaravadivel N, Bennett J, Khush GS: Pyramiding of bacterial blight resistance genes in rice: marker-assisted selection using RFLP and PCR. *Theor Appl Genet* 1997, 95: 313–320. 10.1007/s001220050565

14. Ikeda R, Khush G, Tabien R: A new resistance gene to bacterial blight derived from *O. longistaminata* . *Jpn J Breed* 1990, 40: 280–281.

15. Jennings PR: Plant type as a rice breeding objective. *Crop Sci* 1964, 4: 13–15. 10.2135/cropsci1964.0011183X000400010005x

16. Kauffman H, Reddy A, Hsieh S, Merca S: An improved technique for evaluating resistance of rice varieties to *Xanthomonas oryzae* . *Plant Dis Rep* 1973, 57: 537–541.

17. Larkin PD, Park WD: Transcript accumulation and utilization of alternate and non-consensus splice sites in rice granule-bound starch synthase are temperature-sensitive and controlled by a single-nucleotide polymorphism. *Plant Mol Biol* 1999, 40: 719–727. 10.1023/A:1006298608408

18. Luo Y, Yin Z: Marker-assisted breeding of Thai fragrance rice for semi-dwarf phenotype, submergence tolerance and disease resistance to rice blast and bacterial blight. *Mol Breed* 2013, 32: 709–721.

19. Luo Y, Sangha JS, Wang S, Li Z, Yang J, Yin Z: Marker-assisted breeding of Xa4, Xa21 and Xa27 in the restorer lines of hybrid rice for broad-spectrum and enhanced disease resistance to bacterial blight. *Mol Breed* 2012, 30: 1601–1610. 10.1007/s11032-012-9742-7

20. Mew TW, Alvarez AM, Leach JE, Swings J: Focus on bacterial blight of rice. *Plant Dis* 1993, 77: 5–12. 10.1094/PD-77-0005

21. Sano Y: Differential regulation of Waxy gene expression in rice endosperm. *Theor Appl Genet* 1984, 68: 467–473. 10.1007/BF00254822

22. Sasaki A, Ashikari M, Ueguchi-Tanaka M, Itoh H, Nishimura A, Swapan D, Ishiyama K, Saito T, Kobayashi M, Khush G: A mutant gibberellin-synthesis gene in rice. *Nature* 2002, 416: 701–702. 10.1038/416701a

23. Singh S, Sidhu JS, Huang N, Vikal Y, Li Z, Brar DS, Dhaliwal HS, Khush GS: Pyramiding three bacterial blight resistance genes (*xa5* , *xa13*and *Xa21*) using marker-assisted selection into indica rice cultivar PR106. *Theor Appl Genet* 2001, 102: 1011–1015. 10.1007/s001220000495

24. Song W-Y, Wang G-L, Chen L-L, Kim H-S, Pi L-Y, Holsten T, Gardner J, Wang B, Zhai W-X, Zhu L-H, Fauquet C, Ronald PC: A receptor kinase-like protein encoded by the rice disease resistance gene, Xa21. *Science* 1995, 270: 1804–1806. 10.1126/science.270.5243.1804

25. Spielmeyer W, Ellis MH, Chandler PM: Semidwarf (*sd-1*), "green revolution" rice, contains a defective gibberellin 20-oxidase gene. *Proc Natl Acad Sci USA* 2002, 99: 9043–9048. 10.1073/pnas.132266399

26. Su Y, Rao Y, Hu S, Yang Y, Gao Z, Zhang G, Liu J, Hu J, Yan M, Dong G: Map-based cloning proves qGC-6, a major QTL for gel consistency of japonica/indica cross, responds by Waxy in rice (*Oryza sativa* L.). *Theor Appl Genet* 2011, 123: 859–867. 10.1007/s00122-011-1632-6

27. Suh JP, Jeung JU, Noh TH, Cho YC, Park SH, Park HS, Shin MS, Kim CK, Jena KK: Development of breeding lines with three pyramided resistance genes that confer broad-spectrum bacterial blight resistance and their molecular analysis in rice. *Rice* 2013, 6: 1–11. 10.1186/1939-8433-6-5

28. Suhartini T (1991) Preliminary Evaluation of Performances and Yields of Rice Varieties and Lines at Lebak area. Paper presented at the Proceedings of a seminar of research outcomes of Bogor Research Institute of food crops, Bogor (Indonesia), 11–12 March, 1991 Suhartini T (1991) Preliminary Evaluation of Performances and Yields of Rice Varieties and Lines at Lebak area. Paper presented at the Proceedings of a seminar of research outcomes of Bogor Research Institute of food crops, Bogor (Indonesia), 11–12 March, 1991

29. Sun X, Yang Z, Wang S, Zhang Q: Identification of a 47-kb DNA fragment containing *Xa4* , a locus for bacterial blight resistance in rice.*Theor Appl Genet* 2003, 106: 683–687.

30. Tian Z, Qian Q, Liu Q, Yan M, Liu X, Yan C, Liu G, Gao Z, Tang S, Zeng D: Allelic diversities in rice starch biosynthesis lead to a diverse array of rice eating and cooking qualities. *Proc Natl Acad of Sci USA* 2009, 106: 21760–21765. 10.1073/pnas.0912396106

31. Wang ZY, Zheng FQ, Shen GZ, Gao JP, Snustad DP, Li MG, Zhang JL, Hong MM: The amylose content in rice endosperm is related to the posttranscriptional regulation of the waxy gene. *Plant J* 1995, 7: 613–622. 10.1046/j.1365-313X.1995.7040613.x

32. Wang G-L, Song W-Y, Ruan D-L, Sideris S, Ronald PC: The cloned gene, Xa21, confers resistance to multiple *Xanthomonas oryzae* pv.*oryzae* isolates in transgenic plants. *Mol Plant-Microbe Interact* 1996, 9: 850–855. 10.1094/MPMI-9-0850

33. Zhang Q: Genetics and improvement of bacterial blight resistance of hybrid rice in China. *Rice Sci* 2009, 16: 83–92. 10.1016/S1672-6308(08)60062-1

34. Zhang J, Li X, Jiang G, Xu Y, He Y: Pyramiding of *Xa7* and *Xa21* for the improvement of disease resistance to bacterial blight in hybrid rice.*Plant Breed* 2006, 125: 600–605. 10.1111/j.1439-0523.2006.01281.x

35. Zhu C, Shen W, Zhao H, Wan J: Advances in researches of the application of low-amylose content rice gene for breeding. *Sci Agri Sinica*2003, 37: 157–162.

Chapter 5

Integration of Genomics into Rice Breeding

Shuichi Fukuoka , Kaworu Ebana , Toshio Yamamoto ,Masahiro Yano

QTL Genomics Research Center, National Institute of Agrobiological Sciences, Kannondai 2-1-2, Tsukuba, Ibaraki 305-8602, Japan

ABSTRACT

One of the major challenges in genetics has been to identify the nucleotide polymorphisms responsible for phenotypic variation. Through intensive analysis, several major quantitative trait loci (QTLs) for agronomic traits in rice have been identified and the underlying candidate genes have been delimited. Advanced mapping populations, including chromosome segment substitution lines, have enhanced the power of genetic analysis to detect QTL alleles, even those with minor effects. Recent examples of marker-assisted selection have proven the potential of this strategy for crop improvement. The genome-wide discovery of single nucleotide polymorphisms (SNPs), even among closely related cultivars, has enhanced the power of allele mining in a wide range of rice breeding materials. An array-based SNP genotyping system can be used to visualize pedigree haplotypes in breeding materials, including landraces and modern cultivars. All of these technologies are accelerating the genetic dissection of complex agronomic traits and further improvement of rice.

Keywords: Quantitative trait locus Genetic mapping Marker-assisted selection Single-nucleotide polymorphism Haplotype

INTRODUCTION

Rice genomics has enhanced our understanding of the genetic control of agronomic traits and provided tools for crop improvement. Most agronomic traits are controlled by multiple genes and show complex, quantitative inheritance. Therefore, efforts have been concentrated on identification of quantitative trait loci (QTLs) associated with beneficial traits. QTL alleles with economic value have been tagged by DNA markers and introduced into elite cultivars by marker-assisted selection (MAS). The complete sequence of the rice genome (IRGSP 2005) has supported sophisticated procedures to facilitate genetic analysis of traits of interest, and several genes for agronomic traits have been cloned (Yamamoto et al. 2009). These studies have contributed to our understanding both of the genetic control of several important traits and of the molecular mechanisms of trait expression. Additionally, recent advances in genomics technology have enabled large-scale sequencing and genotyping in rice and other organisms. These innovations have opened up possibilities for genomics-assisted breeding, a new paradigm of selection strategy in plant breeding. This paper introduces recent progress in the genetic dissection of complex traits, marker-assisted introgression of agronomic traits, and whole-genome selection strategies that accelerate the genomics-assisted breeding of rice.

UNCOVERING NATURALLY OCCURRING VARIATION IN COMPLEX AGRONOMIC TRAITS

Naturally occurring variation includes QTL alleles useful for diverse agricultural practices. In contrast to induced mutations, which often have an abnormal phenotype, QTL alleles are a source of genetic variation that is less likely to confer a penalty on agricultural productivity. In the last two decades, much attention has been paid to the genetic and molecular dissection of several important agronomic traits. Around 8,000 QTLs have been detected so far and summarized in the Gramene-QTL database (Jaiswal et al. 2002). Several QTLs with larger effects have been cloned by map-based strategies (Yamamoto et al. 2009); however, precise mapping of QTL alleles with small effects has been a challenge.

An excellent example of uncovering genetic control of a useful trait is the analysis of heading date, a key determinant of the adaptation of rice to different cultivation areas and cropping seasons. Several QTLs have been mapped as single Mendelian factors (Yano et al. 2001; Lin et al.2002, 2003; Doi et al. 2004), and map-based cloning of the underlying genes has pinpointed the nucleotide polymorphisms responsible for phenotypic variation (Yano et al. 2000; Takahashi et al. 2001; Kojima et al. 2002; Doi et al. 2004; Xue et al. 2008). These studies suggest both functional similarity and divergence of genetic control between rice and *Arabidopsis* (Izawa 2007; Tsuji et al. 2008).

Traits associated with grain productivity are a major target for crop improvement. A QTL for semi-dwarf plant height, *sd1*, contributed to the "green revolution" in rice production and was found to contain a defective gibberellin 20-oxidase gene (Sasaki et al. 2002). Another QTL, *Grain number 1a* (*Gn1a*), was identified as encoding cytokinin oxidase/dehydrogenase (Ashikari et al. 2005). These studies illustrate the important role of plant hormones in growth control and in natural trait variation. Recent studies of grain size and grain filling have identified previously uncharacterized genes for which specific alleles have been selected during crop domestication (Shomura et al. 2008; Wang et al. 2008).

Tolerance to environmental stress is an important factor for cultivated crops, enabling them to adapt to specific environments. The rice*Submergence 1* (*Sub1*) gene encodes a putative ethylene-responsive factor that allow rice to survive complete submergence for more than 2 weeks (Xu et al. 2006). *SNORKEL1* and *SNORKEL2* encode transcription factors for ethylene signaling. These factors allow deep-water cultivars to elongate rapidly in response to submergence, helping the plant keep its leaves above water (Hattori et al. 2009). These finding suggest that rice has two different strategies to adapt to submergence due to flooding, both of which are regulated by accumulated ethylene inside the submerged plant. These flooding-tolerant alleles have been found in limited groups of rice, supporting the idea that they may have been lost during breeding selection. This analysis offers the possibility of reintroducing such beneficial QTL alleles into rice.

Durable disease resistance has been a breeding target for many years, because most resistance genes identified so far show complete but race-specific resistance. Blast-resistance QTL allele *pi21*, identified from a durably resistant cultivar, encodes a proline-rich protein structurally different from previously identified resistance proteins (Fukuoka and Okuno 2001; Fukuoka et al. 2009). These studies suggest that a proline-rich motif that may have a role in protein–protein interaction has a crucial role in the protein's function. The allele enhances the defense response and confers non-race-specific resistance, although the response against pathogen attack is not as strong as the typical resistance (*R*-gene) response. This type of nonspecific, weaker resistance might contribute to decreased selective pressure for pathogens to overcome host resistance (Fukuoka et al. 2009).

MARKER-ASSISTED ENHANCEMENT OF COMPLEX AGRONOMIC TRAITS

MAS is a promising strategy to enhance traits with economic and agricultural value in rice (Yamamoto et al. 2009). DNA markers for genes and QTLs for agronomic traits allow breeders to precisely select plants with beneficial traits in breeding programs. Recent examples of MAS address important issues in the enhancement of agronomic traits under complicated genetic control.

In one example, MAS enhanced the cropping potential of an elite cultivar by enabling the development of versions of the cultivar with diverse heading dates. Four QTLs for heading date—*Hd6*, *Hd1*, *Hd4*, and *Hd5*—were introgressed from 'Kasalath' into 'Koshihikari', a leading cultivar in Japan (Takeuchi et al. 2006). The size of the introgressed chromosomal segments in those lines was very small: 300-600 kb in nearly isogenic lines (NILs) for *Hd1*, *Hd6*, and *Hd5*. The resulting Koshihikari NILs provide versions of the cultivar with early and late heading dates but without significant differences in agronomic performance from the original cultivar. In another example, a major QTL allele of *Sub1* that confers improved submergence tolerance was backcrossed into several cultivars, providing good performance in yield and other agronomic traits as well as tolerance to submergence (Neeraja et al. 2007; Septiningsih et al. 2009). The results of these backcrosses confirmed the usefulness of this allele in submergence-prone areas in Southeast Asia.

An effective application of MAS in breeding is in the elimination of linkage drag, the co-introduction of beneficial and undesirable agricultural traits from exotic genetic resources. The durable blast resistance gene *pi21* was found to be linked with gene(s) associated with inferior eating quality within a 40-kb distance (Fukuoka et al. 2009). In that study, desirable recombinants between *pi21* and the genes conferring inferior eating quality were successfully selected from a large breeding population by using DNA markers for the region around *pi21*. The availability of DNA markers closely linked to the genes of interest enabled a breakthrough in the development of durably resistant cultivars that had not been achieved during 80 years of conventional breeding in Japan. This is a clear example of selection in breeding that could not have been achieved without the use of DNA markers tightly linked to the genes and DNA regions of interest.

MAS also allow us to combine beneficial QTL alleles in one plant in a relatively short time (Ashikari and Matsuoka 2006). This concept, gene pyramiding, was successfully used to develop a new high-yield line with lodging resistance in an elite genetic background. This was done by crossing an NIL containing a gene for semi-dwarfism with one containing a gene/QTL encoding increased grain number (Ashikari et al. 2005). Gene pyramiding is also useful to improve traits controlled by multiple QTLs.

ALLELE MINING FOR COMPLEX AGRONOMIC TRAITS WITH CHROMOSOME SEGMENT SUBSTITUTION LINES

A key factor for success in the identification and use of QTLs for agronomic traits is the magnitude of the genetic effect of an allele relative to the effect of the environment. In some cases, a QTL has an effect large enough to be useful as a single gene. For example, the major QTL*Grain number 1a* was cloned by using a map-based strategy; its identification contributed both to our understanding of the genetic control of spikelet development in rice and to the improvement of grain number per panicle (Ashikari et al. 2005). However, in general, a single major QTL allele is not enough to improve a trait, and additional alleles must be identified to combine with it. Information on "validated" minor QTL alleles is limited even for traits that have been characterized by many researchers (Yamamoto et al. 2009). A major reason for the difficulty in dealing with minor QTLs is the low detection power of QTL analysis using F_2 and recombinant inbred lines (RILs), owing to their highly heterogeneous genetic backgrounds.

To overcome problems associated with the limited detection power of conventional genetic mapping populations, researchers have developed chromosome segment substitution lines (CSSLs; Ebitani et al. 2005; Takai et al. 2007; Ando et al. 2008; Hori et al. 2010). In these novel plant materials, a particular chromosomal region in the recurrent parent is replaced by the corresponding segment from a donor genotype. CSSLs allow one to detect QTL alleles in the substituted region with high detection power and to efficiently survey QTLs within that part of the donor chromosome.

Genetic analysis using CSSLs provides a systematic research flow from the detection of QTLs for useful traits to their cloning and molecular breeding (Fig. 1). Once a phenotypic difference between a CSSL and the recurrent parent is identified, researchers can immediately identify the chromosomal region of that QTL, because each CSSL normally contains a single chromosomal segment from the donor. Comparison of the position of donor chromosome regions among a set of CSSLs for that chromosome enables us to further delineate the candidate chromosomal region (substitution mapping). Mapping populations and NILs for further genetic analysis of the QTL without background noise (i.e., the effect of other segregating QTLs) can be obtained readily from the progeny of a single backcross with that CSSL. Such genetic materials allow us to determine the genetic effect of a given QTL allele, even one with a small effect (Miura et al. 2001; Sato et al. 2003; Ueda et al. 2004; Ebitani et al. 2005; Ando et al. 2008; Hori et al. 2010). After confirmation of genetic effect of a QTL, map-based cloning can be started quickly from these materials, by developing populations for high-resolution mapping.

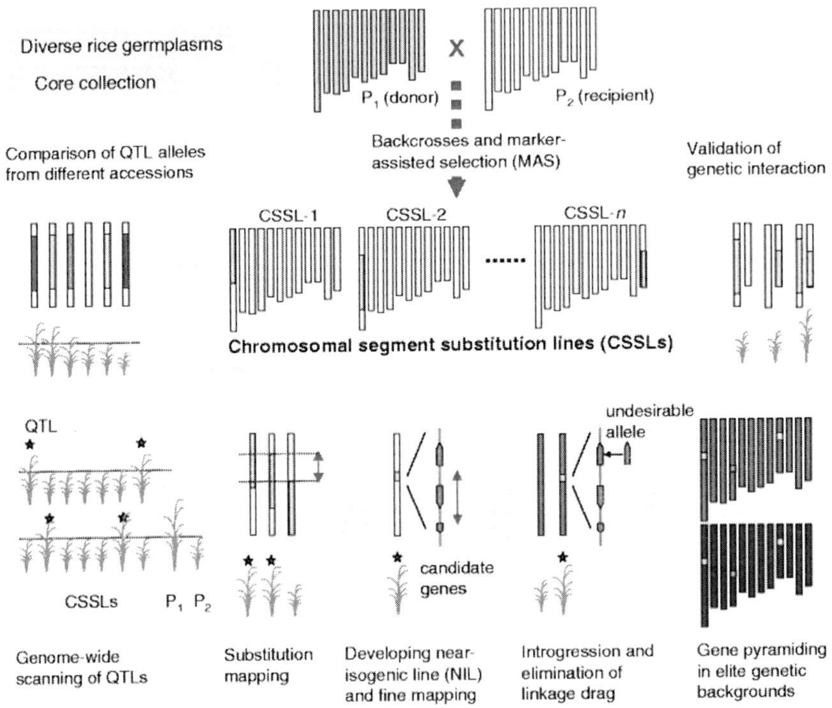

Figure 1. Research flow from QTL mining to molecular breeding. See text for details.

CSSLs are a powerful tool to determine phenotypic differences conferred by various QTL alleles. Sequence analysis of the *pi21* blast resistance gene from a number of cultivars identified 12 alleles (based on insertion–deletion polymorphisms in the region) that might be involved with the protein's function (Fukuoka et al. 2009). However, it was sometimes difficult to infer the effect of each mutation from the sequence only. Developing CSSLs for the chromosome region of interest by using diverse germplasm as the donor material provided a solution to this problem. Evaluation of this series of CSSLs, each possessing one of the *pi21* alleles in the genetic background of a susceptible cultivar, clearly demonstrated that all except one were susceptible, and suggested that two deletions in the allele are required to confer resistance (Fukuoka et al. 2009). The analysis also revealed that the resistance allele was present in some strains of *japonica* rice, suggesting that it could improve blast resistance of rice worldwide. This example illustrates the usefulness of information on the distribution of QTL alleles when a breeding strategy is being planned.

Another application of CSSLs is the analysis of genetic interaction. Although the power of detection of epistatic interactions in CSSLs is not as high as that in primary mapping populations such as F_2s and RILs, CSSLs provide unique opportunities to detect interactions among chromosomal regions involved in target traits within a highly homogeneous genetic background. In previous studies, NILs containing single QTLs have been developed and then crossed to combine two or three QTLs into a single genetic background so as to clarify the epistatic interactions among the QTLs (Lin et al. 2000; Yamamoto et al. 2000; Lin et al. 2003). However, it takes much time and effort to develop NILs for all pairs of putatively interacting QTLs. In contrast, CSSLs already developed for allele mining or other purposes can be used to validate a putative epistatic interaction relatively quickly by using progeny from a cross between CSSLs containing the QTLs of interest.

The power of research using CSSLs depends on the identification and availability of donor cultivars. We analyzed the genetic structure of cultivated rice by genome-wide RFLP analysis and selected core collections that represent the genetic diversity (Kojima et al. 2005). We are now developing CSSLs from a wide range of accessions selected from our core collection as donors and Koshihikari as the recurrent parent. These materials will enhance the potential of systematic research for allele mining and explaining complex agronomic traits in rice.

GENOME-WIDE DISCOVERY OF SINGLE NUCLEOTIDE POLYMORPHISMS AND DEVELOPMENT OF GENOTYPING ARRAYS

Until now, simple sequence repeats have been the major source of DNA markers used for genetic analysis and rice breeding. We believe that single nucleotide polymorphisms (SNPs) will become the main source of genetic markers, because the SNP is the most frequent type of DNA polymorphism. After release of the detailed rice genome sequence by the International Rice Genome Sequencing Project (IRGSP 2005), SNPs between *japonica* 'Nipponbare' and each of two *indica* lines (93-11 and GLA3) were found throughout the genome (Han and Xue 2003; Feltus et al. 2004; Shen et al. 2005; Wang et al. 2009). With the technological innovations of high-throughput sequencing (Schuster 2008), SNP information from various rice cultivars will be discovered rapidly and cost-effectively. Recently, a genome-wide re-sequencing analysis of 20 diverse *indica* and *japonica* cultivars produced about 160,000 SNPs (McNally et al. 2009), which enabled us to introgress a large number of QTLs at high resolution in our rice breeding program. However, it is still unrealistic to determine SNP genotypes by whole-genome sequencing of diverse accessions, improved cultivars, and segregating progeny of the crosses used for rice breeding. Genome-wide SNP typing arrays can be used to solve this problem (Shen et al. 2005; Gupta et al. 2008), simultaneously genotyping thousands of SNPs. The use of SNPs will help to validate the effectiveness of genome-wide association mapping (GWA; Breseghello and Sorrells 2006) for exploiting natural genetic resources with economic value.

In one application of SNP typing arrays to crop breeding, genotypes of improved barley germplasms were surveyed at 1,524 SNP loci (Rostoks et al. 2006). This study examined the extent of linkage disequilibrium (LD) and the potential of GWA among European elite barley cultivars, illustrating the challenge of associating haplotypes with phenotypes resulting from artificial selection at a genome-wide level. Although the application of GWA may be relied on the use of appropriate population and reliable trait evaluation, this study demonstrated a potential power of the connection between haplotype definition and selection in breeding program. In rice, whole-genome genotyping of rice based on SNPs between closely related elite *japonica* cultivars enabled the definition of 18 consensus haplotype blocks that have been inherited from ancestral landraces, and indicated that haplotype diversity has decreased during the history of rice breeding (Yamamoto et al.2010). The accumulated haplotype information will help us to plan breeding programs and to monitor recombination events between haplotype blocks. Eventually, whole-genome SNP typing assays will facilitate efficient rice improvement as they become integrated into current MAS systems.

To obtain sufficient numbers of SNPs in the populations that breeders actually deal with, it will be necessary to evaluate the frequency of the SNPs in these populations. Our recommended strategy is (1) to conduct whole-genome sequencing of several representative cultivars in a set of accessions (selected from ecotypes identified by population structure analysis or from genetically similar breeding lines) and (2) to detect the genotypes of all the lines in the population by using SNP typing arrays. However, we do not yet know the appropriate density of SNPs to discriminate all the recombination events in breeding populations at minimum expense. A comprehensive haplotype database based on genome-wide SNPs will be needed before the typing arrays can be designed.

FROM GENE SELECTION TO GENOME SELECTION

A strategy for the integration of genomics-oriented research activities into rice breeding is summarized in Fig. 2. Currently, marker-assisted breeding is conducted by direct or indirect selection of a target allele by using DNA markers associated with the locus (upper part of Fig. 2), but a trait of interest is not always determined by a single locus. Rather, several loci with minor phenotypic effects may epistatically or cumulatively determine phenotypes. It has been difficult to deal with such genes with minor effects in MAS, but genome-wide SNP analysis will provide new solutions and tools for this problem. By using SNP analysis, we can define haplotype blocks (lower part of Fig. 2). When a haplotype includes more than one valuable allele, it is important to define a set of DNA markers corresponding to chromosomal blocks large enough to confer the desired phenotypic effect. When a haplotype causes both preferable and inferior phenotypes because of linkage drag, it needs to be split by recombination to create a new desirable haplotype. A large number of available SNPs will help to illustrate the genetic structure of haplotypes in detail. However, our ability to define such haplotypes needs to be improved. The decay length of LD calculated among cultivars used in the analysis can offer a clue to the size of the haplotype.

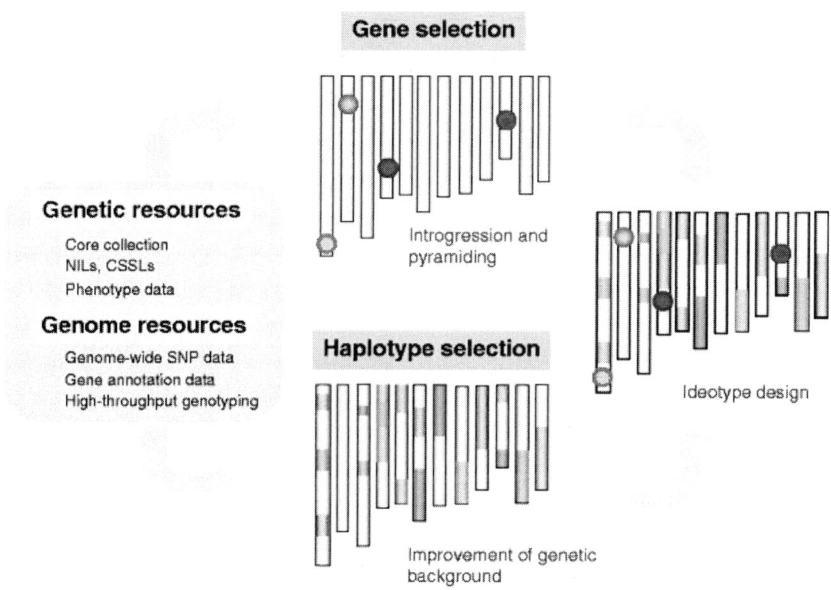

Figure 2. Genome-assisted breeding: integration of gene selection and haplotype selection. See text for details.

To fully utilize haplotypes in breeding, additional progress will be needed in methods to associate a given haplotype with phenotypic performance. Currently, there is no systematic method to link a certain haplotype with a specific phenotype. Although GWA and genomic selection (Zhong et al. 2009) have been proposed as tools for plant genetics and breeding, the power for detecting useful alleles in such approaches and their effectiveness in breeding remain to be clarified. From the view of experimental materials, the validation of nested association mapping populations (McMullen et al. 2009) and multi-parent advanced generation inter-cross populations (Cavanagh et al. 2008) has been performed in maize and *Arabidopsis*, respectively. These approaches need to be practiced in rice. Those efforts will provide new criteria for selection. Finally, the major QTL alleles identified by any map-based strategy must be introgressed into preferable genetic backgrounds selected by haplotype analysis (Fig. 2). Future progress in genotype-driven rice breeding could be realized not only by gene (allele) replacement, but also by background (haplotype) improvement.

FUTURE PERSPECTIVES

MAS, developed as an armchair theory about 20 years ago, have been successful at targeting several QTLs with major effects. Technological improvements in genomics and the establishment of materials such as CSSLs for allele mining will accelerate the accumulation of information on beneficial QTL alleles and linked DNA markers. However, rice breeders need additional QTL alleles, including those with minor effects, for further improvement of this crop. With this goal in mind, the workability of a given strategy largely depends on how we deal with diverse germplasm and how we reliably phenotype materials and identify useful genes with minor effects, to integrate such information into our research flow. In the near future, genotyping will be facilitated by automated and low-cost procedures. Researchers will spend more effort on developing plant materials suitable for phenotyping and on the improvement of phenotyping skills and technologies. The combination of all these resources and information will facilitate the dissection of complex traits with agricultural importance, such as yield performance, drought tolerance, and grain quality. The output from such efforts will further enhance genomics-assisted breeding to more effectively meet future human needs.

DECLARATIONS

Acknowledgements

We thank members of the QTL Genomics Research Center (QTL-GRC) at NIAS for helpful discussions and suggestions in the preparation of this manuscript. Work in QTL-GRC is supported by funds (Integrated Research Project for Plant, Insect and Animals Using Genome Technology and Genomics for Agricultural Innovation, QTL-2001, NVR-0001, NVR-0002) from the Ministry of Agriculture, Forestry and Fisheries, Japan.

REFERENCES

1. Ando T, Yamamoto T, Shimizu T, Ma XF, Shomura A, Takeuchi Y, et al. Genetic dissection and pyramiding of quantitative traits for panicle architecture by using chromosomal segment substitution lines in rice. Theor Appl Genet. 2008;116:881–90.

2. Ashikari M, Matsuoka M. Identification, isolation and pyramiding of quantitative trait loci for rice breeding. Trends Plant Sci. 2006;11:344–50.

3. Ashikari M, Sakakibara H, Lin S, Yamamoto T, Takashi T, Nishimura A, et al. Cytokinin oxidase regulates rice grain production. Science. 2005;309:741–5.

4. Breseghello F, Sorrells ME. Association mapping of kernel size and milling quality in wheat (*Triticum aestivum* L.) cultivars. Genetics. 2006;172:1165–77.

5. Cavanagh C, Morell M, Mackay I, Powell W. From mutations to MAGIC: resources for gene discovery, validation and delivery in crop plants. Curr Opin Plant Biol. 2008;11:215–21.

6. Doi K, Izawa T, Fuse T, Yamanouchi U, Kubo T, Shimatani Z, et al. *Ehd1*, a B-type response regulator in rice, confers short-day promotion of flowering and controls *FT-like* gene expression independently of *Hd1*. Genes Dev. 2004;18:926–36

7. Ebitani T, Takeuchi Y, Nonoue Y, Yamamoto T, Takeuchi K, Yano M. Construction and evaluation of chromosome segment substitution lines carrying overlapping chromosome segments of indica rice cultivar 'Kasalath' in a genetic background of japonica elite cultivar 'Koshihikari'. Breed Sci. 2005;55:65–73.

8. Feltus FA, Wan J, Schulze SR, Estill JC, Jiang N, Paterson AH. An SNP resource for rice genetics and breeding based on subspecies*indica* and *japonica* genome alignments. Genome Res. 2004;14:1812–9.

9. Fukuoka S, Okuno K. QTL analysis and mapping of *pi21*, a recessive gene for field resistance to rice blast in Japanese upland rice. Theor Appl Genet. 2001;103:185–90.

10. Fukuoka S, Saka N, Koga H, Ono K, Shimizu T, Ebana K, et al. Loss of function of a proline-containing protein confers durable disease resistance in rice. Science. 2009;325:998–1001.

11. Gupta PK, Rustgi S, Mir RR. Array-based high-throughput DNA markers for crop improvement. Heredity. 2008;101:5–18.

12. Han B, Xue Y. Genome-wide intraspecific DNA-sequence variations in rice. Curr Opin Plant Biol. 2003;6:134–8.

13. Hattori Y, Nagai K, Furukawa S, Song XJ, Kawano R, Sakakibara H, et al. The ethylene response factors *SNORKEL1* and *SNORKEL2*allow rice to adapt to deep water. Nature. 2009;460:1026–30.

14. Hori K, Sugimoto K, Nonoue Y, Ono N, Matsubara K, Yamanouchi U, et al. Detection of quantitative trait loci controlling pre-harvest sprouting resistance by using backcrossed populations of japonica rice cultivars. Theor Appl Genet. 2010;120:1547–57.

15. IRGSP. The map-based sequence of the rice genome. Nature. 2005;436:793–800.

16. Izawa T. Day length measurements by rice plants in photoperiodic short-day flowering. Int Rev Cytol. 2007;256:191–222.

17. Jaiswal P, Ware D, Ni J, Chang K, Zhao W, Schmidt S, et al. Gramene: development and integration of trait and gene ontologies for rice. Comp Funct Gen. 2002;3:132–6.

18. Kojima S, Takahashi Y, Kobayashi Y, Monna L, Sasaki T, Araki T, et al. *Hd3a*, a rice ortholog of the *Arabidopsis FT* gene, promotes transition to flowering downstream of *Hd1* under short-day conditions. Plant Cell Physiol. 2002;43:1096–105.

19. Kojima Y, Ebana K, Fukuoka S, Nagamine T, Kawase M. Development of an RFLP-based rice diversity research set of germplasm. Breed Sci. 2005;55:431–40.

20. Lin HX, Yamamoto T, Sasaki T, Yano M. Characterization and detection of epistatic interactions of 3 QTLs, *Hd1*, *Hd2*, and *Hd3*, controlling heading date in rice using nearly isogenic lines. Theor Appl Genet. 2000;101:1021–8

21. Lin HX, Ashikari M, Yamanouchi U, Sasaki T, Yano M. Identification and characterization of a quantitative trait locus, *Hd9*, controlling heading date in rice. Breed Sci. 2002;52:35–41.

22. Lin HX, Liang ZW, Sasaki T, Yano M. Identification and characterization of a quantitative trait locus, *Hd4* and *Hd5*, controlling heading date in rice. Breed Sci. 2003;53:51–9.

23. McMullen MD, Kresovich S, Villeda HS, Bradbury P, Li H, Sun Q, et al. Genetic properties of the maize nested association mapping population. Science. 2009;325:737–40.

24. McNally KL, Childs KL, Bohnert R, Davidson RM, Zhao K, Ulat VJ, et al. Genomewide SNP variation reveals relationships among landraces and modern varieties of rice. Proc Natl Acad Sci USA. 2009;106:12273–8.

25. Miura K, Lin SY, Yano M, Nagamine T. Mapping quantitative trait loci controlling low temperature germinability in rice (*Oryza sativa* L.). Breed Sci. 2001;51:293–9.**View Article**

26. Neeraja CN, Maghirang-Rodriguez R, Pamplona A, Heuer S, Collard BC, Septiningsih EM, et al. A marker-assisted backcross approach for developing submergence-tolerant rice cultivars. Theor Appl Genet. 2007;115:767–76.**PubMedView Article**

27. Rostoks N, Ramsay L, MacKenzie K, Cardle L, Bhat PR, Roose ML, et al. Recent history of artificial outcrossing facilitates whole-genome association mapping in elite inbred crop varieties. Proc Natl Acad Sci USA. 2006;103:18656–61.

28. Sasaki A, Ashikari M, Ueguchi-Tanaka M, Itoh H, Nishimura A, Swapan D, et al. Green revolution: a mutant gibberellin-synthesis gene in rice. Nature. 2002;416:701–2.

29. Sato T, Ueda T, Fukuta Y, Kumagai T, Yano M. Mapping of quantitative trait loci associated with ultraviolet-B resistance in rice (*Oryza sativa* L.). Theor Appl Genet. 2003;107:1003–8.

30. Schuster SC. Next-generation sequencing transforms today's biology. Nat Methods. 2008;5:16–8.

31. Septiningsih EM, Pamplona AM, Sanchez DL, Neeraja CN, Vergara GV, Heuer S, et al. Development of submergence-tolerant rice cultivars: the *Sub1* locus and beyond. Ann Bot. 2009;103:151–60.

32. Shen R, Fan JB, Campbell D, Chang W, Chen J, Doucet D, et al. High-throughput SNP genotyping on universal bead arrays. Mutat Res. 2005;573:70–82.

33. Shomura A, Izawa T, Ebana K, Ebitani T, Kanegae H, Konishi S, et al. Deletion in a gene associated with grain size increased yields during rice domestication. Nat Genet. 2008;40:1023–8.

34. Takahashi Y, Shomura A, Sasaki T, Yano M. *Hd6*, a rice quantitative trait locus involved in photoperiod sensitivity, encodes the alpha subunit of protein kinase CK2. Proc Natl Acad Sci USA. 2001;98:7922–7.

35. Takai T, Nonoue Y, Yamamoto S, Yamanouchi U, Matsubara K, Liang ZW, et al. Development of chromosome segment substitution lines derived from backcross between *indica* donor rice cultivar 'Nona Bokra' and *japonica* recipient cultivar 'Koshihikari'. Breed Sci. 2007;57:257–61.

36. Takeuchi Y, Ebitani T, Yamamoto T, Sato H, Ohta H, Hirabayashi H, et al. Development of isogenic lines of rice cultivar Koshihikari with early

and late heading by marker-assisted selection. Breed Sci. 2006;56:405–13.

37. Tsuji H, Tamaki S, Komiya R, Shimamoto K. Florigen and the photoperiodic control of flowering in rice. Rice. 2008;1:25–35.

38. Ueda T, Sato T, Numa H, Yano M. Delimitation of the chromosomal region for a quantitative trait locus, *qUVR- 10*, conferring resistance to ultraviolet-B radiation in rice (*Oryza sativa* L.). Theor Appl Genet. 2004;108:385–91

39. Wang E, Wang J, Zhu X, Hao W, Wang L, Li Q, et al. Control of rice grain-filling and yield by a gene with a potential signature of domestication. Nat Genet. 2008;40:1370–4.

40. Wang L, Hao L, Li X, Hu S, Ge S, Yu J. SNP deserts of Asian-cultivated rice: genomic regions under domestication. J Evol Biol. 2009;22:751–61.

41. Xu K, Xu X, Fukao T, Canlas P, Maghirang-Rodriguez R, Heuer S, et al. *Sub1A* is an ethylene-response factor-like gene that confers submergence tolerance to rice. Nature. 2006;442:705–8.

42. Xue W, Xing Y, Weng X, Zhao Y, Tang W, Wang L, et al. Natural variation in *Ghd7* is an important regulator of heading date and yield potential in rice. Nat Genet. 2008;40:761–7.

43. Yamamoto T, Lin H, Sasaki T, Yano M. Identification of heading date quantitative trait locus *Hd6* and characterization of its epistatic interactions with *Hd2* in rice using advanced backcross progeny. Genetics. 2000;154:885–91.

44. Yamamoto T, Yonemaru J, Yano M. Towards the understanding of complex traits in rice: substantially or superficially? DNA Res. 2009;16:141–54.

45. Yamamoto T, Nagasaki H, Yonemaru J, Ebana K, Nakajima M, Shibaya T, et al. Fine definition of the pedigree haplotypes of closely related rice cultivars by means of genome-wide discovery of single-nucleotide polymorphisms. BMC Genomics. 2010;11:267.

46. Yano M, Katayose Y, Ashikari M, Yamanouchi U, Monna L, Fuse T, et al. *Hd1*, a major photoperiod sensitivity quantitative trait locus in rice, is closely related to the *Arabidopsis* flowering time gene *CONSTANS*. Plant Cell. 2000;12:2473–84.

47. Yano M, Kojima S, Takahashi Y, Lin H, Sasaki T. Genetic control of flowering time in rice, a short-day plant. Plant Physiol. 2001;127:1425–9.

48. Zhong S, Dekkers JC, Fernando RL, Jannink JL. Factors affecting accuracy from genomic selection in populations derived from multiple inbred lines: a barley case study. Genetics. 2009;182:355–64.

Chapter 6

Increasing Leaf Vein Density by Mutagenesis: Laying the Foundations for C_4 Rice

Aryo B. Feldman , Erik H. Murchie 1 2*, Hei Leung , Marietta Baraoidan , Robert Coe , Su-May Yu , 3 3 4 5 , Shuen-Fang Lo5, William P Quick4

1 School of Biosciences, University of Nottingham Malaysia Campus, emenyih, Selangor Darul Ehsan, Malaysia,

2 School of Biosciences, University of Nottingham Sutton Bonington Campus, Sutton Bonington, Leicestershire, United Kingdom,

3 Plant Breeding, Genetics and Biotechnology, the International Rice Research Institute, Los Ban˜ os,Philippines,

4 The C4 Rice Center, the International Rice Research Institute, Los Ban˜ os, hilippines,

5 Institute of Molecular Biology, Academia Sinica, Nankang, Taipei, Taiwan

ABSTRACT

A high leaf vein density is both an essential feature of C_4 photosynthesis and a foundation trait to C_4 evolution, ensuring the optimal proportion and proximity of mesophyll and bundle sheath cells for permitting the rapid exchange of photosynthates. Two rice mutant populations, a deletion mutant library with a cv. IR64 background (12,470 lines) and a T-DNA insertion mutant library with a cv. Tainung 67 background (10,830 lines), were screened for increases in vein density. A high throughput method with handheld microscopes was developed and its accuracy was supported by more rigorous microscopy analysis. Eight lines with significantly increased leaf vein densities were identified to be used as genetic stock for the global C_4 Rice Consortium. The candidate population was shown to include both shared and independent mutations and so more than one gene controlled the high vein density phenotype. The high vein density trait was found to be linked to a narrow leaf width trait but the linkage was incomplete. The more genetically robust narrow leaf width trait was proposed to be used as a reliable phenotypic marker for finding high vein density variants in rice in future screens.

INTRODUCTION

The yield potential of rice (*Oryza sativa* L.) needs to be increased by at least 50% by 2050 to support the burgeoning human population, which can only be underpinned by improvements in rates of biomass production [1]. Rice productivity could be radically improved by the introduction of C_4 photosynthetic properties. C_4 photosynthesis is characterized by a CO_2concentrating mechanism involving the coordination of metabolism in two cell types, the mesophyll and bundle sheath. It results in the elimination or substantial reduction in photorespiration and consequently an enhancement in the capacity and quantum yield of photosynthesis at high temperatures [2] [3] [4].

One of the key C_4 properties is a high leaf vein density, which has been considered to be a prerequisite to the evolution of the complete suite of C_4 traits in plants [5] [6] [7]. It is needed to ensure the optimal ratio of mesophyll and bundle sheath cells with close contact permitting the rapid exchange of photosynthates. This is achieved via 'Kranz' anatomy which typically shows a single or double layer of mesophyll cells enclosing bundle sheath cells in a concentric fashion. Bundle sheath cells in turn enclose the vascular tissue. This arrangement permits bundle sheath cells and mesophyll cells to occupy similar volumes within the C_4 leaf, whereas the total mesophyll cell volume is greater in the C_3 leaf [8]. A possible source of study for identification of rice genes regulating traits relevant to C_4, like high leaf vein density, is mutagenized populations.

Chemical and physical deletion mutagenesis is an important tool in forward and reverse genetics, in addition to inducing new traits and enabling various allelic expressions of certain genes by turning off suppressor genes. Point mutations, caused by EMS mutagens for example, are the removal of a single base pair in the DNA molecule. A more sizeable amount of DNA can be removed by larger deletions. For example, γ-ray mutagenesis can delete a region from 1–4 kb in size. Deletion mutagenesis is especially useful as it can produce phenotypic variation representing alterations to the entire genome of any genotype in a relatively efficient fashion [9] [10]. The value of deletion mutagenesis for rice breeding is reflected by the size of the stock of cv. IR64 mutants held at the International Rice Research Institute (IRRI): 66,891 M_4 lines [10]. Mutagenesis has been widely used in crop improvement. The most successful mutation in rice breeding has been the *sd1* gene, which reduces plant height by 25% without affecting panicle size to diminish lodging. The gene was a spontaneous monogenic recessive mutation and is now present in over 90% of modern high yielding varieties [11]. Other important traits in rice breeding programs induced by mutagenesis include early flowering, low phytic acid content and giant embryo size [12]. The IR64 mutant library itself has been a source of resistance to diseases [13] [14] [15], high biomass production [Barroiden M, IRRI, personal communication], salt tolerance [16] and drought tolerance [17], as well as the identification of genes involved in leaf development [18].

Deletion mutagenesis has an advantage over other means of inducing new traits as deletions usually knockout multiple genes. This is relevant to developing a C_4 rice as it is possible that more than one gene copy in rice would need to be removed to ensure the loss of expression and to discover gene function [19]. Deletion mutagenesis is also able to induce mutations at high rates and so can produce a set of genome wide saturated mutations in a relatively small population [20].

Insertion mutagenesis is designed to disrupt and/or activate gene expression while tagging the site of insertion. The inserted fragment of DNA can inactivate a gene if it is located in the middle of its coding sequence. The insertional construct may also contain activation vectors that carry strong promoters, which can function in either orientation and at a considerable distance from the coding regions. These can cause transcriptional activation of nearby genes, resulting in dominant gain of function mutations [21]. It has been shown that activation tags have an effect of up to 10 kb left and right of the insertion point [10], so can affect multiple genes by individual insertion events. Insertional mutagenesis includes a variety of insertion elements such as T-DNA, transposable elements like the *Tos17* retrotranspon, and the Ac-Ds and Spm/dSpm maize transposons. The Taiwan Rice Insertion Mutant (TRIM) library at Academia Sinica in Taiwan holds 100,000 T-DNA lines with a cv. Tainung 67 background [22].

Insertion mutagenesis is relevant to engineering a C_4 rice since the different attributes of C_4photosynthesis include both the addition and removal of biochemical and anatomical features found in C_3 plants. Tagged insertion mutants are of special interest because they can represent 'engineered' expression levels of known gene sites, so that correlations between tags and phenotypes can be made. The disadvantage of deletion mutagenesis is that it can produce silent mutations, i.e. changes in the base sequence that have no mutant phenotype (due to gene redundancy, for example). Therefore, by enhancing the expression of genes near the insert in insertional mutagenesis, novel variation may be generated in the C_3 genome that is relevant to C_4 photosynthesis.

Alterations in gene function by mutagenesis of rice can be explored for phenotypic changes relevant to C_4 photosynthesis. The success of this relies upon the hypothesis that the rice genome has sufficient functional polymorphism and genetic plasticity to yield measurable C_4traits that can be revealed by artificially inducing variation in the genome. Mutagenesis of rice provides a source of genetic variation that can cover the entire genome and so can give a good indication of whether this is the case. Following a forward genetics scheme allows the detection of relevant phenotypes whose underlying genetics can be characterized.

This paper takes such an approach and describes the high throughput screening of large numbers of lines within rice mutant populations for leaf vein density and the subsequent genetic analysis of resulting high vein density candidates. Here we conduct a study to test the hypothesis that vein density is a heritable trait in rice and that mutant populations can be used to generate phenotypic variation for the study of traits in rice that are relevant to C_4photosynthesis. We also investigated the close link of leaf vein density with leaf width, which was found in the high vein density mutant lines that had narrow leaf widths in [23]. The possibility of uncoupling the two traits was investigated by elucidating the genetic bases of leaf width and vein density in rice mutants.

MATERIALS AND METHODS

Vein density screening

The primary genetic resource was the IRRI IR64 deletion mutant population derived from a single IR64 plant, IR64-21 [13]. The original IR64-21 parent plant was selected for phenotypic uniformity from breeder seeds grown under field conditions. IR64 mutants used in the screening were either deletions (mainly dry seeds mutagenized with γ-radiation at 250 GY at the International Atomic Energy Agency) or point mutations (mainly pre-soaked seed mutagenized with EMS at 0.4, 0.6, 0.8, 1.0 and 1.6% concentrations at IRRI). Mutagenized lines were advanced every four months from the initial M_1 population by single seed descent until at least the M_3 generation. This was to ensure near homozygous inbred lines, which was essential for quantifying phenotypic characteristics in replicated trials. The screened population, consisting of 12,470 lines, was at the M_4 generation derived from either bulked sibling seeds or single seed descent. The deletion mutant population was divided by leaf width phenotypes: wild-type and narrow leaf width. The first vein density screen included only those mutants with wild-type leaf widths whereas the second screen targeted narrow leaf width mutant lines.

The secondary mutant population screened for variations in vein density was 10,830 insertion lines from the TRIM library, produced by Academia Sinica. The TRIM library consisted of lines that had gene disruptions or activation tags by use of randomly inserted *Agrobacterium tumefaciens* mediated transfer DNA (T-DNA) (average copy number of 1.7) [24].

Ten seeds per line were planted in the deletion mutant vein density screening. However, germination rates were rarely 100% and plants with gross morphologies (e.g. those that were severely pale or stunted) were disregarded for screening. Thus, generally 5–8 plants per line were sampled for screening. Ten wild-type IR64 seeds were planted as controls alongside the planting of mutant plants for every day of measurements (per 170 lines). Veins were counted on-site using a handheld microscope (Readiview handheld microscope, Meade Instruments Corporation, Irvine, USA) at 80× magnification.

A similar number of plants per line were sampled in the insertion mutant population. Here, vein counts were made off-site at IRRI from images that were taken of freshly harvested material. Images of veins in the freshly harvested youngest fully expanded leaf at the fifth to the seventh leaf stage were captured for screening the T_1 insertion mutant population by a small digital microscope (AM7013MT Dino-Lite Premier, AnMo Electronics Corporation, Torrance, California, USA) at 4× magnification. Imaged veins in four × 2 mm wide areas of the leaf (not including the midrib or edges) were counted.

The fifth to the seventh fully developed leaf that emerged on the primary tiller, excluding the seed leaf (leaf five to seven), was used for vein density measurements in both populations of mutants. The number of veins in the widest part of the abaxial side of the leaf was counted. It was important to take care to always count the veins of the widest middle part of the leaf as some vein density variation was observed towards the the base and tip (unpublished observations). Care was also taken to avoid counting veins close to the midrib and close to the margins of the leaves, where there was a tendency for vein density values to increase and so they were not representative of the entire leaf width (unpublished observations). If a high vein density value was recorded, then a further vein density reading was taken in the leaf area of the other side of the midrib. If two consistently high vein densities were found and this was confirmed by microscopy analysis, the plant was considered to be a candidate.

Potential candidates detected from the vein density screening had their vein densities confirmed using a laboratory microscope equipped for digital imaging. Vein numbers of fresh leaf samples were viewed with a light microscope (Olympus BX51 Motorized Research Microscope, Tokyo, Japan, connected to an Olympus DP71 Microscope Digital Camera, Tokyo, Japan) at 4× magnification. Images were taken using Olympus Cell∧P imaging software (from the Olympus Cell* software family).

Crossing. Deletion mutant lines were backcrossed to their parental line, IR64-21, to remove a significant portion of background mutations in an effort to revert all traits to wild-type except for high vein density. Mutations were assumed to be recessive, given their low level of detection and the nature of chemical and physical mutations [Hei Leung, IRRI, personal communication], and so were designated as the female parent. This made the recognition of true BC_1F_1 plants easier as they should have exhibited the wild-type phenotype.

Mutant plants that had begun heading were emasculated in the afternoon (at ~15:00). All spikelets (palea and lemma) on the panicles were cut at an obtuse angle (>45°) with small scissors so that the anthers were exposed and the stigma was left unharmed. Anthers were removed with a vacuum emasculator (Gast Model-1022-V103-G272X, Gast Manufacturing Corporation, Benton Harbor, Michigan, USA). Their panicles were then bagged to make sure that stigmas did not desiccate or were unintentionally fertilized.

The following morning (at ~10:00), after removing the bags from the panicles, forceps were used to remove any remaining anthers if there were any. Remaining naked stigmas were then fertilized with the parent plant, whose anthers protruded outside the spikelet. Pollen was dusted onto the receptive female by tapping the stems of the pollen donor panicles beside the emasculated spikelets. Pollen dusting was repeated the following two mornings to ensure successful pollination. Panicles were then bagged for seed setting.

Growing conditions

The initial T_1 insertion mutant population was grown under field conditions at Academia Sinica (25° 4′ 34.51″ N; 121° 61′ 44.91″ E), Taiwan and screened in the summer 2009 (Oct. 2009). The mean temperature was 28.8°C (range: 19.4–38.6°C), mean relative humidity was 74.3%, and mean daily sunshine duration was 5.0 hours.

Subsequent analysis of the insertion mutant population and of the deletion mutant population took place at the IRRI greenhouses (14° 10' 23.29", 121° 15' 32.44") from mid-2009 to mid-2012. The mean temperature was 26.8°C (range: 19.9–35.0°C), mean relative humidity was 86.3%, and mean daily sunshine duration was 6.4 hours.

Seeds were incubated at 50°C for 72 hr if their dormancy had to be broken, i.e. if seeds were more than two years old or if germination rates were particularly poor (<30%). They were then pre-germinated in petri dishes on filter paper treated with sterilized distilled water and incubated at 35°C until the radical and coleoptile were sufficiently swollen (usually by 24 hr).

In the screening of deletion mutants, 100 seeds were planted in a tray with length width x height dimensions, 36.5 cm × 27.0 cm × 11.5 cm. For both rice populations, soil was sourced from the IRRI upland farm and was treated with 0.09-0.01-0.09 g NPK fertilizer per kg soil as a basal dressing and 0.09 g N fertilizer per kg soil every 2–3 weeks, depending on plant growth rate (determined qualitatively on a relative scale). Irrigation was applied generously to ensure that the soil moisture content was always high.

RESULTS

Vein density screening

Vein densities of the IR64 wild-type were largely (83%) 5 or 5.5 veins mm^{-1} (Figures 1 and 2). They were rarely (0.10%) 6.5 veins mm^{-1} and those instances were confined to one side of the midrib only. In addition, the average vein density of all deletion mutant plants was 5.4 veins mm^{-1}, which was more than two standard deviations (i.e. >0.7) away from 6.5 veins mm^{-1}. Hence, any plant that had a vein density of 6.5 veins mm^{-1} or more on both sides of the midrib was considered to be a potential high vein density candidate.

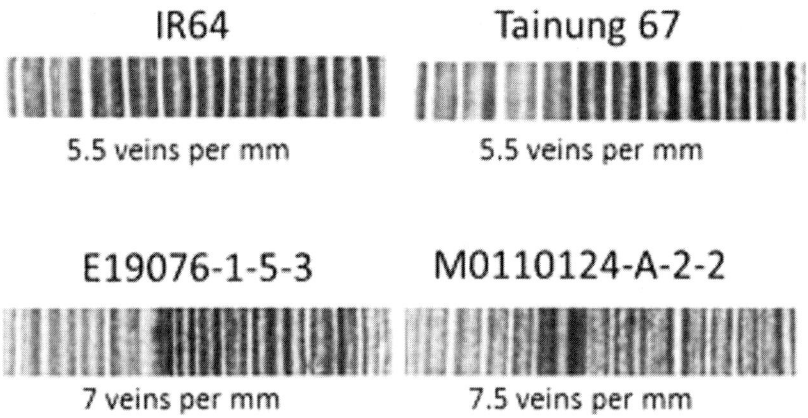

Figure 1. Leaf veins of mutant and wild-type rice lines. Images of veins within a 2 mm leaf width were captured at 2–3 weeks after sowing. Line E19076-1-5-3 is an M_5 IR64 mutant. Line M0110124-A-2-2 is a T_3 Tainung 67 mutant.

Figure 2 shows that there were overlapping frequency distributions of vein density values of the IR64 wild-type line and M_4 deletion mutant plants for vein counts per mm. This meant that mutagenesis did not affect the vein density of most of the lines. Those that showed differences in vein density were at the upper tail end of the bell shaped curve.

The screening of 10,830 T_1 insertion lines showed that leaf vein density frequencies of the Tainung 67 wild-type and insertional mutants were similarly distributed (Figure 3). As was the case in the deletion mutants (Figure 2), this implied that mutagenesis did not affect the vein density in the large majority of insertional lines but rather created enough variation to produce a small fraction of mutants with an increase in vein density. The distribution frequency of insertion lines had a heavier tail compared to the Poisson distribution of the deletion mutants.

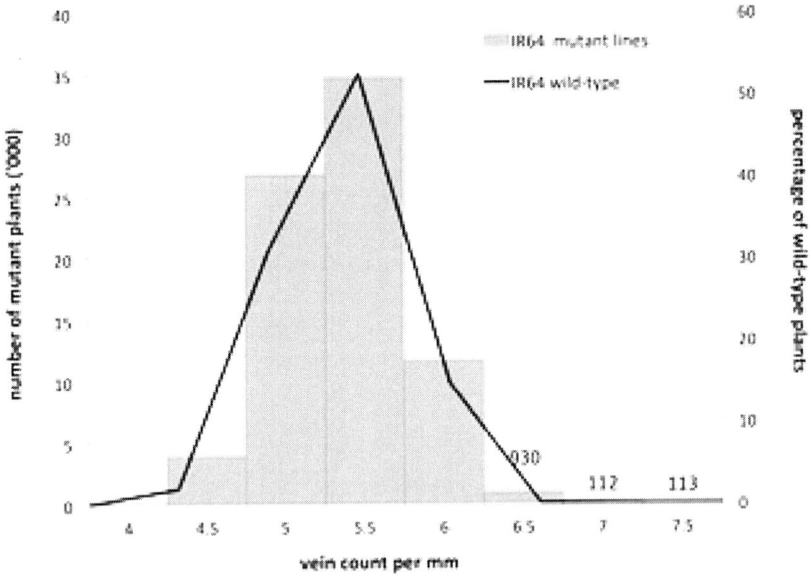

Figure 2. Frequency distribution of vein densities in the deletion mutant population and their control in the screen. Values above chart bars denote the number of mutant plants for each high vein density. Distribution frequencies of deletion mutant and control plants were largely the same. Differences were at the extremity of the binomial curve at higher vein densities (930 (1.2%), 112 (0.14%) and 13 (0.02%) mutant plants for respectively 6.5, 7 and 7.5 veins mm^{-1}) where lines were selected as potential candidate mutants for increased vein density.

Figure 4 compares the frequency distribution of the insertion lines to that of the deletion mutant lines in terms of vein density. Gene activation/disruption created a greater range of vein density values than deletion and point mutations. The increased vein density variation was expected as gene activation was perceived to be a more effective tool in creating phenotypic variation than deletion and point mutagenesis [22] [William P Quick, IRRI, personal communication]. In light of this, vein counts of at least 7 veins mm^{-1} were considered to show an increase in vein density in the insertion mutant population.

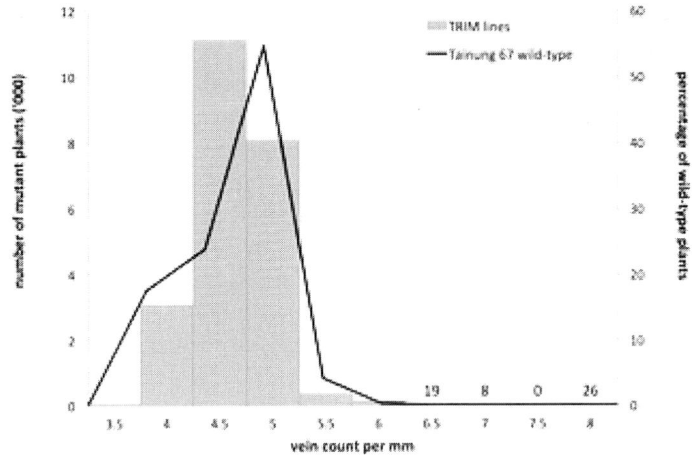

Figure 3. Frequency distribution of vein densities in the T_1–T_3 insertion mutant populations and their controls. Values within the chart denote the number of mutant plants for each of the highest vein density values. Distribution frequencies of insertion mutant and wild-type plants were largely the same. Differences were at the extremity of the binomial curve at higher vein densities (eight (0.04%) and 26 (0.12%) mutant plants for respectively 7 and 8 veins mm^{-1}) where lines were selected as potential candidate mutants for altered vein spacing.

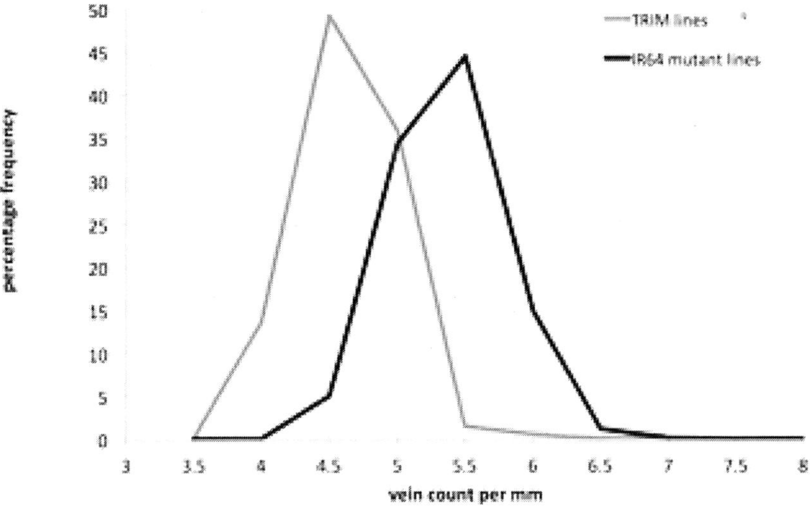

Figure 4. Comparing vein density frequencies of IR64 deletion mutants and Tainung 67 insertion mutants. Frequencies are expressed in percentages of the corresponding mutant population.

There were insertion lines, but no deletion mutants, with vein densities of 3.5 and 8 veins mm^{-1} (0.12% of the insertion mutant population for both vein densities). The deletion mutants peaked at a higher vein count ($p<0.001$): 5–5.5 veins mm^{-1} compared to 4.5–5 veins mm^{-1} in the insertion lines. This is likely to be attributed to the background genotypes as the Tainung 67 parental line has been consistently shown to have a lower vein density than the IR64 parental line. To support this, in a long term field experiment (data not shown) there was a difference ($p<0.001$) in vein densities between IR64 wild-type (overall mean of 5.32 ± 0.06 veins mm^{-1}) and Tainung 67 wild-type (overall mean of 4.63 ± 0.05 veins mm^{-1}). In a long term greenhouse experiment (data not shown), IR64 wild-type had a higher ($p<0.001$) vein density (overall average of 5.06 ± 0.04 veins mm^{-1}) than Tainung 67 wild-type (overall average of 4.77 ± 0.05 veins mm^{-1}).

Inheritance and segregation of vein density mutations

Deletion mutant population.

Selfing 97 potential deletion candidate lines from the first screening of M_4 lines with normal leaf width did not always result in progeny with the increased vein density phenotype. Instead, the majority (78%) of mutant lines regressed to the wild-type vein density phenotype in the subsequent generation. Twenty-one M_5 lines showed transmission of a high vein density trait and did so at varying degrees. Only four M_5 lines showed high inheritance of the phenotype. However, the progeny of these four lines also reverted to the wild-type phenotype. These lines were abandoned and focus turned to a second screening for narrow width lines.

The second screening of vein densities in deletion mutants with the narrow leaf phenotype used a population constructed by both observations of the original M_4 screen and previous phenotypic evaluations made by the Genetics Laboratory, IRRI [Marietta Baraoidan, personal communication]. These lines had leaf widths of 4–6 mm for the widest part of the youngest fully developed leaf at the fifth to seventh leaf stage, compared to 8–9 mm for the IR64 wild-type. From 50 narrow leaf lines, 13 were found to have increased vein density progeny. Five lines out of these 13 (E11068-1-10, E19076-1-5, G558-11-5, E22097-1-3 and E26181-1-1) were selected as being candidates after checking for the stability of the higher vein density trait over developing stages, confirming the transmission of the trait to the following generations and analyzing microscopy data.

Segregation data of these lines at the BC_1F_2 generation gave ratios that were closer to being Mendelian (1:3 for a monogenic recessive mutation) (Table 1) than the non-narrow leaf M_5deletion candidate mutants. The narrow leaf trait segregated at a 1:3 ratio for the BC_1F_2 lines, E11068-1-10/IR64 and E26181-1-

1/IR64 (respectively, χ^2 = 0.88 and 1.52 < 3.84; df =1). The high vein density trait (≥6.5 veins mm^{-1}) segregated at a 1:3 ratio for these two lines, as well as for E19076-1-5/IR64 (respectively, χ^2 = 0.88, 0.28 and 1.65 < 3.84; df =1).

Table 1. Number of BC_1F_2 progeny of narrow leaf high vein density deletion candidate lines with narrow leaves and high vein densities.

BC_1F_2 line	n	n/4	no. progeny with		
			narrow leaves	6.5 veins mm^{-1}	veins mm^{-1}
E11068-1-10/IR64	152	38.00	43	43	40
E19076-1-5/IR64	182	45.50	34*	38	31*
E22097-1-3/IR64	170	42.50	27**	21***	14***
E26181-1-1/IR64	97	24.25	19	22	14*
G558-11-5/IR64	117	29.25	15**	15**	7***

Narrow leaf width: ≤6 mm. *Observed value differed from expected value (n/4) at $p<0.05$. **Observed value differed from expected value (n/4) at $p<0.01$. ***Observed value differed from expected value (n/4) at $p<0.001$.

Advancing some of these high vein density BC_1F_2 progeny did not genetically fix the high vein density. Even though the narrow leafed trait was fixed in BC_1F_3 lines, the ratio of high vein density progeny was at best 3:1 for a monogenic dominant mutation.

A complementation test was performed by inter-mating the candidate mutant lines G558-11-5-2 (♀) and E22097-1-3-1 (♂). The F_1 progeny had a wild-type phenotype, which suggested that the narrow leafed trait and the high vein density trait in the mutant lines were in separate complementation groups (and so were non-allelic). The F_2 progeny descended from a single F_1parent plant showed a range of phenotypes, some of which were not present in either mutant (or the wild-type) (Table 2). This again suggested independence of the mutations, as well as random assortment.

Table 2. Phenotype combinations in the F_2 progeny of double mutant, G558-11-5-2/E22097-1-3-1.

phenotype	genotype	no. progeny
very narrow leaf & very high vein density	$n_1n_1n_2n_2v_1v_1v_2v_2$	1
very narrow leaf & high vein density	$n_1n_1n_2n_2V_{x}\_v_yv_y$	2
narrow leaf & very high vein density	$n_xn_xN_{y}\_v_1v_1v_2v_2$	6
narrow leaf & high vein density	$n_xn_xN_{y}\_V_{x}\_v_yv_y$	6
wild-type leaf width & high vein density	$N\_V_{x}\_v_yv_y$	14
wild-type leaf width & wild-type vein density	$N\_V\_$	50

n_n: recessive, narrow leaf width allele. N_n: dominant, wild-type leaf width allele. v_n: recessive, high vein density allele. V_n: dominant, wild-type vein density allele. Phenotypes were evaluated at the 5th–7th leaf stage except for the vein densities of progeny plants with very narrow leaves. Their vein counts per 2 mm could not be taken at the 5th–7th leaf stage so were instead taken at the late vegetative stage. Wild-type leaf width: 8–9 mm. Narrow leaf width: 4–6 mm. Very narrow leaf width: 1–2 mm. Wild-type vein density: ≥5<6.5 veins mm^{-1}. High vein density: ≥6.5<7.5 veins mm^{-1}. Very high vein density: ≥7.5 veins mm^{-1}. Both candidate mutant lines G558-11-5-2 and E22097-1-3-1 had the phenotypic traits of narrow leaves and high vein density. The data characterizes F_2 progeny descended from a single F_1 plant. The F_1 plants all had wild-type vein density and leaf width.

There were three progeny plants with very narrow leaves (1–2 mm leaf width at the fifth to seventh leaf stage), which appeared to demonstrate an additive or synergistic relationship between the narrow leaf mutations of the two mutant parent lines. There appeared to be another additive effect in the case of the two high vein density mutations in seven progeny plants, which had very high vein densities (≥7.5 veins mm^{-1}).

In the next round of complementation tests, all possible crossing combinations between candidate lines were attempted. Resultant F_1 phenotypes gave some indication as to the independence of the mutations. Firstly, the high vein density and narrow leaf width mutations were not necessarily linked. This was most noticeable in the F_1 plants from E11068-1-10-1/G58-11-5-2 and from E19076-1-5-3/G558-11-5-2. Both crosses generated offspring with wild-type vein density but their leaf widths were narrow (i.e. were of a mutant phenotype) (Table S1).

Secondly, sibling sets of F_1 plants with wild-type phenotypes (in regard to vein density and/or leaf width) that indicated independence of mutations included E11068-1-10-1 x G58-11-5-2 for vein density, E19076-1-5-3 x E22097-1-3-1 for both vein density and leaf width, possibly E19076-1-5-3 x G558-11-5-2 for vein density and possibly E22097-1-3-1 x G58-11-5-2 for both vein density and leaf width (which would back up the results of the first crossing round presented previously). 'Possibly' is relevant in some instances as reciprocal crosses in these mutant combinations gave conflicting results. Disregarding these inconsistencies, it was suggested that the mutations that were independent of each other for vein density were for: E11068-1-10-1 and G558-11-5-2, E19076-1-5-3 and E22097-1-3-1, E19076-1-5-3 and G558-11-5-2, and E22097-1-3-1 and G558-11-5-2.

Given that E11068-1-10-1 and E22097-1-3-1 were considered to share the same mutation for vein density (i.e. their F_1 progeny had wild-type vein densities), it followed from the independence of E11068-1-10-1 and G558-11-5-2 mutations for vein density that E22097-1-3-1 and G558-11-5-2 vein density mutations should also be independent (Table S1 supports this). From the independence of E19076-1-5-3 and E22097-1-3-1 mutations for vein density and the shared mutation between E11068-1-10-1 and E22097-1-3-1 for vein density, it followed that the vein density mutations for E19076-1-5-3 and E22097-1-3-1 should also be independent. From the independence of E22097-1-3-1 and G558-11-5-2 mutations for vein density and the shared mutation between E11068-1-10-1 and E22097-1-3-1 for vein density, it followed that the vein density mutations for E11068-1-10-1 and G558-11-5-2 should also be independent.

In terms of leaf width mutations, E19076-1-5-3 and E22097-1-3-1 mutations were independent and the mutation of E11068-1-10-1 and E22097-1-3-1 were shared. Therefore, E11068-1-10-1 and E19076-1-5-3 leaf width mutations should also be independent. E22097-1-3-1 and G558-11-5-2 also had independent mutations for leaf width. Given that E11068-1-10-1 and E22097-1-3-1 shared a leaf width mutation, E11068-1-10-1 and G558-11-5-2 should also have independent leaf width mutations, which was contradicted by the experimental findings (i.e. progeny from this cross had narrow leaf widths).

One notable finding from the second round of complementation tests was that all crosses with E26181-1-1-2 resulted in progeny with mutant vein densities and leaf widths.

Insertion mutant population.

Thirty-five (0.32%) of the insertion lines had increased vein counts per leaf width when screened in Academia Sinica, Taiwan. Twenty-five T_2 seeds of these lines were grown at IRRI and eight were confirmed for increased vein density. In the T_3 generation, three lines were identified as potential candidate lines for increased vein density. These three lines, M0104656-B, M0105588-A and M0110124-A, transmitted the increased vein density trait over two generations.

The three candidate lines had a range of vein density values in both the T_2 and T_3 generations. Nevertheless, advancing the highest vein density T_2 progeny to the next generation skewed the values away from the control and towards the upper end of the vein count scale. This was especially evident in the M0105588-A insertion line. This line had only one T_2 progeny with a higher vein density than the control. However, when this particular progeny was advanced, the majority of its T_3 progeny had greater vein counts than the control with several (six out of 15) exceeding the threshold for what was considered a candidate. Among these was a progeny plant that averaged 9 veins mm^{-1}, which was the highest vein count measured from the screening of both deletion and insertion mutant populations.

The best T_3 progeny of each candidate line were selected as parents for segregation analysis based upon inheritance data and vein density related phenotypes (Table 3). Segregation ratios were based on using 6.5 veins mm^{-1} as the lower boundary for what was considered to be an increase in vein density, as in the deletion mutant population. M0104656-B-11-5 had a 3:1 mutant to wild-type vein density ratio (χ^2 = 1.06 < 3.84; df =1) and M0110124-A-2-3 had a 1:1 ratio (χ^2 = 0.12 < 3.84; df =1). M0105588-A-4-13 did not segregate at a 3:1 mutant to wild-type vein density ratio by only a very narrow statistical margin (χ^2 = 3.843 > 3.841; df =1).

Table 3. Inheritance data for T_4 insertion candidate lines.

insertion mutant family	vein density segregation			leaf width segregation		
	no. m	no. w-t	ratio	no. m	no. w-t	ratio
M0104656-B-11-5	57	14	3:1	70	1	1:0
M0105588-A-4-13	58	10	3:1*	67	1	1:0
M0110124-A-2-3	38	35	1:1	58	15	3:1

*Observed ratio differed from expected ratio at $p<0.05$. m: mutant phenotype. wt: wild-type phenotype. The control, Tainung 67 wild-type, had an average of 5.02 ± 0.02 veins mm^{-1} and a maximum of 6 veins mm^{-1}. Vein densities of 6.5 veins mm^{-1} were considered to be a mutant phenotype. The Tainung 67 wild-type had two out of eighty-eight plants with <7 mm leaf width, one plant with <6 mm leaf width, and no plants with <5 mm leaf width (and an average leaf width of 9.53 ± 0.08 mm). Leaf widths of 6 mm or less were considered to be a mutant phenotype.

For segregation ratios of leaf width traits, lines M0104656-B-11-5 and M0105588-A-4-13 were fixed (p>0.05 for the line not segregating the expected ratio) for the mutant phenotype (i.e. narrow leaf width) and line M0110124-A-2-3 had a 3:1 mutant to wild-type ratio (p>0.05 for the lines not segregating at the expected ratio). In M0110124-A-2-3, there were two clear distinct phenotypes for both leaf and plant phenotypes: narrow leaves and dwarfed plant stature compared to wild-type leaf widths and plant stature. Plants with the latter phenotype flowered notably early in comparison to both the mutant segregant phenotype and wild-type Tainung 67. As in the deletion candidate lines, there was evident linkage between the narrow leaf width trait and the high vein density trait in the insertion T_4 insertion candidate lines. All of the wild-type leaf width progenies in every line had wild-type vein densities.

DISCUSSION

We have carried out the first study that examines in detail the genetics of leaf vein density in rice, which is considered to be important for the long term goal of inserting the C_4 mechanism into C_3 crops. We have shown variation in vein density within a deletion and insertion mutant population, but this was found to be linked to leaf width. Here we discuss the biological nature of the high vein density trait in rice and the significance of our data.

Evaluating the high throughput vein density screen

The forward genetics scheme was effective in discovering phenotypic alterations relevant to C_4 photosynthesis. The accuracy of scoring in the screening method for identifying mutant lines with increased vein densities was confirmed by microscopy tests and the consistently higher vein densities of candidate mutant lines in different environments. In addition, the vein density screening established sufficient throughput to identify a workable number of putative deletion candidate mutants over time: average of one candidate line per month. Their number was not so high that there would be too many lines to handle for concurrent in-depth measurements and not so low that there would be no lines taken beyond the screen. The rate of detection was approximately one high vein density event in every thousand lines screened. This frequency of detecting a mutation in leaf anatomy is consistent with that observed in other traits in the mutant collection [13]. Also, this was the same frequency as finding IR64 wild-type plants with a high vein density on one side of the midrib.

The screening used two major assumptions. The first was that the strongest indication of a genetic basis for vein density was trait inheritance. The second was that it was sufficient for a single mutation to have some measurable effect on vein density rather than having to find and combine multiple mutant lines. These two assumptions meant that potentially valuable material had been lost.

The poor inheritance of the high vein density trait in some experiments may have involved an environmental effect on vein density [25]. Different rates of high vein densities were counted in the same line on different measuring sessions (separated by months). The effect of the environment on leaf morphology was demonstrated in [26]. The leaf thickness of rice plants increased if they were transferred from a low light to a high light environment 14 days (but not eight days or less) before full leaf extension [26]. Hence, changes in the environment can alter the morphology of the same leaf number of a rice genotype given a sufficient interval (an acclimation period) between observations.

Variation in leaf morphology as determined by the environment was likely to be an issue in BC_1F_3 deletion candidate mutant lines, which should have had a fixed high vein density phenotype after sufficient selfing but instead some of their progeny reverted to wild-type vein densities. The environmental interaction of vein number could have been due to water stress[27], high temperatures [28], and/or variable irradiance, wind and nutrient conditions [29]. This poses an obvious challenge for defining the criteria for the selection of candidates. The environmental effect may indeed have been the cause of the number of false positives in the screening. Both the inheritance of the high vein density phenotype between selfed generations in the narrow leaf lines and the heritability estimate of vein density counters this, however. Narrow sense heritability was reasonably high (h^2 = 0.56), which meant that most of the variance in the vein density trait was genetic.

The genetics of high vein density mutations and their link with the narrow leaf width trait

Backcrossing narrow leaf high vein density deletion candidate mutants was performed to investigate the genetics of the two traits of narrow leaf width and high vein density. Results supported the conclusion that the traits were recessive and also indicated that the traits were recessive for a single, and potentially the same, gene. This was to be expected as loss of function mutations by deletion mutagenesis are generally recessive (as the leftover allele can typically produce sufficient protein to compensate).

On the other hand, segregation ratios of T_4 insertion mutant candidates for increased vein density indicated that the trait was controlled by a single dominant gene. This suggested that gene activations (gain of function mutations) were responsible for the changes in vein density as they behave dominantly [30] [21]. The alternative explanation is that the mutations were caused by somaclonal variation (which occurs at the chromosomal level as in chromosome breakage) [31]. Even though T_4 line, M0110124-A-2-3 segregated at a 1:1 ratio for increased vein density, it may still have been due to a monogenic dominant ratio as one of the alleles in the gene locus may have been masking the other one due to allelic dominance.

The narrow leaf trait appeared to be controlled by separate genes in the insertion mutant candidates as it was fixed in the T_4 lines, M0104656-B-11-5 and M010558-A-4-13 but was only monogenic dominant in T_4 line, M0110124-A-2-3.

The complementation tests made between deletion candidate mutant lines indicated that the candidate population contained both shared and independent high vein density and narrow leaf width mutations. The tests also showed that the mutation(s) of E26181-1-1-2 is/are key upstream regulator(s) of vein density and leaf width as all progeny sets of crosses that included line E26181-1-1-2 as a parent had mutant phenotypes.

The effort to uncouple the two traits of high vein density and narrow leaf width was difficult. The data seemed to suggest that the high vein density trait could not be inherited independently but was instead a consequence of leaves being narrow. This was what was indicated by [23], which linked the two traits of a high vein density and narrow leaf width by small cell sizes.

The very close overlap between segregation for vein density and for leaf width in the BC_1F_2deletion mutant plants suggested complete linkage between the two traits. However, F_2 plants from the double mutant, G558-11-5-2/E22097-1-3-1 appeared to uncouple the narrow leaf and high vein density traits. This was evident from progeny plants that had both wild-type leaf widths and high vein densities. Even though this was also found in BC_1F_2 plants (as well as in the screening and even the wild-type plants), it occurred at a much higher extent in the F_2double mutant progeny: 22% of the progeny plants with wild-type leaf widths had high vein densities compared to <5% in all other mutant families measured. Therefore, even though linkage was present as all the narrow leafed progeny had high vein densities, the linkage was incomplete.

Incomplete linkage was also backed up by the second round of complementation tests. Sets of F_1 progeny from E11068-1-10-1/G558-11-5-2 and from E19076-1-5-3/G558-11-5-2 had wild-type vein densities but also had mutant leaf widths. This result was in contrast to other genetic results, including the phenotypes of G558-11-5-2/E22097-1-3-1 F_2 progeny, which suggested that high vein densities followed on from narrow leaf widths in candidate mutants.

The development of high vein density and the evolution of C_4photosynthesis

A series of ecosystem changes took place to bring about the onset of C_4 evolution. The reduced atmospheric [CO_2] and other environmental changes associated with open habitats should be regarded as laying down the road, as it were, to C_4 photosynthesis. The drop in atmospheric [CO_2] was the precondition that established the early section of the road. However, it was the move of plants out of densely vegetated forests that made them vulnerable to a large host of abiotic and biotic stresses, which first triggered morphological changes within the C_3 photosynthetic model and then evolved C_4 photosynthesis [32].

Crucially, changes in the plant's water relations would have required modifications in stomata, venation and leaf dimensions, such as the narrow leaves and high vein densities found in the candidate mutants of this paper. A more efficient hydraulic system is usually present in narrow leaves regardless of changes in vein density (as there is less lamina surface to travel), which is beneficial for superior CO_2 assimilation rates [33] [34]. In addition, the smaller boundary layer of a narrow leaf also reduces transpiration losses and dissipates excess heat [35], which can be essential in reducing photorespiration losses.

It could be argued that the improved efficiency of CO_2 utilization and the well-developed leaf vein systems seen in C_4 systems no longer necessitated flexible alterations in, for example, leaf area, leaf width, angle of leaf blades, leaf thickness and stomatal density seen in C_3plants [26] [36].

The link found in this study between narrow leaves and high vein density has been previously reported. In [37], it was explained that leaf shape and venation patterns are closely associated (see also [38]) and that both are probably controlled by a common regulator like auxin. In [39], the intimate link between vein formation and leaf width is likened to fracture propagation in a stretched material, which results in auxin-like signals in the new spaces that form new veins. Therefore, the connection between auxin and vascular formation might explain the connection between vein density and leaf width.

On the other hand, the relationship between leaf shape and venation may depend at least in large part to the underlying cell anatomy. We explore this relationship in our following paper.

CONCLUSION

The close linkage of the two traits of high leaf vein density and narrow leaf width was suggested from the results, while the complementation tests showed that it probably was not complete. We conclude that the phenotypically plastic trait of vein density is genetically coupled with leaf width, which itself is a much more genetically robust trait. We hypothesize that a narrow leaf width is a precondition for narrow vein spacing. In the deletion mutant population, a similar finding was found in high biomass mutants, whose high plant height was more genetically robust than increased biomass production [Marietta Barroiden, IRRI, personal communication], which was also found in the evaluation of a range of rice lines in [40]. Therefore, leaf width could be used as a reliable phenotypic marker for finding modifications in vein density, as well as a means of speeding up the throughput of vein density screenings.

The hypothesis that mutagenesis could be used as a tool for creating enough variability latent within the rice genome to uncover measurable increases in leaf vein density was confirmed. The final candidate list included lines containing heritable mutations from all three types of mutagenesis techniques investigated, i.e. point, deletion and insertional mutagenesis. The most significant difference between the techniques was that the former two produced mutations that appeared to be monogenic recessive and the latter produced mutations that appeared to be monogenic dominant.

The study has shown that research in the C_4 Rice Project can utilize mutant populations for the exploration of a C_4 relevant trait in rice. A forward genetics scheme is a viable means of gene discovery for other C_4 traits, given a strong phenotype and a high throughput screening technique. One major advantage of discovering C_4 anatomy related rice mutants is that they can serve as breeding lines in which the counterpart biochemistry can be inserted. Using increased vein density rice as acceptor lines is especially relevant given that leaf anatomy, and vein density in particular, has been hypothesized to be foundational to the evolution from C_3 to C_4 photosynthesis [5] [6] [7].

ACKNOWLEDGMENTS

We're grateful to Dr John Sheehy for advice and guidance with the mutant screen and the following people at IRRI for their expertise and assistance: Abigail Elmido Mabilangan (and her team) and Pedro Reaño.

AUTHOR CONTRIBUTIONS

Conceived and designed the experiments: ABF EHM HL WPQ. Performed the experiments: ABF. Analyzed the data: ABF RC HL. Contributed reagents/materials/analysis tools: HL MB SMY SFL. Wrote the paper: ABF EHM.

REFERENCES

1. Sheehy JE, Ferrer AB, Mitchell PL, Elmido-Mabilangan A, Pablico P, et al. (2007). How the rice crop works and why it needs a new engine. In: Sheehy JE, Mitchell PL, Hardy B, editors. Charting new pathways to C_4 rice. Los Baños: International Rice Research Institute. pp. 3–26.

2. Evans J, von Caemmerer S (2000) Would C_4 rice produce more biomass than C_3 rice? In: Sheehy JE, Mitchell PL, Hardy B, editors. Redesigning Photosynthesis to Increase Yield. Manila: International Rice Research Institute and Amsterdam: Elsevier. pp. 53–71.

3. Mitchell PL, Sheehy JE (2007) The case for C_4 rice. In: Sheehy JE, Mitchell PL, Hardy B, editors. Charting new pathways to C_4 rice. Los Baños: International Rice Research Institute. pp. 27–36.

4. Sage TL, Sage RF (2009) The functional anatomy of rice leaves: implications for refixation of photorespiratory CO_2 and efforts to engineer C_4 photosynthesis into rice. Plant & Cell Physiology 50(4): 756–772.

5. McKown AD, Dengler NG (2007) Key innovations in the evolution of Kranz anatomy and C_4 vein pattern in *Flaveria* (Asteraceae). American Journal of Botany 94(3): 382–399.

6. McKown AD, Dengler NG (2009) Shifts in leaf vein density through accelerated vein formation in C_4 *Flaveria* (Asteraceae). Annals of Botany 104: 1085–1098.

7. Kajala K, Covshoff S, Karki S, Woodfield H, Tolley BJ, et al. (2011) Strategies for engineering a two-celled C_4 photosynthetic pathway into rice. Journal of Experimental Botany 62(9): 3001–3010.

8. Gowik U, Westhoff P (2011) The path from C_3 to C_4 Photosynthesis. Plant Physiology 155: 56–63.

9. Hirochika H, Guiderdoni E, An G, Hsing Y-i, Eun MY, et al. (2004) Rice Mutant Resources for Gene Discovery. Plant Molecular Biology 54: 325–334.

10. Krishnan A, Guiderdoni E, An G, Hsing Y-iC, Han C-d, et al. (2009) Mutant resources in rice for functional genomics of the grasses. Plant Physiology 140: 165–170.

11. Neeraja CN, Vemireddy LR, Malathi S, Siddiq EA (2009) Identification of alternate dwarfing gene sources to widely used Dee-GeeWoo-Gen allele of *sd1* gene by molecular and biochemical assays in rice (*Oryza sativa* L.). Electronic Journal of Biotechnology 12(3): 1–9. doi: 10.2225/vol12-issue3-full11

12. Rutger JN (2009) The induced sd1 mutant and other useful mutant genes in modern rice varieties. In: Shu QY, editor. Induced Plant Mutations in the Genomics Era. Rome: Food and Agriculture Organization of the United Nations. pp. 44–47.

13. Wu J-L, Wu C, Lei C, Baraoidan M, Bordeos A, et al. (2005) Chemical- and irradiation-induced mutants of indica rice IR64 for forward and reverse genetics. Plant Molecular Biology 59: 85–97.

14. Wu C, Bordeos A, Madamba MRS, Baraoidan M, Ramos M, et al. (2008) Rice lesion mimic mutants with enhanced resistance to diseases. Molecular Genetics and Genomics 279(6): 605–619.

15. Madamba MRS, Sugiyama N, Bordeos A, Mauleon R, Satoh K, et al. (2009) A recessive mutation in rice conferring non-race- specific resistance to bacterial blight and blast. Rice 2(2–3): 104–114.

16. Nakhoda B, Leung H, Mendioro MS, Mohammadi-nehad G, Ismail AM (2012) Isolation, characterization, and field evaluation of rice (*Oryza sativa* L., Var. IR64) mutants with altered responses to salt stress. Field Crops Research 127(27): 191–202.

17. Cairns JE, Acuna BTL, Simborio FA, Dimayuga G, Praba LM, et al. (2009) Identification of deletion mutants with improved performance under water-limited environments in rice (*Oryza sativa* L.). Field Crops Research 114(1): 159–168.

18. Shi YF, Chen J, Liu WQ, Huang QN, Shen B, et al. (2009) Genetic analysis and gene mapping of a new rolled-leaf mutant in rice (*Oryza sativa* L.). Science in China Series C: life sciences 52(9): 885–890.

19. Murchie EH, Horton P (2007) Toward C_4 rice: learning from the acclimation of photosynthesis in the C_3 leaf. In: Sheehy JE, Mitchell PL, Hardy B, editors. Charting new pathways to C_4 rice. Los Baños: International Rice Research Institute. pp. 333–350.

20. Bhat RS, Upadhyaya NM, Chaudhury A, Raghavan C, Qiu F, et al. (2007) Chemical- and irradiation-induced mutants and TILLING. In: Upadhyaya NM, editor. Rice Functional Genomics: challenges,

progress and prospects. New York: Springer Science+Business Media, LLC. pp. 149–180.

21. Wan S, Wu J, Zhang Z, Sun X, Lv Y, et al. (2009) Activation, tagging, an efficient tool for functional analysis of the rice genome. Plant Molecular Biology 69: 69–80.

22. Hsing Y-I, Chern C-G, Fan M-J, Lu P-C, Chen K-T, et al. (2007) A rice gene activation/knockout mutant resource for high throughput functional genomics. Plant Molecular Biology 63: 351–364.

23. Smillie IRA, Pyke KA, Murchie EH (2012) Variation in vein density and mesophyll cell architecture in a rice deletion mutant population. Journal of Experimental Botany 63(12): 4563–4570. doi: 10.1093/jxb/ers142

24. Chern C-G, Fan M-J, Yu S-M, Hour A-L, Lu P-C, et al. (2007) A rice phenomics study- phenotype scoring and seed propagation of a T-DNA insertion-induced rice mutant population. Plant Molecular Biology 65: 427–438.

25. Noblin X, Mahadevan L, Coomaraswamy IA, Weitz DA, Holbrook NM, et al. (2008) Optimal vein density in artificial and real leaves. Proceedings of the National Academy of Sciences, USA 105: 9140–9144.

26. Murchie EH, Hubbart S, Peng S, Horton P (2005) Acclimation of photosynthesis to high irradiance in rice: gene expression and interactions with leaf development. Journal of Experimental Botany 56(411): 449–460.

27. Klich MG (2000) Leaf variations in *Elaeagnus angustifolia* related to environmental heterogeneity. Environmental and Experimental Botany 44: 171–183.

28. Uhl D, Mosbrugger V (1999) Leaf venation density as a climate and/or environmental proxy; a critical review and new data. Palaeogeography, Palaeoclimatology, Palaeoecology 149: 17–30.

29. Roth-Nebelsick A, Uhl D, Mosbrugger V, Kerp H (2001) Evolution and function of leaf venation architecture: a review. Annals of Botany 87: 553–566.

30. Kuromori T, Takahashi S, Kondou Y, Shinozaki K, Matsui M (2009) Phenome analysis in plant species using loss-of-function and gain-of-function mutants. Plant & Cell Physiology 50(7): 1215–1231.

31. Chen CX, Sun JS (1994) Somaclone variation in plants. Journal of Biological Sciences 3: 4–6.

32. Keeley JE, Rundel PW (2003) Evolution of CAM and C_4 carbon-concentrating echanisms. International Journal of Plant Science 163(3 Supplement): S55–S77.

33. Sack L, Frole K (2006) Leaf structural diversity is related to hydraulic capacity in tropical rain forest trees. Ecology 87: 483–491.

34. Brodribb TJ, Field TS, Jordan GJ (2007) Leaf maximum photosynthetic rate and venation are linked by hydraulics. Plant Physiology 144: 1890–1898.

35. Givnish TJ (1987) Comparative studies of leaf form: assessing the relative roles of selective pressures and phylogenetic constraints. New Phytologist 106(Supplement): 131–160.

36. Oguro H, Hinata K, Tsunoda S (1985) Comparative anatomy and morphology of leaves between C_3 and C_4 species in *Panicum*. Annals of botany 55(6): 859–867.

37. Dengler NG, Kang J (2001) Vascular patterning and leaf shape. Current Opinion in Plant Biology 4: 50–56.

38. .Nelson T, Dengler NG (1997) Leaf vascular pattern formation. The Plant Cell 9: 1121–1135.

39. Runions A, Fuhrer M, Lane B, Federl P, Rolland–Lagan A-G, et al. (2005) Modeling and visualization of leaf venation patterns. ACM Transactions on Graphics 24(3): 702–711.

40. Jahn CE, McKay JK, Mauleon R, Stephens J, McNally KL, et al. (2011) Genetic variation in biomass traits among 20 diverse rice varieties. Plant Physiology 155: 157–168.

Chapter 7

Characterization of DNA Repair Deficient Strains ofChlamydomonas reinhardtii Generated by Insertional Mutagenesis

Andrea Plecenikova1,2¤, Miroslava Slaninova2, Karel Riha1,3*

[1] *Gregor Mendel Institute, Austrian Academy of Sciences, Vienna Biocenter (VBC), Vienna, Austria,*

[2] *Department of Genetics, Faculty of Natural Sciences, Comenius University in Bratislava, Bratislava, Slovakia,*

[3] *Central European Institute of Technology (CEITEC), Masaryk University, Brno, Czech Republic*

ABSTRACT

While the mechanisms governing DNA damage response and repair are fundamentally conserved, cross-kingdom comparisons indicate that they differ in many aspects due to differences in life-styles and developmental strategies. In photosynthetic organisms these differences have not been fully explored because gene-discovery approaches are mainly based on homology searches with known DDR/DNA repair proteins. Here we performed a forward genetic screen in the green algae *Chlamydomonas reinhardtii* to identify genes deficient in DDR/DNA repair. We isolated five insertional mutants that were sensitive to various genotoxic insults and two of them exhibited altered efficiency of transgene integration. To identify genomic regions disrupted in these mutants, we established a novel adaptor-ligation strategy for the efficient recovery of the insertion flanking sites. Four mutants harbored deletions that involved known DNA repair factors, DNA Pol zeta, DNA Pol theta, SAE2/COM1, and two neighbouring genes encoding ERCC1 and RAD17. Deletion in the last mutant spanned two*Chlamydomonas*-specific genes with unknown function, demonstrating the utility of this approach for discovering novel factors involved in genome maintenance.

INTRODUCTION

DNA double strand breaks (DSBs) pose a serious threat to genome integrity as their erroneous repair may lead to chromosomal rearrangements with potentially lethal consequences for an organism. The response to DSBs elicits a highly organized and complex cellular program, called the DNA damage response (DDR), which sets in motion processes that mitigate the adverse effects of DNA damage and facilitate DNA repair. Broken DNA is usually repaired by one of two mechanistically distinct pathways: homologous recombination (HR) and non-homologous end joining (NHEJ). While HR uses a homologous DNA strand as a template for error-free repair, NHEJ is inherently error-prone and does not rely on sequence homology. The preferred mode of repair and cellular consequences of DDR varies between organisms and is also dependant on cell type and cell cycle context [1], [2]. For example, while HR is the preferred mode of repair in many unicellular organisms such as budding and fission yeast, NHEJ is the prevalent pathway in plants and animals.

In many aspects, plants seem to respond differently to DNA insults than do animals. The constant risk of tumor formation in animals has led to evolution of DDR that assures precise genome maintenance, often resulting in apoptotic death of damaged cells. The lack of such a strong selective constraint presumably permitted evolution of a more relaxed DDR in plants, making them more tolerant to genome damage [3], [4]. Furthermore, plant cells are exposed to high levels of genotoxic stress resulting from long-term exposure to solar ultraviolet (UV) irradiation, photosynthesis and extended periods of desiccation [5], [6]. Thus, some features of plant DDR and DSB repair may deviate from models primarily established from studies in yeasts and mammals. Functional characterization of plant DDR and DSB repair is mainly limited to a few model organisms including *Arabidopsis thaliana*, rice, maize, and the moss*Physcomytrella patens* [6], [7], [8]. The key experimental strategy for DDR gene discovery in plants is based on homology searches for conserved DNA repair proteins known from yeast and animals [9], [10]. Forward genetic screens have a much higher potential to identify unknown genes with novel DDR and DSB repair functions, and they have been successfully applied for dissecting the mechanism of meiotic recombination [11], [12]. However, only a few screens were performed with the aim to identify novel components of somatic DSB repair [13],[14], [15]. This is partially due to the complicated nature of these screens, as the scored phenotype is often represented by sensitivity to a genotoxic treatment and hence the death of the plant that carries the desired mutation.

In this study we exploited the potential of the unicellular green algae *Chlamydomonas reinhardtii* in discovery of genes related to DDR and DSB repair in the plant kingdom. *C. reinhardtii* is an established model that is traditionally used for studying photosynthesis and cell motility. Interest in this organism has also been sparked by prospects of using algae as source of hydrogen or lipids for biofuel production. In addition, sharing a common ancestor 1.1 billion years ago, *C. reinhardtii* provides an excellent complementary model to higher plants for comparative gene analyses within the plant kingdom [16]. Whereas unicellular *C. reinhardtii*appears to be a relatively simple organism, in regards to genome complexity and mechanisms that govern genome maintenance and expression it is comparable with higher eukaryotes. The draft genome sequence revealed more than 15,000 genes that contain a large number of introns and numerous repetitive DNA sequences and transposable elements [17]. The presence of DNA methylation, siRNA and microRNAs indicates that epigenetic and post-transcriptional regulation of gene expression is more complex than in many other unicellular models [18], [19]. In addition, *C. reinhardtii* displays an extremely low efficiency of HR, which is unusual for a unicellular organism [20] and similar to higher plants. *C. reinhardtii* therefore offers the unique opportunity to study some of the cellular processes that are featured in higher eukaryotes while providing the advantages of a unicellular experimental system.

Here, we performed a forward genetic screen of a population of *C. reinhardtii* mutagenized by insertional mutagenesis for sensitivity to genotoxic stress and describe an efficient strategy for identification of disrupted loci. We characterized five mutants and determined that they encompass deletions in known DNA repair genes, which validated our approach, as well as in genes with unknown functions, demonstrating the potential of this approach in identification of novel DDR related genes.

MATERIALS AND METHODS

Strain and growth media

The *C. reinhardtii* strain *cw15-302 arg2* [21] was obtained from the laboratory of Dr. Christoph Beck (University of Freiburg, Germany). Cells were grown under constant light at 22°C in Tris-acetate phosphate media (TAP) supplemented with L-arginine (100 mg/L) and with 1% agar when solid medium was required [22].

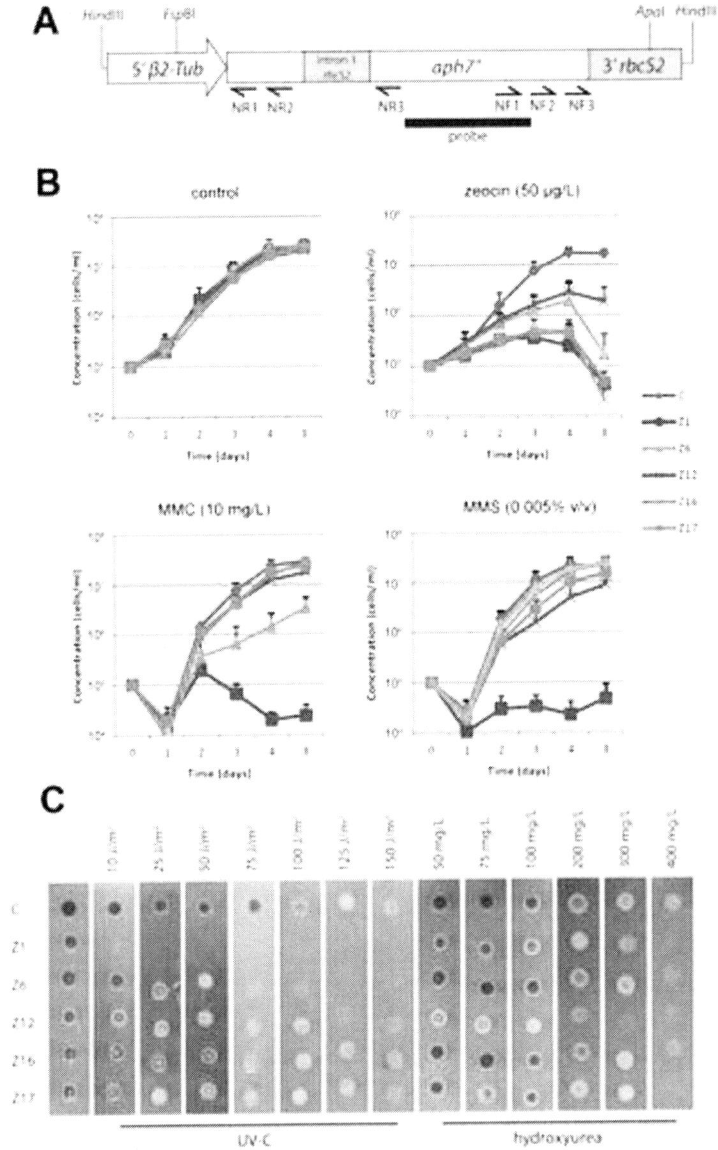

Figure 1. Sensitivity of *C. reinhardtii* mutant strains to genotoxic treatments. (A) Structure of the *aph7"* construct used for the insertional mutagenesis. Primers (arrows), restriction sites and the region used as a probe for Southern hybridization are indicated. (B) Growth curves of mutant and the parental *cw15-302 arg2* strains (denoted as C) in TAP liquid media supplemented with zeocin, MMC or MMS. (C) Growth of mutant strains on TAP plates exposed to increasing dose of UV-C or HU. Pictures were taken five days after inoculation.

GENETIC SCREEN

Plasmid pHyg3 [23] was digested with *Hin*dIII and the linear fragment containing the chimeric*aph7"* selection marker gene was used as the insertional cassette (Figure 1A). Insertional mutagenesis of *C. reinhardtii* cells was performed by the glass-bead transformation method[24]. Immediately after transformation, cells were spread onto selective TAP plates containing 10 mg/L of hygromycin B (Calbiochem) and grown for 1 week. Transformed colonies resistant to hygromycin B were resuspended in sterile distilled water and replicated to one set of replica TAP plates with or without 300 μg/L of zeocin (Duchefa Biochemie) and to the second set of replica plates, one of which was irradiated with UV-C (70 J/m^2) immediately after plating (Stratalinker UV Crosslinker 2400, Stratagene). For the UV-C sensitivity screen, the plates were incubated for one day in the dark to avoid photoreactivation. Colonies sensitive to the selection conditions were maintained on TAP plates and in the subsequent 3–6 months re-screened at least twice for the sensitivity to zeocin and UV-C irradiation.

Genotoxicity tests

Sensitivity to hydroxyurea (HU) was assessed by inoculating cells on TAP plates supplemented with 50–400 mg/L HU (Sigma-Aldrich). Sensitivity to UV-C irradiation was examined by scoring cell growth on TAP media exposed to 10 J/m^2–150 J/m^2 UV-C (Stratalinker UV Crosslinker 2400, Stratagene). Sensitivity to zeocin (50 μg/L), mitomycin C (MMC; 10 mg/L; Sigma-Aldrich) and methyl methanesulfonate (MMS; 0,005%) was tested by determining growth curves in TAP liquid media. Mutant and control strains were first grown for 3 days on TAP plates with L-arginine (100 mg/L) and hygromycin B (10 mg/L) under continuous light and then transferred to liquid TAP media supplemented with L-arginine and the tested drug. Cell density was determined in a Bürker chamber after 3 days (time point 0) and then again at the 24, 48, 72, 96 and 192 hours. Data from three independent experiments were collected and analyzed.

Nucleic acid isolation and analysis

Total genomic DNA was isolated using a modified protocol according to [25]. 50 ml of *C. reinhardtii* culture grown for 3 days (~2.10^7 cells. ml^{-1}) was pelleted by centrifugation, resuspended in 2×CTAB buffer [1.4 M NaCl, 0.1 M Tris ph 8.0, 0.02 M EDTA, 2% hexadecyltrimethylammoniumbromide (CTAB, Sigma)] and incubated for 1 h at 65°C. DNA was extracted with phenol:chloroform:isoamyl alcohol (25:24:1, v/v) and precipitated by isopropanol. RNA was removed by 30 min treatment with RNase (164 mg/L) at 37°C that was subsequently removed by phenol and chloroform:isoamyl alcohol (24:1, v/v) extractions. DNA was purified by ethanol precipitation. Genomic DNA was subjected to Southern hybridization analysis according to standard protocols [26]. A PCR fragment amplified from the pHyg3 plasmid with primers HygR_PR1_F616 5′-GAGAGCACCAACCCCGTACTGG-3′ and HygR_PR1_R1196 5′-GTGAAGTCGACGA TCCCGGT-3′ was radioactively labeled with [α ^{32}P] dCTP and used as a probe.

Isolation of insert-flanking regions by hairpin-PCR

2 μg of genomic DNA was digested with a restriction endonuclease according to supplier's specifications and isopropanol precipitated. 1 μg of digested DNA was ligated to a hairpin adaptor using T4 DNA ligase (Thermo Fisher Scientific) for 16 h at laboratory temperature. The hairpin adaptor [27] compatible with the DNA ends generated by restriction digest was used in a final concentration of 1.6 μM. DNA was isopropanol precipitated and 40–80 ng were used as the template in PCR. Genomic DNA flanking the site of insertion was amplified by two rounds of PCR with the primer PETRA-B (5′-CTCTAGACTGTGAGACTTGGAGATG-3′) and nested insert specific primers complementary to the *aph7″* gene (Figure 1A): NR1 (5′-CCAGTGCTCGCCGAACAGCT-3′), NR2 (5′-TCGTTCCGCAGGCTCGCGTA-3′), NF1 (5′-GAGACTCCCGCTACAGCCTG-3′) and NF2 (5′-CTGCACGACTTCGAGGTGTT CG-3′). PCR mixture consisted of 1×GoTaq Reaction Buffer (Promega), 0.2 mM dNTP, 0.5 μM PETRA-B, 0.5 μM NR1 or NF1, 40 ng–80 ng DNA, 3% DMSO, Go Taq DNA polymerase (Promega) and the reaction was performed with 35 cycles of 45 s at 94°C, 30 s at 64°C and 2 min 30 s at 72°C. 0.5 μl of PCR amplification product was used as the template for nested PCR with PETRA-B and NR2 or NF2 primers. PCR products were cloned into pGEM-T Easy Vector (Promega) or to pCR2.1-TOPO Vector (Invitrogen) and transformed to *E. coli* DH5α competent cells and sequenced using a standard procedure.

Insert integration assay

C. reinhardtii strains were transformed by the glass-bead method with plasmids pUCARG7.8 and pUCBM20ΔARG [21] and transformation efficiency and HR/NHEJ ratios were calculated as previously described [28].

RESULTS

We generated a library of *C. reinhardtii* insertional mutants by transforming the cell-wall deficient strain *cw15-302 arg2* with the 1.7 kb fragment of the pHyg3 plasmid containing the chimeric *aph7″* gene that confers resistance to hygromycin B [23] (Figure 1A). We used a glass-bead transformation technique and a low amount of DNA (500 ng) per transformation to decrease the chance of multiple insertions per genome [29]. In total, we generated 4588 transformants that were subsequently screened for growth on agar plates supplemented with zeocin (300 mg/L), an antibiotic that induces DSB, and for survival on plates exposed to UV-C irradiation (70 J/m^2). After two rounds of screening we obtained 19 clones sensitive to zeocin, 52 clones sensitive to UV-C, and 9 clones sensitive to both treatments. However, only five mutant clones retained the phenotype after 3–6 months of subculturing. All five mutant strains, which we named Z1, Z6, Z12, Z16 and Z17, were originally selected for their inability to grow on agar plates supplemented with zeocin, and this phenotype was also confirmed by cultivation in liquid media (Figure 1B). Z1, Z16 and Z17 exhibited the highest growth retardation in the presence of zeocin, while Z6 showed intermediate sensitivity and the growth of Z12 mutants was least affected (Figure 1B and Table 1). The growth of clones Z1 and Z6 was also impaired by UV-C irradiation (Figure 1C, Table 1).

Table 1. Relative sensitivity to genotoxic treatments in respect to other mutants and the control strain *cw15-302 arg2* (+++ high sensitivity, ++ intermediate sensitivity, + mild sensitivity, −no sensitivity).

Mutant	Zeocin	UV	MMC	MMS	HU
Z1	+++	+++	+++	+++	+
Z6	++	++	+	–	–
Z12	+	+	–	++	+++
Z16	+++	–	–	–	–
Z17	+++	–	–	+	–

We next tested sensitivity of the mutants to the genotoxic drugs MMC, methyl MMS and HU. These drugs induce DNA damage by different mechanisms: MMC causes DNA interstrand cross-linking, MMS treatment methylates DNA which is believed to lead to replication fork stalling, and HU impairs production of deoxyribonucleotides, which inhibits DNA replication and results in stalled replication forks. MMC almost fully suppressed growth of the Z1 mutant, while proliferation of Z6 was impaired to lesser extent (Figure 1B, Table 1). Other strains were unaffected. Z1 also exhibited the highest sensitivity to MMS. A slight inhibition of growth by MMS was detected in Z12 and Z17 (Figure 1B, Table 1). HU had the strongest effect on Z12, exhibiting a discernable effect already at the lowest concentration of HU. A slight inhibition was also detected in the Z1 mutant (Figure 1C, Table 1). The observation that almost all mutants were sensitive to at least two independent genotoxic treatments validated the screening approach and suggested that the mutants we obtained are indeed impaired in DDR/DSB repair. Moreover, the differential sensitivity of the 5 mutant strains to particular treatments indicated a deficiency in different aspects of DDR and DNA repair.

Insertion of exogenous DNA into a genome during transformation is governed by processes involved in DSB repair. To examine whether the efficiency of DNA integration via NHEJ or homology driven repair is altered in the isolated strains, we took advantage of the missense mutation *arg7-8* in the *ARG7* gene that encodes argininosuccinate lyase (ASL) and confers arginine auxotrophy of the *cw15-302 arg2* strain [21]. We transformed the mutant strains with plasmids carrying either the entire *ARG7* gene (pUCARG7.8), or a truncated version that lacks the 5′ region encoding the promoter and N-terminal portion of ASL (pUCBM20ΔARG) [21], [28]. Since the transformants were selected for complementation of ARG7 function by their ability to grow on plates without arginine, prototrophy of the pUCBM20ΔARG transformants was anticipated to be caused by homology driven repair that corrected the *arg7-8* mutation [21]. While transformation efficiency with the pUCARG7.8 construct was unaffected in Z1, Z6 and Z17 mutants as compared to the *cw15-302 arg2* strain, a 10-fold decrease was detected in the Z16 background suggesting deficiency in NHEJ ($P=0.063$, two-tailed Student's t test; Figure 2A). Interestingly, occurrence of Arg prototrophic clones increased by one order of magnitude in Z12. Homology driven integration in *C. reinhardtii* has been reported to be 1.5–4 orders of magnitude lower than random integration and depends on the transformation method and constitution of transformed DNA [20], [30], [31], [32]. In our conditions, the frequency of pUCBM20ΔARG transformants in the *cw15-302 arg2* strain was approximately 1000-times lower than of pUCARG7.8. Similar ratios were detected in the Z1, Z6, and Z16 strains while a slight decline in the ratio was detected in Z17, indicating a deficiency in HR ($P=0.088$, two-tailed Student's t test; Figure 2B).

The high frequency of Arg prototrophs recovered from the Z12 background was unusual and warranted further verifications. We found that prototrophic colonies appeared in a transformation-independent manner and hence, represented spontaneous phenotypic reversions. These reversions are specific to the Z12 strain and were never observed in the parental or other mutant strains. The Arg prototrophic phenotype of Z12 revertant clones was stable, but they appeared to grow slower in liquid media without arginine than *ARG7*complemented *cw15-302 arg2* cells (Figure 2C). Sequencing of the *ARG7* region in several Z12 revertants excluded the possibility that the phenotypic reversions were caused by spontaneous repair of the *arg7-8* mutation (data not shown). Thus, we conclude that Z12 cells are predisposed to activate an adaptive mechanism enabling them to overcome *arg7-8* deficiency mediated arginine auxotrophy.

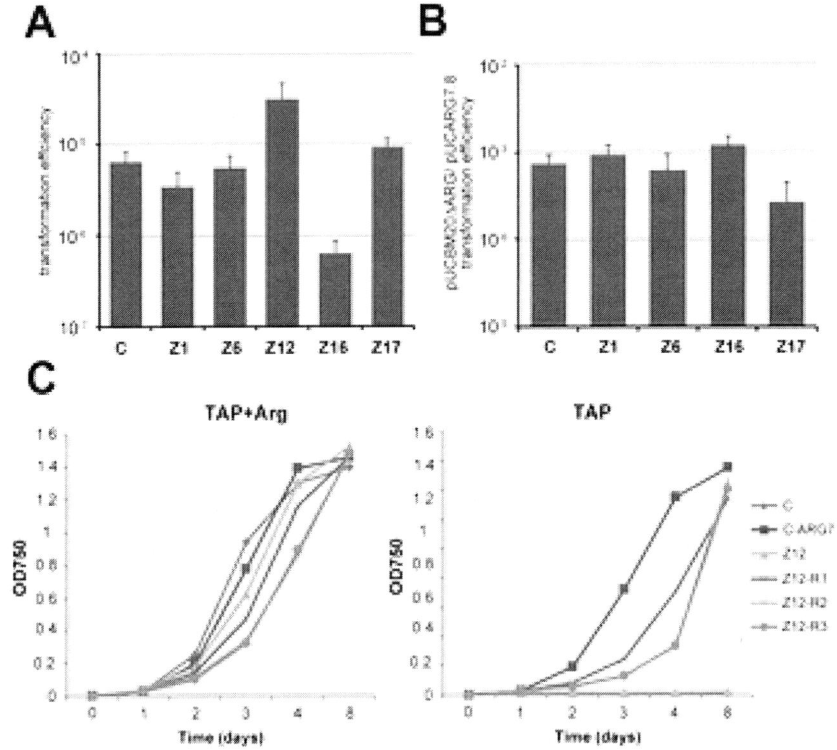

Figure 2. Transformation efficiency of *C. reinhardtii* mutants. (A) Transformation efficiency calculated as the frequency of Arg-prototrophic transformants per a total number of transformed cells normalized to 1 pmol of pUCARG7.8 construct used for transformation. (B) Efficiency of homology-driven integration estimated as a ratio between the transformation efficiencies obtained with the pUCBM20ΔARG and pUCARG7.8 constructs. (A, B) C=*cw15-302 arg2*; standard errors of three independent experiments are indicated (N=3). (C) Growth curves of the Z12 strain and selected Z12 Arg-prototrophic revertants in TAP liquid media with or without Arg. *cw15-302 arg2* cells complemented with the *ARG7* construct were used as an Arg-prototrophic control (C-ARG7).

Next we sought to identify the loci disrupted by insertional mutagenesis. First, we determined the number of independent insertions in the genomes of the mutant strains by Southern analysis with a probe specific for the insertional cassette (Figure 1A, 3A). Restriction analysis with enzymes that digest outside of the construct (*Pst*I, *Pvu*II) revealed the presence of a single insertion in all mutants (Figure 3A). Multiple bands that appeared in the Z1 mutant after digestion with enzymes that cleave within the *aph7"* gene (*Fsp*BI, *Apa*I; Figure

1A, 3A) were suggestive of three copies of the insert in a single site. Other mutants appeared to harbour only a single copy of *aph7"* in their genome.

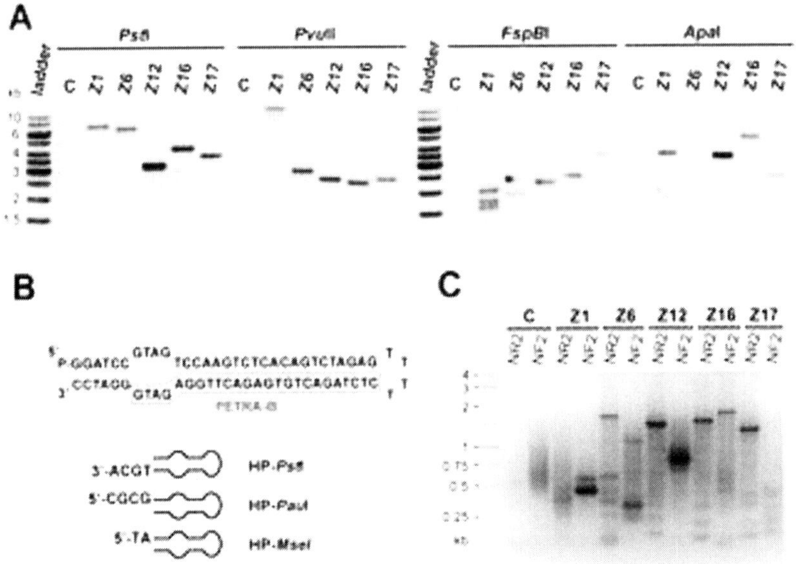

Figure 3. Molecular characterization of *aph7"* insertions. (A) Southern analysis of *aph7"* insertions in individual mutants. The restriction enzymes used for each experiment are indicated; positions of the restriction sites within the*aph7"* construct probe used for hybridization are depicted in Figure 1A. (B) Structure of the hairpin adaptor with a blunt end; examples of hairpins with ends compatible with selected restriction enzymes are indicated. (C) PCR products, separated in agarose gels and stained with ethidium bromide, that were generated after two rounds of nested PCR from a genomic library produced by *Pst*I digest.

To localize the insertion sites, we developed a PCR-based strategy utilizing ligation of DNA hairpin adaptors to digested genomic fragments (Figure 3B). We have originally designed these adaptors to specifically amplify blunt-ended telomeres in *A. thaliana* [27]. The end of the hairpin can be designed to contain complementarity to DNA ends generated by any restriction enzyme (Figure 3B). The adaptor further harbours a short stretch of non-complementary sequence forming a bubble near the open end of the hairpin. This bubble corresponds to the sequence of a PCR primer (PETRA-B; Figure 3B) that does not amplify ligated adaptor dimers, but can only anneal to DNA strands arising from the extension of a primer specific to a gene of interest. To amplify sequences flanking the insertion site, ligation of the adaptor to digested genomic DNA was followed by two rounds of nested PCR with the PETRA-B and insert specific primers (Figures 1A and 3C). Several adaptor-ligated libraries generated by different restriction enzymes were usually tested for each mutant. The most suitable restriction enzymes for this purpose were *Pst*I, *Pau*I, *Pvu*II, *Nsb*I and *Mse*I. PCR products were sequenced and aligned to the scaffold of the *C. reinhardtii* genome. Using this approach we identified both flanking sites in Z6, Z12, Z16 and Z17 mutants. Insertions in these strains were accompanied by large deletions ranging from 11 to 30 kb that encompassed two to five genes (Figure 4). Deletions in Z6, Z16 and Z17 disrupted genes coding for proteins homologous to the known DNA repair factors DNA Pol zeta, DNA Pol theta and SAE2/COM1 nuclease, respectively. An 11 kb deletion in Z12 affected two proteins of unknown function, Cre10.g441650 and Cre10.g441700, which show no extensive sequence homology in protein databases and appear to be specific to *Chlamydomonas* and closely related genomes, e.g. *Volvox carteri*. Only one flanking site with homology to the nuclear genome was retrieved in Z1 mutants. The other insertion border contained chloroplast DNA (Figure 4). The Z1 insertion was localized to the vicinity of genes coding for the ERCC1 endonuclease and the RAD17 DNA damage checkpoint protein. PCR analyses demonstrated that both genes are absent in the Z1 mutant, indicating that a large deletion occurred at the insertion site in this strain. We attempted to complement the zeocin sensitivity phenotypes by transforming the mutant strains with BAC clones spanning the deleted regions. However, despite extensive efforts, these experiments failed due to technical difficulties transforming *C. reinhardtii* with large DNA fragments. Therefore, we could not ascertain that the DNA damage sensitivity of mutant strains was indeed caused by disruption of genes at the insertion loci.

DISCUSSION

The use of forward genetic screens for discovery of genes involved DNA repair and DNA damage response has been limited in plants; most plant DDR/DNA repair proteins were identified based on sequence homology with their yeast and mammalian counterparts. The major caveat of this homology driven approach is its inability to identify genes and processes that may have evolved specifically in the plant lineage to deal with genomic insults. The importance of unbiased approaches is documented by a screen in *A. thaliana* that revealed novel genes linking DDR with epigenetic regulation and stress signalling pathways [13], [14]. Similarly, SOG1, a transcription factor governing DDR is a plant specific protein that was discovered in a screen to identify suppressors of the radiosensitivity of a nucleotide excision repair mutant [33].

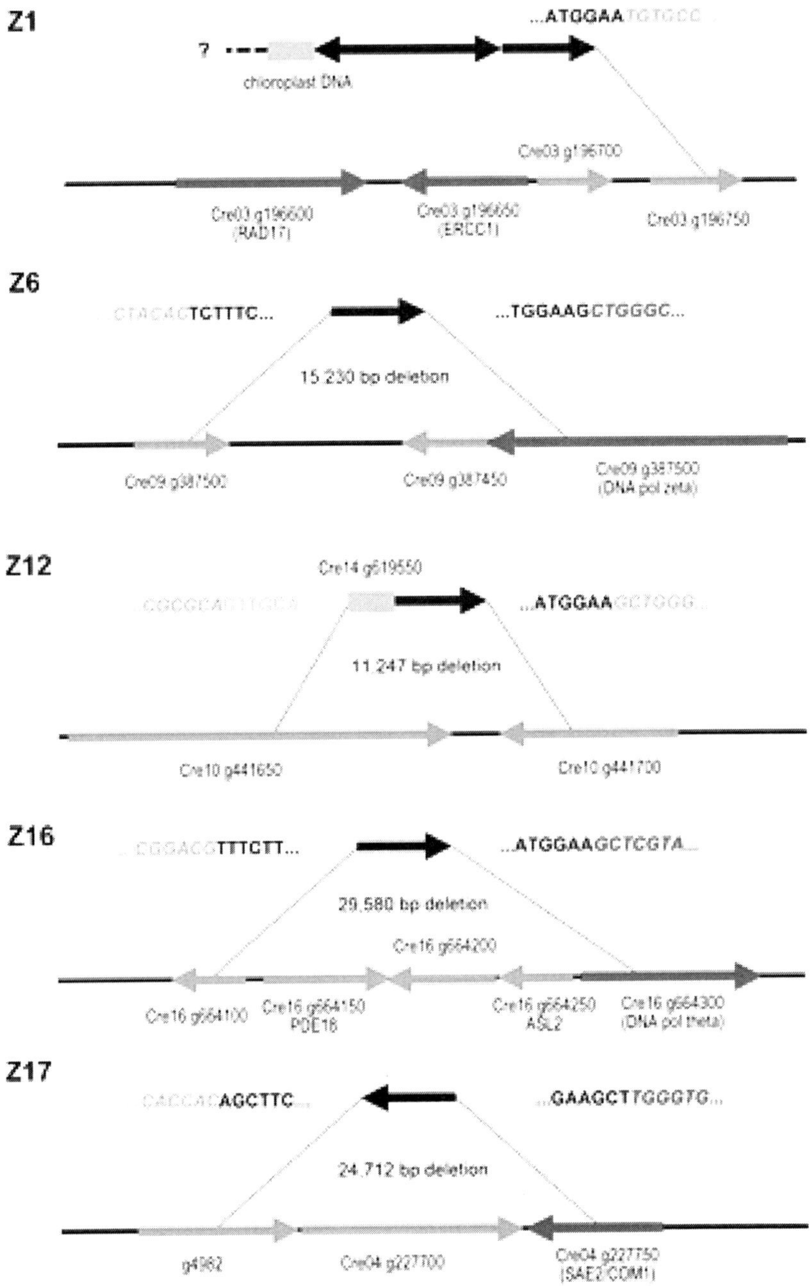

Figure 4. Structure of insertion sites. Black arrows represent the aph7″ insert, red arrows indicate homologues to known DNA repair genes, blue arrows represent genes unrelated to DNA repair, and green boxes depict DNA fragments that were co-

transformed to the insertion sites from ectopic locations. Sequences at the insert borders are indicated.

Here we exploited the unicellular green algae *C. reinhardtii* in a forward genetic screen for the discovery of DDR/DNA repair genes. By surveying a mutagenized population of *C. reinhardtii*, we have identified five transformants that exhibited stable sensitivity to the radiomimetic drug zeocin and, with exception of the mutant Z16, to at least one other genotoxic treatment. To identify the loci disrupted in these mutants, we have developed a novel PCR-mediated strategy based on ligation of hairpin adapters to genomic fragments generated by restriction endonucleases cutting within G/C-rich sequences. Inverse PCR and TAIL-PCR are the usual methods of choice for cloning sequences flanking insertion sites, but the high GC content of the *C. reinhardtii* genome renders the success of these techniques highly variable [34], [35]. With the hairpin-ligation strategy we were able to readily obtain sequences flanking both insertion sites for all mutants; in fact, in some mutants we obtained these sequences several times independently using different restriction enzymes (data not shown). The forward genetic screen in *C. reinhardtii* was extremely time efficient in comparison to other plant models as the entire procedure from making the mutant library to the identification of disrupted loci took about 9 months. Validation of the DNA-damage sensitivity proved to be the most critical and time consuming step as a large proportion of isolated mutants lost the phenotype after 3 to 6 months of subculturing. Lack of tools for efficient validation of genes identified in the screen was the major hurdle we faced in this study. Insertional mutagenesis led to deletions spanning several genes. As *C. reinhardtii* is not amenable to gene targeting, genetic complementation appears to be a method of choice for confirming association between scored phenotypes and disrupted loci. However, also this approach can be in *C. reinhardtii* troublesome. *C. reinhardtii*genes can span tens of kbs which in combination with very high CG content impedes classical cloning and PCR based approaches for making complementing constructs. For these reasons, we were unable to complement mutant phenotypes and formally confirm identity of causative genes. We anticipate that advances in genome-editing technologies utilizing TALENs or CRISPR/Cas9 will eventually solve this issue and facilitate gene discovery in *C. reinhardtii*. Nevertheless, the fact that four out of five insertion-tagged loci contained disruption of known DNA repair genes validates suitability of the experimental approach in discovery of genes involved in plant DDR.

In the case of the Z1 insertion we were able to map only one border to the nuclear genome, while the other flanking sequence consisted of a chloroplast DNA fragment. The insertion event led to deletion of genes homologous to *ERCC1* and *RAD17*. ERCC1 forms, together with XPF, a structure specific nuclease discovered for its role in nucleotide excision repair [36]. It is also involved in processing of DNA interstrand crosslink lesions and homologous recombination intermediates. ERCC1 deficiency in *A. thaliana* has been reported to cause sensitivity to MMC, MMS and UV and gamma irradiation [15]. RAD17 is a conserved replication checkpoint protein important for genome stability and its inactivation in *A. thaliana*renders plants sensitive to bleocin and MMC [37]. Thus, the sensitivity of *C. reinhardtii* Z1 mutants to a broad range of genotoxic treatments is likely caused by the combined absence of these DNA repair proteins.

Characterization of insertion sites in the Z6 and Z16 mutants revealed deletions disrupting genes coding for translesion synthesis DNA polymerases zeta and theta, respectively. DNA Pol zeta can bypass a range of DNA lesions including adducts induced by cisplatin, UV and apurinic/apyrimidinic sites [38] and there is evidence suggesting its function in interstrand crosslink repair and HR [39], [40]. This is consistent with the relatively mild, but broad sensitivity of the Z6 strain to genotoxic insults that include zeocin, UV and MMC; a similar spectrum of sensitivity was also observed in the *A. thaliana* DNA pol zeta mutant [41]. In contrast, the Z16 strain carrying a deletion in DNA pol theta (also known as POLQ) exhibits strong sensitivity to zeocin, but not to the other treatments. Interestingly, *A. thaliana* pol theta deficient plants display developmental defects, constitutive DNA damage response and sensitivity to MMC and MMS [42]. Thus, this polymerase may have acquired more fundamental role in DNA repair in higher plants. Mouse and human cell lines impaired in DNA pol theta are sensitive to bleomycin and ionizing radiation, implicating its function in some aspects of DSB repair [43], [44]. Studies in *Drosophila melanogaster* indicated that DNA Pol theta contributes to DSB repair by facilitating microhomology-mediated DNA end-joining [45], which is in accordance with our observation of a 10-fold decrease in the transformation efficiency of the Z16 mutant.

The Z17 strain harbours a deletion that disrupts SAE2/COM1/CtIP, a protein that co-operates with the MRE11-RAD50-NBS1/Xrs2 nuclease to initiate resection of DSB during HR [46], [47]. This complex also plays a central role in the repair of stalled or collapsed replication forks, and mutants described in a number of organisms are sensitive to a wide range of DNA damaging agents including MMC, ionizing radiation, hydroxyurea, bleomycin or MMS [48], [49], [50]. The conserved HR function of SAE2/COM1 in *C. reinhardtii* is also indicated by decreased efficiency of homology-driven integration of the pUCBM20ΔARG construct into Z17 genome. In higher plants, COM1 appears to participate in interstrand crosslink repair as *A. thaliana com1*mutant displayed sensitivity to MMC [51].

Identification of several well characterized DNA repair factors at the insertion sites in Z1, Z6, Z16 and Z17 strains demonstrated the ability of the screen to uncover genes involved in DDR and genome integrity. Nevertheless, the ultimate goal of forward genetic screens is to identify unknown genes or unanticipated links to the processes under study. Two previous studies reported on the identification of novel genes involved in DNA repair in *C. reinhardtii* by combining insertional mutagenesis with a screen for sensitivity to genotoxic treatments [52],[53]. The disruption in the Z12 mutant also affects two genes encoding proteins of unknown function that appear to be specific to *C. reinhardtii* and closely related species. The Z12 mutant exhibits the strongest sensitivity to HU and MMS among all the mutants recovered in our screen, suggesting the impairment of processes associated with DNA replication. Interestingly, we observed that in addition to DNA damage sensitivity, Z12 cells are able to overcome arginine auxotrophy caused by the *arg7-8* mutation in the *ARG7* gene. Mutations in the *ARG7* have been used for decades as auxotrophic markers in *C. reinhardtii* transformation experiments for their stability and rare reversions [54], [55]. However, the spontaneous phenotypic reversions to arginine prototrophy occur in Z12 at an extremely high frequency and are not accompanied by correction of the causative mutation in the *ARG7* gene. The adaptive mechanism that is activated in Z12 revertants is unknown. It is also unclear whether the propensity of Z12 cells to switch to Arg prototrophy is linked to the DNA repair deficiency phenotype and whether it associates with either of the disrupted genes. The high rate of the phenotypic reversion, along with its stability, evokes parallels with de-repression of an epigenetically suppressed mechanism. Spontaneous changes in epigenetic states can be stably inherited and occur at much higher frequencies than genetic mutations. Furthermore, malfunction of numerous DNA replication and repair factors has been linked to de-repression of epigenetically silent loci, and, *vice versa*, deficiency in chromatin regulators may lead to impaired DNA repair [13], [56], [57], [58]. Thus, we speculate that the phenotypic switch to arginine prototrophy in Z12 cells is conditioned by a mutation in a gene that is required for both genome integrity and epigenetic inheritance. While further in depth mechanistic studies are required to decipher this phenomenon, the example of the Z12 mutant illustrates the power of forward genetics in *C. reinhardtii* in uncovering novel genes and processes involved in genome maintenance.

ACKNOWLEDGMENTS

We thank to Sona Valuchova and Matt Watson for helpful discussions.

AUTHOR CONTRIBUTIONS

Conceived and designed the experiments: AP MS KR. Performed the experiments: AP. Analyzed the data: AP MS KR. Contributed to the writing of the manuscript: AP KR.

REFERENCES

1. Nagaria P, Robert C, Rassool FV (2013) DNA double-strand break response in stem cells: mechanisms to maintain genomic integrity. Biochim Biophys Acta 1830: 2345–2353.

2. Helle F (2012) Germ cell DNA-repair systems-possible tools in cancer research? Cancer Gene Ther 19: 299–302.

3. Riha K, McKnight TD, Griffing LR, Shippen DE (2001) Living with genome instability: plant responses to telomere dysfunction. Science 291: 1797–1800.

4. Doonan JH, Sablowski R (2010) Walls around tumours - why plants do not develop cancer. Nat Rev Cancer 10: 794–802.

5. Bray CM, West CE (2005) DNA repair mechanisms in plants: crucial sensors and effectors for the maintenance of genome integrity. New Phytol 168: 511–528.

6. Waterworth WM, Drury GE, Bray CM, West CE (2011) Repairing breaks in the plant genome: the importance of keeping it together. New Phytol 192: 805–822.

7. Mannuss A, Trapp O, Puchta H (2012) Gene regulation in response to DNA damage. Biochim Biophys Acta 1819: 154–165.

8. Schaefer DG (2001) Gene targeting in Physcomitrella patens. Curr Opin Plant Biol 4: 143–150.

9. Singh SK, Roy S, Choudhury SR, Sengupta DN (2010) DNA repair and recombination in higher plants: insights from comparative genomics of Arabidopsis and rice. BMC Genomics 11: 443.

10. Vlcek D, Sevcovicova A, Sviezena B, Galova E, Miadokova E (2008) Chlamydomonas reinhardtii: a convenient model system for the study of DNA repair in photoautotrophic eukaryotes. Curr Genet 53: 1–22.

11. De Muyt A, Pereira L, Vezon D, Chelysheva L, Gendrot G, et al. (2009) A high throughput genetic screen identifies new early meiotic recombination functions in Arabidopsis thaliana. PLoS Genet 5: e1000654.

12. Timofejeva L, Skibbe DS, Lee S, Golubovskaya I, Wang R, et al. (2013) Cytological characterization and allelism testing of anther developmental mutants identified in a screen of maize male sterile lines. G3 (Bethesda) 3: 231–249.

13. Takeda S, Tadele Z, Hofmann I, Probst AV, Angelis KJ, et al. (2004) BRU1, a novel link between responses to DNA damage and epigenetic gene silencing in Arabidopsis. Genes Dev 18: 782–793.

14. Ulm R, Revenkova E, di Sansebastiano GP, Bechtold N, Paszkowski J (2001) Mitogen-activated protein kinase phosphatase is required for genotoxic stress relief in Arabidopsis. Genes Dev 15: 699–709.

15. Hefner E, Preuss SB, Britt AB (2003) Arabidopsis mutants sensitive to gamma radiation include the homologue of the human repair gene ERCC1. J Exp Bot 54: 669–680.

16. Gutman BL, Niyogi KK (2004) Chlamydomonas and Arabidopsis. A dynamic duo. Plant Physiol 135: 607–610.

17. Merchant SS, Prochnik SE, Vallon O, Harris EH, Karpowicz SJ, et al. (2007) The Chlamydomonas genome reveals the evolution of key animal and plant functions. Science 318: 245–250.

18. Molnar A, Schwach F, Studholme DJ, Thuenemann EC, Baulcombe DC (2007) miRNAs control gene expression in the single-cell alga Chlamydomonas reinhardtii. Nature 447: 1126–1129.

19. Feng S, Cokus SJ, Zhang X, Chen PY, Bostick M, et al. (2010) Conservation and divergence of methylation patterning in plants and animals. Proc Natl Acad Sci U S A 107: 8689–8694.

20. Sodeinde OA, Kindle KL (1993) Homologous recombination in the nuclear genome of Chlamydomonas reinhardtii. Proc Natl Acad Sci U S A 90: 9199–9203.

21. Mages W, Heinrich O, Treuner G, Vlcek D, Daubnerova I, et al. (2007) Complementation of the Chlamydomonas reinhardtii arg7-8 (arg2) point mutation by recombination with a truncated nonfunctional ARG7 gene. Protist 158: 435–446.

22. Gorman DS, Levine RP (1965) Cytochrome f and plastocyanin: their sequence in the photosynthetic electron transport chain of Chlamydomonas reinhardi. Proc Natl Acad Sci U S A 54: 1665–1669.

23. Berthold P, Schmitt R, Mages W (2002) An engineered Streptomyces hygroscopicus aph7" gene mediates dominant resistance against hygromycin B in Chlamydomonas reinhardtii. Protist 153: 401–412.

24. Kindle KL (1990) High-frequency nuclear transformation of Chlamydomonas reinhardtii. Proc Natl Acad Sci U S A 87: 1228–1232.

25. Borevitz JO, Liang D, Plouffe D, Chang HS, Zhu T, et al. (2003) Large-scale identification of single-feature polymorphisms in complex genomes. Genome Res 13: 513–523.

26. Sambrook J, Russell D (2001) Molecular Cloning: A Laboratory Manual. New Yoprk: CSHL Press.

27. Kazda A, Zellinger B, Rossler M, Derboven E, Kusenda B, et al. (2012) Chromosome end protection by blunt-ended telomeres. Genes Dev 26: 1703–1713.

28. Plecenikova A, Mages W, Andresson OS, Hrossova D, Valuchova S, et al. (2013) Studies on recombination processes in two Chlamydomonas reinhardtii endogenous genes, NIT1 and ARG7. Protist 164: 570–582.

29. Matsuo T, Okamoto K, Onai K, Niwa Y, Shimogawara K, et al. (2008) A systematic forward genetic analysis identified components of the Chlamydomonas circadian system. Genes Dev 22: 918–930.

30. Nelson JA, Lefebvre PA (1995) Targeted disruption of the NIT8 gene in Chlamydomonas reinhardtii. Mol Cell Biol 15: 5762–5769.

31. Zorin B, Hegemann P, Sizova I (2005) Nuclear-gene targeting by using single-stranded DNA avoids illegitimate DNA integration in Chlamydomonas reinhardtii. Eukaryot Cell 4: 1264–1272.

32. Zorin B, Lu Y, Sizova I, Hegemann P (2009) Nuclear gene targeting in Chlamydomonas as exemplified by disruption of the PHOT gene. Gene 432: 91–96.

33. Yoshiyama K, Conklin PA, Huefner ND, Britt AB (2009) Suppressor of gamma response 1 (SOG1) encodes a putative transcription factor governing multiple responses to DNA damage. Proc Natl Acad Sci U S A 106: 12843–12848.

34. Tuncay H, Findinier J, Duchene T, Cogez V, Cousin C, et al. (2013) A forward genetic approach in Chlamydomonas reinhardtii as a strategy for exploring starch catabolism. PLoS One 8: e74763.

35. Dent RM, Haglund CM, Chin BL, Kobayashi MC, Niyogi KK (2005) Functional genomics of eukaryotic photosynthesis using insertional mutagenesis of Chlamydomonas reinhardtii. Plant Physiol 137: 545–556.

36. Davies AA, Friedberg EC, Tomkinson AE, Wood RD, West SC (1995) Role of the Rad1 and Rad10 proteins in nucleotide excision repair and recombination. J Biol Chem 270: 24638–24641.

37. Heitzeberg F, Chen IP, Hartung F, Orel N, Angelis KJ, et al. (2004) The Rad17 homologue of Arabidopsis is involved in the regulation of DNA damage repair and homologous recombination. Plant J 38: 954–968.

38. Shachar S, Ziv O, Avkin S, Adar S, Wittschieben J, et al. (2009) Two-polymerase mechanisms dictate error-free and error-prone translesion DNA synthesis in mammals. EMBO J 28: 383–393.

39. Shen X, Jun S, O'Neal LE, Sonoda E, Bemark M, et al. (2006) REV3 and REV1 play major roles in recombination-independent repair of DNA

interstrand cross-links mediated by monoubiquitinated proliferating cell nuclear antigen (PCNA). J Biol Chem 281: 13869–13872. doi: 10.1074/jbc.c600071200

40. Wu F, Lin X, Okuda T, Howell SB (2004) DNA polymerase zeta regulates cisplatin cytotoxicity, mutagenicity, and the rate of development of cisplatin resistance. Cancer Res 64: 8029–8035.

41. Sakamoto A, Lan VT, Hase Y, Shikazono N, Matsunaga T, et al. (2003) Disruption of the AtREV3 gene causes hypersensitivity to ultraviolet B light and gamma-rays in Arabidopsis: implication of the presence of a translesion synthesis mechanism in plants. Plant Cell 15: 2042–2057.

42. Inagaki S, Suzuki T, Ohto MA, Urawa H, Horiuchi T, et al. (2006) Arabidopsis TEBICHI, with helicase and DNA polymerase domains, is required for regulated cell division and differentiation in meristems. Plant Cell 18: 879–892.

43. Goff JP, Shields DS, Seki M, Choi S, Epperly MW, et al. (2009) Lack of DNA polymerase theta (POLQ) radiosensitizes bone marrow stromal cells in vitro and increases reticulocyte micronuclei after total-body irradiation. Radiat Res 172: 165–174.

44. Higgins GS, Prevo R, Lee YF, Helleday T, Muschel RJ, et al. (2010) A small interfering RNA screen of genes involved in DNA repair identifies tumor-specific radiosensitization by POLQ knockdown. Cancer Res 70: 2984–2993.

45. Chan SH, Yu AM, McVey M (2010) Dual roles for DNA polymerase theta in alternative end-joining repair of double-strand breaks in Drosophila. PLoS Genet 6: e1001005.

46. Mimitou EP, Symington LS (2008) Sae2, Exo1 and Sgs1 collaborate in DNA double-strand break processing. Nature 455: 770–774.

47. Nicolette ML, Lee K, Guo Z, Rani M, Chow JM, et al. (2010) Mre11-Rad50-Xrs2 and Sae2 promote 5′ strand resection of DNA double-strand breaks. Nat Struct Mol Biol 17: 1478–1485.

48. Prinz S, Amon A, Klein F (1997) Isolation of COM1, a new gene required to complete meiotic double-strand break-induced recombination in Saccharomyces cerevisiae. Genetics 146: 781–795.

49. Sartori AA, Lukas C, Coates J, Mistrik M, Fu S, et al. (2007) Human CtIP promotes DNA end resection. Nature 450: 509–514.

50. Limbo O, Chahwan C, Yamada Y, de Bruin RA, Wittenberg C, et al. (2007) Ctp1 is a cell-cycle-regulated protein that functions with Mre11 complex to control double-strand break repair by homologous recombination. Mol Cell 28: 134–146.

51. Uanschou C, Siwiec T, Pedrosa-Harand A, Kerzendorfer C, Sanchez-Moran E, et al. (2007) A novel plant gene essential for meiosis is related to the human CtIP and the yeast COM1/SAE2 gene. EMBO J 26: 5061–5070.

52. Sarkar N, Lemaire S, Wu-Scharf D, Issakidis-Bourguet E, Cerutti H (2005) Functional specialization of Chlamydomonas reinhardtii cytosolic thioredoxin h1 in the response to alkylation-induced DNA damage. Eukaryot Cell 4: 262–273.

53. Cenkci B, Petersen JL, Small GD (2003) REX1, a novel gene required for DNA repair. J Biol Chem 278: 22574–22577.

54. Debuchy R, Purton S, Rochaix JD (1989) The argininosuccinate lyase gene of Chlamydomonas reinhardtii: an important tool for nuclear transformation and for correlating the genetic and molecular maps of the ARG7 locus. EMBO J 8: 2803–2809.

55. Purton S, Rochaix JD (1995) Characterisation of the ARG7 gene of Chlamydomonas reinhardtii and its application to nuclear transformation. Eur J Phycol 30: 141–148.

56. Liu J, Ren X, Yin H, Wang Y, Xia R, et al. (2010) Mutation in the catalytic subunit of DNA polymerase alpha influences transcriptional gene silencing and homologous recombination in Arabidopsis. Plant J 61: 36–45.

57. Barrero JM, Gonzalez-Bayon R, del Pozo JC, Ponce MR, Micol JL (2007) INCURVATA2 encodes the catalytic subunit of DNA Polymerase alpha and interacts with genes involved in chromatin-mediated cellular memory in Arabidopsis thaliana. Plant Cell 19: 2822–2838.

58. Kirik A, Pecinka A, Wendeler E, Reiss B (2006) The chromatin assembly factor subunit FASCIATA1 is involved in homologous recombination in plants. Plant Cell 18: 2431–2442.

Chapter 8

Induced Mutations Unleash the Potentials of Plant Genetic Resources for Food and Agriculture

Chikelu Mba

Plant Genetic Resources and Seeds Team, Plant Production and Protection Division, Food and Agriculture Organization of the United Nations, Viale delle Terme di Caracalla, Rome, 00153, Italy;

ABSTRACT

The options for increasing food production by at least 70% over the next four decades so as to keep pace with a rapidly increasing human population are bedeviled by erratic climatic conditions, depleted arable lands, dwindling water resources and by the significant environmental and health costs for increasing the use of agrochemicals. Enhanced productivities through "smart" crop varieties that yield more with fewer inputs is a viable option. However, the genetic similarities amongst crop varieties—which render entire cropping systems vulnerable to the same stresses—coupled with unvarying parental materials limit the possibilities for uncovering novel alleles of genes and, hence, assembling new gene combinations to break yield plateaux and enhance resilience. Induced mutation unmasks novel alleles that are harnessed to breed superior crop varieties. The historical antecedents, theoretical and practical considerations, and the successes of induced mutations in crop improvement are reviewed along with how induced mutagenesis underpins plant functional genomics. The roles of cell and molecular biology techniques in enhancing the efficiencies for the induction, detection and deployment of mutation events are also reviewed. Also, the integration of phenomics into induced mutagenesis and the use of pre-breeding for facilitating the incorporation of mutants into crop improvement are advocated.

Keywords: induced mutations; mutagens; mutant; crop improvement; genomics; cell biology; TILLING; phenomics; pre-breeding

1. ADDRESSING THE CHALLENGE OF FEEDING THE 21ST CENTURY WORLD

A most critical challenge to crop production is the attainment of enhanced productivities in farmer's fields, especially in developing countries. The Food and Agriculture Organization of the United Nations (FAO) estimated that 70% more food is needed to feed adequately the over nine billion people expected to inhabit planet earth by 2050 [1]. The task of meeting this target, which requires a 37% increase over the historical annual incremental rates [2], is clearly as daunting as the scope of the envisaged increases is unprecedented. The difficulties are compounded by the finite nature of arable lands and agricultural water resources. Indeed, in most parts of the world—including the critically food insecure regions of the developing world [3,4,5,6], these resources are either stagnant or are dwindling on accounts of additional demands for them by the demographic and economic pressures of urbanization and the competing needs for producing livestock feeds, bioenergy production and several other industries [1]. The effects of climate change and variations are also rendering significant areas of arable lands unsuitable for crop production [7,8,9]; this is a major driver for food insecurity.

1.1. Enhanced Productivities on Farmers' Fields Are Critical

It is obvious therefore that a significant portion of the required increases in food production cannot be attained by the further deployment of additional land and water resources. The increased use of agrochemicals is also not a sustainable option on account of its deleterious impacts on health and the environment. Simply, more food must be produced with fewer inputs. The admixture of complementary solutions being adduced for feeding the world's teeming population with fewer agricultural inputs and with minimal ecological footprints constitute the "greener", ecosystem-based and knowledge-intensive paradigm that is commonly referred to as sustainable crop production intensification [10,11].

FAO [11] recommended that, for realizing the imperative of the low-input agriculture being proffered for the 21st Century, farmers require a suite of improved crop varieties that are genetically diverse, climate change resilient, input use-efficient, high yielding, have enhanced nutritional and other quality attributes and have been bred for adaptation to a range of agroecosystems and farming practices. But, the envisaged genetically diverse portfolio of suitable crop varieties are neither available to farmers [2,12,13] nor do the current breeding strategies hold promise for delivering them [14]. The extremely narrow genetic base of the available varieties of crops and the parental lines for breeding new ones nullify efforts to enhance productivities in farmers' fields, increase vulnerabilities and thereby imperil food security.

1.2. The Unmasking of New Alleles Increases Chances for Success

Clearly, the devising and adoption of winning strategies for addressing these dire scenarios necessitate a profound re-evaluation of all aspects of the crop production value chain including the suitability of the crop varieties that farmers grow. There are several mechanisms for facilitating the enhanced harnessing of the inherent potentials encoded into the genetic blueprints of crops so as to make available wider sources of heritable variations to crop improvement. Mba *et al.* [14] have suggested that, in addition to pre-breeding strategies that involve the increased incorporation of traits from non-adapted genetic resources including landraces and crop wild relatives in crop improvement, putative parental materials can also be induced to mutate as means for unleashing new alleles of genes that control the traits desired for the "smart" crop varieties of the 21st Century. We review the exploitations of spontaneous and induced mutations in crop husbandry and genetic improvement. The advances made in the uses of contemporary scientific and technological methodologies to enhance the efficiency of the induction, detection and deployment of induced mutations are also explored. We also provide perspectives for facilitating the integration of induced mutations in plant breeding programs.

2. MUTATION

Mutation is the heritable change to the genetic material; individuals that manifest modified characteristics on account of heritable changes are known as mutants. In nature, the combinations of the errors that occur during the replication of deoxyribonucleic acid (DNA) and damages to this hereditary material on account of an individual's exposure to sunlight and ultra-violet (UV) radiation and diseases lead to heritable changes to the genetic blueprint. Other types of mutations include duplications whereby whole parts of the genome are doubled and translocations whereby a part of a chromosome is transferred to a non-homologous chromosome; these occur as a result of errors during meiotic cell division. Their effects depend on whether or not, and how, genes are affected in such mutations. The accumulations of such aberrations have been the primer for evolution and speciation.

Stretches of DNA make up the basic unit of heredity, the gene. The subtle changes to DNA sequences are the most useful for crop improvement as the more gross alterations (at the chromosomal and cellular levels) invariably confer some levels of unfitness and even lethality. One way for classifying mutations at the DNA sequence level is to categorize the mutation on the basis of how the modification affects the gene's ability to synthesize the protein that it is responsible for. In general, the mutations that are important in crop improvement usually involve single bases and may or may not affect protein synthesis.

2.1. Common Types of Mutations Relevant to Crop Improvement

The common types of mutations in DNA sequences include:

2.1.1. Single Base Substitutions or Point Mutations

One base is replaced by another could be either transitions or transversions. In a transition, one purine (A or G) or pyrimidine (C or T) is replaced by the other. In situations where a purine is replaced by a pyrimidine or vice-versa, this is known as transversion. Point mutations include:

- Samesense, silent, or wild-type mutation occurs when the substitution does not affect the amino acid that is encoded by a codon; this is because many amino acids are encoded by different codons. The mutation in this case can only be detected through sequencing (or some other appropriate molecular assays) as there is no detectable phenotypic change.
- Missense mutation refers to the situation where the new nucleotide modifies the codon such that an altered amino acid is encoded for in the protein. This results in detectable phenotypic changes.
- Nonsense mutation describes the substitution that results in a codon that instead of encoding an amino acid is modified into one of the stop codons. This results in the premature termination (or truncation) of the translation of codons to amino acid sequences in the proteins leading to a protein with shortened polypeptides that might be inactive. Severely

truncated proteins, *i.e.*, those with the truncation sites early in the gene sequence are inactive.

2.1.2. Insertions and Deletions (Indels)

Mutations involving situations where extra base pairs, ranging from one to thousands, are inserted into (insertion) or deleted from (deletion) the sequence of a gene are referred to as indels. Indels result in shifts in the reading frame of the gene with the rearrangement of the nucleotides into a completely new set of codons. This has grave consequences for protein synthesis as the amino acid sequences are manifestly different from the wild-type. Indels of one or two base pairs (or multiples) result in frameshifts while those of three base pairs (or multiples) are more innocuous as the reading frame tends to be preserved, *i.e.*, the original triplet codons are not disturbed.

2.2. Spontaneous Mutations

Spontaneous mutants are aberrant types that are found in nature and for which no deliberate intervention by man has engendered the novel phenotypes. The domestication of off-type plants, *i.e.*, spontaneous mutants that manifested traits suitable for crop husbandry, about 10 millennia ago contributed immensely to the origins of agriculture and hence, the current relatively sedentary lifestyles and permanent human settlements.

The loss of certain "wild" characteristics in crops through spontaneous mutations was the principal driver in the founding of agriculture in the Fertile Crescent of West Asia and independently in northern and southern China, Africa's Sahel, New Guinea, the Andes and several regions of the Americas. Notable examples of spontaneous crop mutants whose heritable spontaneously altered characteristics led to domestication and ease of cultivation and hence ready availability for human consumption include peas, wheat and barley. The accumulation of spontaneous mutations that abolished pod or head shattering (natural seed dispersal mechanisms) and the thinning of seed coats resulting in reductions in lengths of seed dormancy periods made these plants suitable for cultivation at will. Another important example of a heritable permanent change that, in distinguishing mutants from their wild-types, made them suitable for human consumption is the loss of bitterness in almonds, lima beans, watermelons, potatoes, egg-plants, cabbages and several types of nuts. Other spontaneous mutations that have become advantageous include parthenocarpy in bananas and grapes. Table 1 provides a list of some spontaneous mutations that have facilitated the processes of crop domestication over time and hence settled agriculture.

Table 1. Spontaneous mutations and domestication of crops.

Mutation that facilitated domestication	Examples of plant
Abolishment of bitterness and toxicity	almonds, lima beans, watermelons, potatoes, egg-plants, cabbages nuts
Abolishment of the need for sexual reproduction (seedlessness or parthenocarpy)	bananas, grapes, oranges, and pineapples
Loss of natural seed dispersal mechanism—shattering of pods and heads	peas, wheat, barley
Loss of the hard seed coat and other germination inhibitors (dormancy)	wheat, barley, peas
Facility for self-compatible hermaphroditism	grapes, papaya, *etc.*

More recently, the green revolution, a term used to describe the huge increases in grain yields from the 1960s, was made possible through the introgression into wheat and rice of spontaneous mutant alleles of genes that, in conferring reduced plant height (and stiffer lodging-resistant straws), are known as dwarfing genes. Unprecedented stable high yields were achieved for dwarf wheat varieties and marked an epochal defining moment in international agricultural research and development. Mutation induction in crops therefore represents man's successful attempt at reproducing and scaling up the frequency of a naturally occurring phenomenon so that it can be exploited advantageously.

2.3. Induced Mutations

Agents that are used to induce hereditary changes are broadly divided into physical and chemical mutagens. Crops are induced to mutate through the exposure of their propagules to physical and chemical mutagenic agents. For seed propagated crops, botanical seeds are treated with the mutagens while for vegetatively propagated plants, other plant parts used for propagation such as stem cuttings, twigs, buds and tubers are exposed to the mutagenic agent. More recently, the induction of mutations in vegetatively propagated plants is becoming more efficient as scientists take advantage of totipotency, *i.e.*, the inherent ability of individual plant cells to regenerate into whole plants, to use single cells and other forms of *in vitro* cultured plant tissues as starting materials for the induction of mutations.

2.3.1. Physical Mutagens

Man's ability to deliberately induce mutations in plants derives directly from the discoveries of X-rays by Roentgen in 1895; radioactivity by Becquerel in 1896; and radioactive elements by Marie and Pierre Curie in 1898. For these achievements, the Nobel Prize for Physics was awarded to Roentgen in 1901 and to Becquerel, Marie and Pierre Curie in 1903 [15]. Deriving directly from these discoveries were the seminal demonstrations soon afterwards that radiation (X-rays) caused alterations to the genetic make-up of fruit flies [16] and in the crop plants—maize and barley [17,18,19,20,21]—respectively. These proofs of concept proved to be the watershed moments in induced mutagenesis as they provided the impetus for the subsequent widespread adoption of this mimicry of nature in crop improvement and more recently as a strategy to discover genes and elucidate their functions.

Ionizing radiations constitute the most commonly used physical mutagens [22,23]. These are parts of the electromagnetic (EM) spectrum that, on account of their relatively high energy levels, are capable of dislodging electrons from the nuclear orbits of the atoms that they impact upon. The impacted atoms therefore become ions hence the term ionizing radiation. These ionizing components of the EM include cosmic, gamma (γ) and X-rays. Ultra violet (UV) light, though non-ionizing, is capable of some level of tissue penetrability and has been used in inducing mutations as well. The mutagenicity of UV derives from its ability to react with DNA and other biological molecules as its wavelengths are preferentially absorbed by bases in DNA molecules and by the aromatic amino acids of proteins.

In addition to the ionizing radiations, the other commonly used physical mutagens are the high energy ionizing particles, alpha (α) and beta (β) particles and neutrons. The properties of physical agents that are used in inducing crops to mutate are summarized in Table 2. The mutagenicity of these agents derives from a combination of their ability to produce dimmers and reactive ions which in turn cause damage to living organisms; the damage caused ranges from aberrations at the DNA level (breaking the chemical bonds in DNA molecules, deleting or adding nucleotides and by substituting one nucleotide for the other) to gross chromosomal breakages and rearrangements.

Gamma and X-rays are the most commonly used physical mutagens [22,23]. Gamma rays are emitted in the process of the decay of the radioisotopes cobalt-60 (^{60}Co), cesium-137 (^{137}Cs) and to a less extent, plutonium-239 (^{239}Pu). Gamma sources containing one of these radioisotopes are typically installed as gammacell irradiators. A gammacell is used mostly for acute irradiation (i.e. for short periods). For chronic irradiations, *i.e.* the exposure of plants to irradiation for extended periods of times, irradiators are installed in specially designed gamma rooms (or chambers), greenhouses or fields. In general, the irradiators are sealed sources which are normally

Table 2. Some commonly used physical mutagens and their properties.

Mutagen	Typical Frequency (s^{-1})	Typical Energy (kJ/mol)	Typical Photon Energy (eV)
Particles			
α-particles		4.1×10^{8}	
β-particles		1.5×10^{7}	
Electromagnetic radiation			
cosmic rays	$6 \times 10^{21}\ s^{-1}$	2.4×10^{9}	
γ-rays	$3 \times 10^{2}\ s^{-1}$	1.2×10^{8}	1 MeV
x-rays	$3 \times 10^{17}\ s^{-1}$	1.2×10^{5}	100 keV
ultraviolet	$3 \times 10^{15}\ s^{-1}$	1200	4 eV

encapsulated within a stainless steel casing as a safety measure against unintended irradiation. In all cases, only adequately trained and, depending on national statutory requirements, certified personnel operate the sources. Also, special precautions, including strictly controlled access to the facility, are imposed in order to prevent accidents.

The implantation of ion beams is rapidly gaining currency in the induction of mutations in crops since the pioneering work in China and Japan in the early and mid-1990s, respectively [24,25]. Ion beams differ from aforementioned physical mutagens in that in addition to energy transfer, mass deposition and charge exchange play vital roles in their mutagenicity. It is postulated that these additional properties predispose the treated plant materials to the production of spectra of variations that are distinct from those produced through exposure to other mutagens. Ion beams are usually generated by particle accelerators, e.g., cyclotrons using ^{20}Ne, ^{14}N, ^{12}C, ^{7}Li, ^{40}Ar, or ^{56}Fe as radioisotope sources.

Cosmic radiation, the collective term for high energy particles that originate from outer space (mostly protons), has also been demonstrated to be mutagenic to crops [26,27,28]. With China playing the leading role, strong cosmic radiation in conjunction with the super vacuum and microgravity that characterize the outer space environment, have been used to induce useful mutations in rice, wheat, cotton, sweet pepper, tomato and sesame [29] and in maize [27]. Induced mutations, exploitable in crop breeding, have thus been achieved by transporting plant propagules, mostly botanical seeds, in recoverable space orbiting satellites. It is plausible, though, to infer that the highly specialised nature, expense and start up infrastructure make the use of cosmic radiation the most improbable choice for most scientists.

2.3.2. Chemical Mutagens

It would be about two decades after the demonstrations of the mutagenicity of physical agents that nitrogen mustard (component of poisonous mustard gas used in World Wars I and II) would be shown to cause mutations in cells [30,31,32,33,34,35,36,37,38]. This would pave the way for the identification of several other chemical mutagens; these include base analogues, alkylating and intercalating agents and chemicals that modify DNA structure. Their effects on DNA molecules manifest in deamination, the induction of transitions and insertions, the stoppage of transcription and replication and even strand breaks. The properties of the chemical agents commonly used for inducing mutations in crops are shown inTable 3.

Table 3. Some commonly used chemical mutagens and their modes of action.

Chemical mutagenic agent	Mode of action
Base analogues e.g., 5-bromouracil (BU), 5-bromodeoxyuridine, 2-aminopurine (2AP)	Incorporates into DNA in place of the normal bases during DNA replication thereby causing transitions (purine to purine or pyrimidine to pyrimidine); and tautomerization (existing in two forms between which they interconvert e.g., guanine can exist in keto or enol forms).
Nitrous acid	Acts through deamination, the replacement of cytosine by uracil which can pair with adenine and thus from subsequent cycles of replication lead to transitions.
Alkylating agents such as: sulfonates e.g., ethylmethane sulfonate (EMS), diethyl sulfonate (DES); Sulphur mustards e.g., ethyl-2-chloroethyl sulphide; Nitrogen mustards e.g., 2-chloroethyl-dimethyl amine; and Epoxides e.g., ethylene oxide Others are ethyleneimine, hydroxylamine (NH_2OH), *N*-methyl-*N'*-nitro-*N*-nitrosoguanidine (MNNG), sodium azide and diazomethane.	They react with bases and add methyl or ethyl groups and, depending on the affected atom, the alkylated base may then degrade to yield a baseless site, which is mutagenic and recombinogenic, or mispair to result in mutations upon DNA replication.
Intercalating agents such as acridine orange, proflavin, ethidium bromide	They insert between bases of DNA thereby causing a "stretching" of the DNA duplex and the DNA polymerase in turn recognizes this stretch as an additional base and inserts an extra base opposite this stretched (intercalated) molecule. This results in frameshifts *i.e.*, an alteration of the reading frame since codons are groups of three nucleotides.
Miscellaneous group of agents; Large molecules referred to as "bulky" lesions (e.g., *N*-acetoxy-*N*-2-acetyl-aminofluorine—NAAAF).	They bind to bases in DNA and cause them to be noncoding thereby preventing transcription and DNA replication; They cause intra- and inter-strand crosslinks (e.g., Psoralens); They also cause DNA strand breaks (e.g., peroxides).

3. AN INDUCED MUTAGENESIS PROGRAM

The generalized scheme for induced crop mutagenesis—as a crop improvement strategy—is straightforward and involves the sequential steps of the exposure of plant propagules to pre-determined doses of a mutagen, the identification of stable mutants amongst the progeny and the incorporation of the desirable mutants into breeding programs or the use of mutant stocks for the identification of genes and the elucidation of their functions. To be successful, the researcher would have to make informed decisions that range from the choice of mutagen through the doses to be administered to the handling of the putative mutants.

3.1. Choice of Mutagen

A long held notion, which probably has been influencing the choice of mutagen types (chemical *vs.* physical), is that chemical mutagens preferentially induce point mutations while physical mutagens induce gross lesions, such as chromosomal breakages and rearrangements. This view has probably been aided by the relative ease for procuring chemical mutagens as well as simpler set-up requirements. Current empirical data do not support this conclusion; rather the more plausible inference from available information is that the frequency and types of mutations are direct results of the dosage and rate of exposure or administration of the mutagen rather than its type. This implies therefore that the extended immersion of plant propagules in high concentrations of EMS or NaN_3 would be equally as deleterious or even lethal as extended exposures to high doses of ionizing radiation. In all cases, optimal doses of mutagens should always be determined prior to large-scale induction of mutations. It should also be pointed out that chemical mutagens are very toxic poisons that usually require elaborate detoxification of laboratory ware and disposal of used reagents. Relatively, the risks to health posed by physical mutagens are far less. In the end, the choice of a mutagen will be based on the particular researcher's circumstances such as the availability of the mutagens, associated costs and available infrastructure.

3.2. Practical Considerations in Induced Crop Mutagenesis

Mba *et al.* [39] provided the protocols for the induction of mutations in seed and vegetatively propagated plants, based on procedures validated for rice and cassava, respectively. The protocols were for the physical mutagen (gamma rays generated by a cobalt-60 source) and the chemical mutagen (ethyl methane sulfonate, EMS). The authors [39] listed the required equipment and reagents and described the sequential procedures, the pre-treatment handling of the propagules, the mutagenic treatments, post-treatment handling of the materials; and the methodologies for the collection and analysis of data. In general, the steps to the induction and detection of mutants vary between seed and vegetatively propagated crops but would nonetheless involve the following considerations:

- The setting of realistic targets matched with the availability of the requisite appropriate human and material resources and the prior determination that induced crop mutagenesis offers a comparative advantage over other available technologies and strategies.
- A clear understanding of the genetics of the traits to be improved; for instance, polygenic traits (*i.e.*, characters under the influence of many genes) have less chance of being modified with induced mutations than monogenic traits (i.e. characters under the influence of single genes).
- A knowledge of the reproductive biology of the crop, *i.e.*, seed *vs.* vegetative propagation; if seed propagated, self- *vs.*cross-fertilization; and if vegetatively propagated, which propagation methods are possible (*in vivo vs. in vitro*).
- The determination of the plant propagule to be treated *i.e.* botanical seeds or gametes for seed propagated crops and nodal segments, stem cuttings, twigs, buds, or different types of *in vitro* cultures for vegetatively propagated crops and the assemblage of sufficient quantities.
- A knowledge of the ploidy levels of the crop especially when ploidy-related hybridization barriers could impact on the envisaged utility of the induced mutants.
- A determination of the genetic background of the target materials for induction of mutations. Usually, the best genetic background to be induced to mutate should in general be the best all-round genotype that is deficient in a single trait (the breeding objective); and is homozygous.
- An informed decision on which induction method, *i.e.*, physical *vs.* chemical mutagens, and the appropriate doses (including considerations of concentrations and durations for chemical mutagens). A pilot assay is usually necessary in order to determine the optimal dosage prior to the large-scale bulk treatment of the propagules.
- A clear definition of the plans, mechanisms and the identification of the requisite facilities for handling (including for planting and evaluating) the mutant population. This would also involve an understanding of, and planning for, the adequate population sizes.
- The scheduling, as the case may be, for the use of the identified infrastructure (irradiator, screen house, fields, laboratories, *etc.*) and other required inputs.
- As may be necessary, a definition of methodologies for the identification of the mutation events and subsequent selection of desired mutants. Options may include molecular biology assays with techniques that lend

themselves to high throughput set-ups being preferred since they generally obviate the drudgery associated with handling large mutant populations, the majority of which will be jettisoned.

- Methodologies for the dissociation of chimeras and the production of stable mutants must also be clearly thought through. For putative mutants destined for plant breeding programs, a clear definition, along with relevant counterparts, of the means for incorporating the stable mutants into hybridization schemes is critical to success.

The important experimental factors that influence the outcomes of chemical mutagenesis include the concentration of mutagen; volume of mutagen relative to size of samples; treatment duration; temperature; pre-soaking of seeds; pH; use of catalytic agents; and the post-treatment handling of the putative mutants [39]. For physical mutagens, the authors [39] identified the presence of elements such as oxygen and chemical mutagenic agents; moisture content; impurities such as dust and fibers; and viral and bacterial infectious agents as being critical to the outcome. The determination of the optimal doses of mutagens and the dissociation of chimeras in putative mutants are probably the most critically important of factors for the success of induced mutations.

3.3. Optimal Doses for Mutagenic Treatment

For crop breeding purposes (as against generating mutant stocks for functional genomics, for instance), the aim should be to obtain a reasonable number of desired mutations for a trait of interest while inflicting the least unintended disruption to the genotypic integrity of the crop. This ensures that the fitness of the desirable induced mutant is not otherwise compromised by the presence of unintended induced deleterious alleles which would require additional interventions in time and resources for, e.g., backcrossing to the elite starting genotype in order to break linkage drags.

The universally adopted norm is to select a dosage that results in reductions of 30 to 50 or 40 to 60 percent in growth or survival rates, respectively of the first generation mutant (M_1) seedlings compared to the seedlings of untreated seeds. Reductions in germination rate, seedling height, survival rate, number of tillers, seed set, and fertility test in the M_2generation as well as chlorophyll mutation are the main parameters measured in the sensitivity tests for determining the optimal doses. The reductions are plotted against the mutagen doses and the dose corresponding to the desired level of reduction is read off the graph. This could also be determined through regression analysis. A sample graph, Figure 1, shows that a 50% reduction in seedling plant height is caused by gamma irradiation at 300 Gy. Doses corresponding to other percentage reductions can be read off the gradient of the graph. Mba *et al.* [39,40] described the procedures for using the radiosensitivity to the physical mutagen, gamma rays, and the reactions to EMS to ascertain their optimal doses for inducing mutation events in both seed and vegetatively propagated plants.

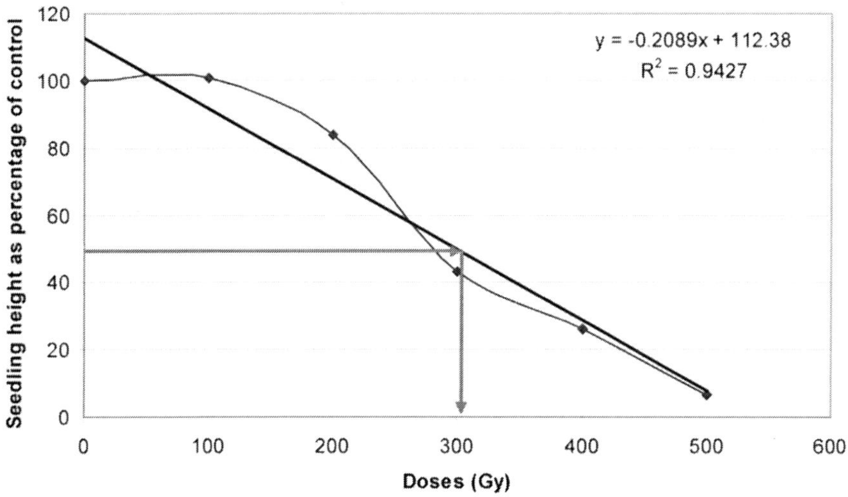

Figure 1. Percentage of reduction in plant height of seedlings germinated from seeds exposed to a mutagen (compared with seedlings from untreated seeds), plotted against mutagen dosage. The red arrows indicate that seedling height reduction of 50% is caused by 300 Gy.

3.4. Incidence of Chimeras

Usually, multi-cellular tissues are used as the starting material for mutation induction. Following induced mutagenesis, the individual cells of putative mutants typically vary amongst themselves in terms of the induced mutations that they harbor. This is chimerism, *i.e.*, sectoral differences in a mutant, whereby cells of different genotypes exist side-by-side in the tissues of the same individual plant. In essence, the plant consists of genetically distinct somatic tissues. Chimeras are significantly dissociated in seed propagated crops in one to two generations of selfing (M_2 to M_3) while for vegetatively propagated crops, several generations of vegetative propagation are required to produce solid homohistonts. Along with the absence of meiotic sieves, the fixation of deleterious alleles, the transmission of pathogens to subsequent generations, the incidence of chimeras are considered the most serious drawbacks to the induction of mutations in vegetatively propagated plants [40,41]. Considerable efforts are currently being invested in the determination of strategies for exploiting totipotency,*i.e.*, the ability of individual single plant cells to regenerate a whole organism, to quicken the pace of chimera dissociation in induced mutants. Some of the cell and tissue biology techniques that enable this are discussed in subsequent sections.

3.5. Induced Mutants as Raw Materials for Crop Improvement

Following the demonstrations of the mutagenicity of X-rays on fruit flies and maize by Muller [16] and Stadler [18], respectively, plant scientists devoted significant efforts to establishing the experimental protocols for the use of radiation to generate heritable variations that could be used in crop improvement. Within three decades, mutation breeding had become an established plant breeding strategy and in a little less than a century is credited with the development of over 3200 crop varieties that are being grown all over the world [42]. The success of plant breeding depends largely on the availability of utilizable heritable variations. When desirable variations are easily discernible from well characterized germplasm collections, the plant breeder's task is fairly straightforward. In instances where such variations are either unavailable to the breeder or are observed only in materials with otherwise undesirable genetic backgrounds, induced mutations may be the only way to generate desirable variations for use in breeding superior crop varieties.

The Joint Division of the Food and Agriculture Organization of the United Nations and the International Atomic Energy Agency (Joint FAO/IAEA) in Vienna, Austria which maintains a Mutant Varieties Database (MVD) for these crop varieties [42] introduced some newly developed mutant rice varieties in China and Bangladesh on its website [43]. For China, these were the rice variety with modified starch characteristics suitable for managing diabetes and obesity; one with high contents of zinc; and yet another with round grains of reduced size that is suitable for formulating baby food. The newly released high yielding rice variety in Bangladesh, BINA Dhan-7, matured earlier than other varieties of comparable yield by about one month, was flood and disease resistant and also amenable to intercropping. These varied induced variations are reflective of the other over 3200 crop varieties that have been developed through induced mutation-mediated strategies in the over 80 years since the demonstration of the feasibility of induced mutations for crop improvement [42].

Most, about 77%, of these mutant crop varieties are seed propagated probably underscoring the relative ease for the induction of mutations using botanical seeds as against stem cuttings, tubers and other vegetative propagules, as starting materials for induced mutagenesis. It might also be a reflection of the preponderance of seed propagated crops (cereals and pulses) as the major food crops of the world. Almost half (48%) of all mutant crop varieties recorded in the MVD are cereals. Rice is the crop with the highest number of mutants and accounts for 53% of the mutant cereals under cultivation followed by barley which makes up 20% of all cereal mutant varieties globally. Figure 2, Figure 3 show the relative distributions of these mutants by the continent where they were developed and crop usage type, respectively. More than half (60%) of the mutant crop varieties have been released in Asia. China alone accounts for more than 25% of all mutant varieties that have been officially released globally. Interestingly, China along with two other Asian countries, Japan and India are the top three countries with the most officially released mutant crop varieties. These trends are largely unchanged from the observations of [44] who chronicled the over 2200 officially released mutant varieties by the year 2000 while [45] provided an in-depth review of the contributions of both spontaneous and induced semi-dwarfness to the genetic improvement of the cereals, rice, bread and durum wheats and barley, especially in the context of the Green Revolution.

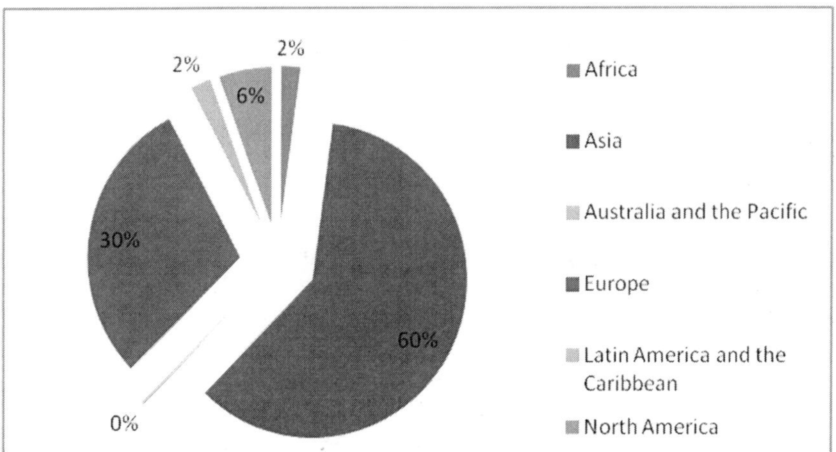

Figure 2. Distribution of mutant crop varieties reported on the Joint Division of the Food and Agriculture Organization of the United Nations and the International Atomic Energy Agency (Joint FAO/IAEA) Mutant Varieties Database [42] by continents of official release.

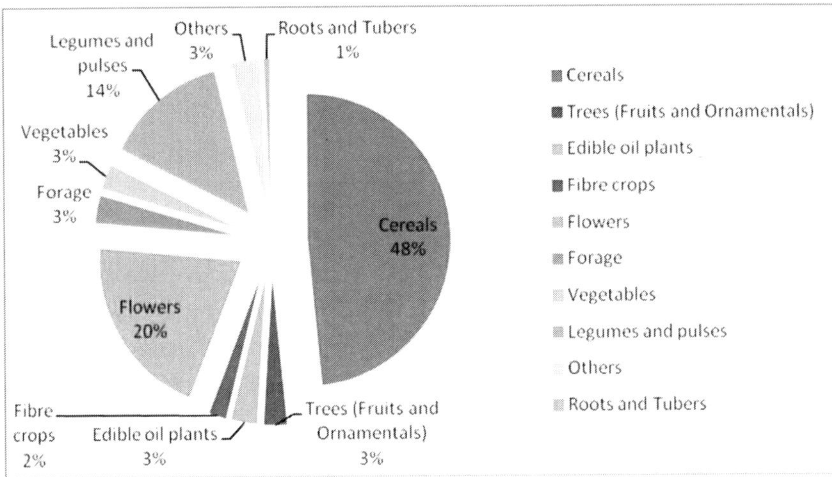

Figure 3. Distribution of officially released mutant crop varieties reported on the Joint FAO/IAEA Mutant Varieties Database [42] by crop type and usage.

Ahloowalia *et al.* [46] surveyed the economic importance of some of the outstanding mutant crop varieties globally based on such indices as annual acreage; estimated reduction in the use of agricultural inputs (on account of improved hardiness); synergistic increases in yield; enhanced market values (on account of improved quality traits); increase in export earnings for the farmers; reduction in food importations; *etc.* These included varieties of rice in Australia, China, India, Pakistan and Thailand; cotton and wheat in Pakistan; pear in Japan; grapefruit, sunflower and peppermint in the USA; barley in several countries of Europe; durum wheat in Italy; sorghum in Mali; groundnut and pulses in India; and several ornamental plants in India, the Netherlands and Germany. Taking barley for instance, the authors [46] reported that the high-yielding and dwarf mutant cultivars of barley "Diamant" and "Golden Promise" had jointly added billions of dollars in additional income to the brewing and malting industries. According to the authors [46], more than 150 leading barley cultivars in several countries in Europe, North America and Asia were derived from crosses involving Diamant. Over 20 million US dollars in additional income accrue to farmers per annum on account of growing these two barley varieties and their progenies making them therefore the mainstay of Europe's malting industry.

Ahloowalia *et al.* [46] provided other examples of the benefits accruing from the cultivation of other notable mutant crop varieties; these included:

- Most varieties of durum wheat grown in Italy and used in making pasta that is marketed worldwide are induced mutants; their cultivation generates tens of millions of dollars in additional income to farmers and the seed industry per annum;

- The Rio Star variety of grape fruit in the USA is a mutant that accounts for 75% of the US grapefruit industry;
- The cultivation of the mutant Japanese pear variety, "Gold Nijesseiki", contributes US$30m in additional income to farmers annually
- A mutant cotton variety, NIAB78, grown widely in Pakistan accounts for over US$20m in additional income to farmers annually
- Most of the rice grown in Asia (especially in China, Japan. Vietnam and India) and Australia are mutants. The additional income to farmers for growing and marketing these rice varieties is estimated in billions of US$ annually
- Mutant barley varieties with enhanced adaptation to extremely harsh environmental conditions in high altitudes are literally extending the frontiers of arable lands in Peru. Their cultivation provides much needed employment for resource-poor farmers.

More recently, Kharkwal and Shu [47] reviewed the contributions of induced crop mutants to global food security by highlighting some of the important varieties developed through induced mutagenesis mediated-strategies, their annual acreages in the countries of cultivation and some other indices that underscore their importance in food security and income generation. The crops covered in this review included rice in China, Thailand, Vietnam, and the USA; barley in European countries and Peru; durum wheat in Bulgaria and Italy; wheat in China; soybean in China and Vietnam; and some other food legumes in India and Pakistan. They concluded that there was considerable evidence that mutant crop varieties would continue to contribute significantly to addressing the food and nutritional securities of many countries especially in view of the potentials for harnessing novel traits in enhancing the adaptabilities of crops to climate change and variations.

A crop variety with significant economic potentials that has been developed recently through induced mutations is linola, the trade name for the edible linseed crop, developed from the non-edible linseed flax, *Linum usitatissimum,* by the Australian Commonwealth Scientific and Industrial Research Organisation [48,49]. With mutations to the fatty acid synthesis pathway, the alpha-linolenic acid (ALA) content was reduced to 2% (as against 50% in the wildtype). Oil produced from the wildtype becomes rancid quite rapidly on account of this high content while linola oil is comparable to the oils from sunflower and canola. Linola is now being produced in Australia, Canada, the United Kingdom and the United States. Interestingly, canola was developed exploiting spontaneous mutations that lowered the levels of erucic acid and glucosinalates, both anti-nutritional factors found in the wildtype rapeseed or field mustard.

Mutant crop varieties therefore form integral parts of daily diets all over the world; are raw materials in industries; and contribute billions of dollars in additional income to farmers. The traits that confer superiority on the mutants over other cultivars include synergistic higher yields due to enhanced resistances to biotic and abiotic factors; improved nutrient use efficiencies; and nutritional and other quality traits. In being hardier and more nutrient use efficient and therefore requiring less agricultural inputs, several mutant crop varieties are more economical to grow and contribute to more environment friendly agriculture.

3.6. Induced Mutants as Tools for Understanding and Harnessing Heredity

The use of characterized mutants in plant breeding has long been an established practice. But interestingly, the observed resurgence in induced mutagenesis in plants (after a lull that commenced in the 1980's in the wake of rapid epochal advances in recombinant DNA technologies), seems to be mostly on account of increased investments in the production of genetic resources for functional genomics, *i.e.* the identification genes and the elucidation of their functions. With the ever increasing ease for obtaining DNA sequence information and even sequencing whole genomes, massive amounts of information on the genomes of many organisms have become publicly available. The deciphering of the information coded into the sequences has not been as routine, however.

Taking advantage of these massive troves of genomic information that is increasingly available for a number of plant species, induced mutants are used in identifying and ascribing functions to genes. The genomic region(s) implicated in the expression of a trait, *i.e.*, the gene, is identified by studying a series of induced mutants vis-à-vis the normal or wildtype variants. Adopting the deductive process of identifying which traits have been modified and relating the modifications to changes in genomic regions of the induced mutants, as compared to the normal types, the gene(s) controlling a trait may be identified.

A significant part of the current efforts and resources invested in induced mutagenesis in plants (including the systematic molecular and morphological characterizations of mutant collections) is devoted to functional genomics. For a majority of the model plants and other intensively researched plant species, including major crop plants, huge amounts of genotypic and phenotypic data are being generated and housed in publicly accessible searchable databases. Typically, researchers rely on

Table 4. Some characterized mutant stocks of crops and the host institutions.

Crop	Host institution	Reference
Maize	The Maize Genetics Cooperation Stock Centre. University of Illinois. Urbana/Champaign. IL. USA	[51]
Arabidopsis	European Arabidopsis Stock Centre (or Nottingham Arabidopsis Stock Centre. NASC). University of Nottingham. Sutton Bonington Campus. UK	[52]
	Arabidopsis Biological Resource Centre. (ABRC) Ohio State University. OH. USA	[53]
Tomato	CM Rick Tomato Genetics Resource Centre. University of California at Davis. CA. USA	[54]
Cucurbits (cucumber. melon. *Cucurbita*. and watermelon)	Cucurbit Genetics Cooperative (CGC) North Carolina State University Raleigh. NC. USA	[55]
Rice	The Oryzabase of the National BioResource Project—Rice National Institute of Genetics. Japan	[56]
	IR64 Rice Mutant Database of the International Rice Functional Genomics. International Rice Research Institute. Manila. Philippines	[57]
	Plant Functional Genomics Lab.. POSTECH BIOTECH CENTER. San 31 Hyoja-dong. Nam-gu Pohang. Kyoungbuk. Korea	[58]
Barley and wheat	Barley mutants. Scottish Crop Research Institute. Dundee. Scotland	[59]
	Barley and Wheat Genetic Stock of the USDA-ARS	[60]
	USDA-ARS Cereal Crops Research Unit. Fargo. ND. USA	[61]
	Wheat Genetics Resource Center. Kansas State University. Manhattan. KS. USA	[62]
	Wheat Genetic Resources Database of the Japanese National BioResource Project	[63]
Pea	Pea mutants. John Innes Centre. Norwich. UK	[64]

such information to request the characterized mutants for scientific investigations. Examples of plant mutant stocks that are publicly available to requestors include the model plant, *Arabidopsis,* maize, rice, barley, peas, cucurbits and tomato (Table 4). The U.S. Department of Agriculture, Agricultural Research Service (USDA/ARS) has a number of germplasm collections that hold publicly available characterized mutant stocks under the National Plant Germplasm System (GRIN) [50]. A number of public germplasm collections in other countries also have substantial number of mutant accessions that constitute a critical resource for the continually increasing plant functional genomics research community. Some of these mutant stocks are listed in Table 4.

3.7. Enhancing the Efficiency for Induced Mutagenesis

Major drawbacks to induced mutagenesis include the requirement for generating large mutant populations, the incidence of chimeras and the heterozygousity of the mutated loci. By integrating cell and tissue biology techniques and molecular biology strategies in the induction and detection of mutations, these constraints are mitigated without compromising the imperative of generating large mutant population sizes. The judicious integration of these novel biotechnologies permits the rapid generation of large mutant populations with desired genetic backgrounds; the targeted querying of specific regions of the genome that control a trait of interest for induced alterations. The following are some molecular biology and cell biology techniques with demonstrated potentials for enhancing the efficiency of induced crop mutagenesis.

3.7.1. Reverse Genetics

The classical genetics approach that relies on the comparing of the phenotypic expressions of variants (spontaneous or induced mutants) with those of the wildtypes in order to identify and characterize the causal genes is known as forward genetics. With the current advances in genetics—the discovery of DNA, the polymerase chain reaction and genome sequencing, for instance—it is now possible to alter or disrupt a specific gene or gene product (whose effect may be previously unknown) and then determine the phenotypic effect of such a modification. In essence, rather than working from the phenotype to the genotype, scientists can now work backwards, from the genotype to the phenotype; this is known as reverse genetics. The reverse genetic technique, TILLING (Targeting Induced Local Lesions IN Genomes [65], that was aptly characterized "Traditional Mutagenesis Meets Functional Genomics" [66], permits the high throughput querying of putative mutants for point mutation events in specific genomic regions. With TILLING, the mutation events are identified by enzymatic cleavage of mismatches between a mutated strand and a wild type one. It therefore has huge potentials for removing the major bottleneck to the routine application of induced mutagenesis: the need to produce, handle and query large putative mutant populations of sufficient mutation density for low frequency recessive events. Since the proof of concept about twelve years ago [65], TILLING has been successfully applied in the detection of mutations in several plant species. These include *Arabidopsis* [67,68,69,70]; rice [71,72,73,74]; maize [75]; wheat [76,77]; sugar beet [78]; barley [79];

soybean [80]; pea [81]; beans [82,83]; tomato [84] and in the vegetatively propagated banana [85]. The related technique, Ecotilling, is also a robust method for identifying spontaneous mutants and hence useful for characterizing germplasm in general [86]. Initially optimized with the chemical mutagen, EMS, TILLING has also been effective in identifying point mutations induced by physical mutagen [87]. The TILLING innovators [65] used denaturing HPLC for detecting the mutations. This methodology lacked the high throughput functionality needed for handling large mutant populations. To redress this, a more robust and high throughput platform that ran on LI-COR gel analyzer system (Lincoln, NE, USA) with an analytical software was described by Colbert *et al.* [68]. This method became widely applicable and was the platform for using TILLING to identify mutants in the above plant species but more recently, Tsai *et al.* [88] have described an alternative, TILLING by sequencing. Rather than use endonuclease to cut DNA strands at heteroduplexes produced by the mismatch between wildtype and mutant strands, the next generation sequencer Illumina is used to sequence target genes from pooled DNA samples. Till *et al.* [89] surveyed some ongoing TILLING projects and their activities; a number of these facilities also provided mutation discovery services to requestors. Table 5 shows some of the facilities providing TILLING services.

Table 5. Some research and development facilities offering TILLING services.

Plant	TILLING Facility	Reference
Rice, tomato, *Arabidopsis*	The UC Davis TILLING Core, University of California, Davis, CA, USA	[90]
Tomato	Metapontum Agrobios Centre, Metaponto di Bernalda, MT, Italy	[91]
Maize	The Maize TILLING Project, Purdue University, West Lafayette, IN, USA	[92]
Brassica napus	CAN-TILL, University of British Columbia.	[93]
Lotus, Medicago, Brassica rapa	RevGenUK, John Innes Centre, Norwich Research Park, Colney, Norwich, NR4 7UH, UK	[94]
Arabidopsis	Seattle TILLING Project, Fred Hutchinson Cancer Research Center, North Seattle, WA, USA	[95]
Tomato, Rapeseed, Pea, Medicago and Melon	Plant Genomic Research, INRA/CNRS—URGV, Evry cedex, France	[96]

3.7.2. Cell and Tissue Biology

The very low levels of efficiency for the induction and detection of mutation events and the production and evaluation of large putative mutant populations constitute significant hindrances to the routine application of induced mutations in plants for crop improvement and functional genomics.

Large populations of induced mutants must be produced in order to enhance the chances for identifying the low frequency and largely recessive induced mutations. Equally confounding efforts in induced mutagenesis is the imperative of dissociating chimeras, a problem that is more pronounced with vegetatively propagated plants. Also problematic is the need for homozygousity of the mutated loci so that the induced mutation, usually recessive, could translate to a phenotype. The efficacies of some *in vitro* techniques in mitigating these constraints significantly have been demonstrated. Some proven methodologies are the incorporations of cell suspension cultures, including somatic embryogenesis; doubled haploidy and rapid *in vitro* multiplication techniques into the processes for the induction, detection and evaluation of mutation events.

3.7.2.1. Cell suspension cultures

The incidence of chimeras in putative mutants is a major problem in induced mutagenesis especially with vegetatively propagated plants on account of the multicellular vegetative propagule which lacks the corrective effects of the meiotic sieve. Chimeras are dissociated by subjecting the putative mutants to several cycles of vegetative regeneration (both *in vitro*and *in vivo*). This, of course, introduces added expense in time and resources. The use of cell suspension cultures as starting materials for inducing mutations such that single cells are mutagenized would be a means to avoid the additional expense and investments in time and labor. The efficacy of cell suspension cultures derives from the potential of each plant cell to regenerate into a whole plant; this is the phenomenon of totipotency. For cell suspension cultures, cell lines are produced from callus followed by the regeneration of plantlets through somatic embryogenesis. In practice, cell suspension involves the culturing *in vitro* of single cells and small cell aggregates; these proliferate and complete a growth cycle while suspended in liquid medium. Ever since the pioneering work of Nickell [97] with *Phaseolus vulgaris*, the production of cell suspension cultures has become routine for many plant species. It holds great promise therefore for induced mutagenesis in that the incidence of chimeras in plants arising from mutagenized single cells would be quite minimal. This would result in significant savings in the time that would ordinarily be invested in the several cycles of regenerations that would ordinarily be required to dissociate chimeras. The pilot testing of this concept with *Musa* has been promising [98].

3.7.2.2. Doubled Haploidy

Another bottleneck to the exploitation of induced crop mutations is the predominantly recessive nature of mutations. The consequence of this is that the expression of the desired phenotype is masked if the alleles are in a heterozygous state. Attaining homozygousity at the mutated alleles so that the desired phenotypes can manifest requires several cycles of selfing and hence added expense. Again, totipotency is exploited to produce doubled haploids (DHs), *i.e.*, whole plant that arise from the doubling of the chromosome number of gametic cells, pollens and egg cells, prior to their regeneration [99,100].

There is considerable promise for the routine incorporation of doubled haploidy into induced mutagenesis by treating these gametic cells prior to regeneration of the DHs [101,102,103]. With the availability of *in vitro* mechanisms that permit the spontaneous and/or induced doubling of the haploid chromosomes, homozygous individuals could be produced quite routinely [104]. This would be a very rapid route to homozygousity [105,106] of the mutant's genome and hence the mutated allele. Indeed, the authors [106] concluded that a mutation is captured in a homozygous, pure line in just one generation by using either the haploid or doubled haploid cells as starting materials for mutation treatments. This implies that the gametes of the M_1 plants could be subjected to doubled haploidy or the gametes (haploid cells) of the wildtypes could be induced to mutate and then subsequently subjected to doubled haploidy. Both routes achieve the same purpose of the attainment of homozygousity in one generation. Szarejko and Forster [106] reviewed the productions of DH from M_1materials of *Hordeum vulgare*, *Oryza sativa*, *Solanum nigrum* and *Triticum aestivum*. Mutations had also been induced successfully in haploid cells of different species of *Brassica*, *Datura innoxia*, *Hordeum vulgare*, *Malus x domestica*, different species of *Nicotiana* and *Oryza sativa*. Different physical and chemical mutagens were used. For seed propagated crops therefore, DH strategies provide the fastest method for attaining homozygousity. The savings in time and cost are significant as recessive mutations usually are not detectable till the second (M_2) or later generations of selfing. Reproducible DH protocols are available for over 250 plant species [107] across most plant genera. Technical challenges still remain however with regard to developing genotype-independent protocols, improving the level of callus production by the pollen grains, the differentiation of callus into green plants and the doubling of chromosomes.

4. THE FAO AND IAEA PARTNERSHIP FOR NUCLEAR TECHNIQUES IN AGRICULTURE

FAO and the IAEA have for over 45 years through their Joint FAO/IAEA Programme of Nuclear Techniques in Food and Agriculture headquartered in Vienna, Austria partnered to support countries' efforts to attain food security [108]. The Plant Breeding and Genetics Section of this program assists countries in using radiation induced mutations, facilitated by biotechnologies, to develop superior crop varieties [108]. The Section implements its mandates through a combination of field projects in developing countries; the coordination of collaborative research networks; and a research and development laboratory arm in Seibersdorf, outside Vienna. Currently, there are a total of 86 field projects with themes relating to induced crop mutations (Table 6). The main thrust of the field activities, known as Technical Cooperation Projects (TCP), is technology transfer characterized by the strengthening of human and infrastructural capacities. The irradiation facilities, mostly cobalt-60 sources, in many countries were provided through this TCP mechanism. A more upstream intervention of this partnership is the Coordinated Research Project (CRP) mechanism which fosters global networking between institutions that pool efforts and resources in addressing a mutually important research problem [109]. There are currently six of such CRPs with themes pertaining to induced crop mutations and ancillary biotechnologies [110]. These interventions are complemented by the adaptive research on similar themes carried out its laboratory in Seibersdorf Austria [111]. Training and the provision of irradiation services also form part of the core activities of the laboratory.

Table 6. Ongoing field projects relating to induced crop mutations managed by the Joint FAO/IAEA Program.

Geographical Location	Number of Projects		
	National	Regional	Total
Africa	34	2	36
Asia and the Pacific	29	8	37
Europe	2	2	4
Latin America and the Caribbean	7	2	9
Total	72	14	86

5. EMERGING TRENDS

The case has been made that in order to become an effective and efficient complement to the suite of modern crop improvement and functional genomics strategies, a significant amount of the associated drudgery must be mitigated. This desired efficiency can be achieved by the incorporation of the very powerful novel biotechnologies that have significantly amplified the scopes for addressing biological questions into the processes for the induction, detection and deployment of mutation events. Genomics, the study of the whole genetic makeup of an individual at the molecular level, is particularly suited to the enhancing of efficiencies of the detection and deployment of mutation events. To fully realize the benefits of induced mutagenesis however, there must also be reliable means for generating phenotypic data at high throughputs. A means for effectively injecting the mutants, despite the deleterious alleles that they additionally carry, into breeding programs must also be well thought through. Clearly, the bottlenecks to phenotyping large mutant populations must be removed just as the additional costs for breaking linkage drags must be reduced implying therefore that phenomics and pre-breeding will become increasingly important ancillary aspects to induced mutagenesis.

5.1. Phenomics

Granted, the amenability of genomics platforms to high throughput processes permits the simultaneous generation of very reliable genotypic data from multiple samples and hence abates the major drawback to induced mutations: the imperative of generating and evaluating large numbers of putative mutants in quest of invariably low frequency events. But, for the genotypic data to lead to valid inferences on the variants, there must be reliable mechanisms for generating the complementary phenotypic data. The routine adoption of phenomics, defined by Houle *et al.* [112] as the "the acquisition of high-dimensional phenotypic data on an organism-wide scale" as a component of the detection and deployment of mutation events holds immense promise in this regard. Phenomics facilities typically make use of thermal infra-red, near infra-red, fluorescence and even magnetic resonance imaging equipment to collect large amounts of physiological, morphological and biochemical data from parts of live plants in very short periods of time. Phenomics is permitting a greater understanding of the mechanisms that govern such complex traits as drought and salinity tolerances [113,114]. The set up of phenomics facilities is undoubtedly an expensive endeavor that may be unaffordable for many national agricultural research organizations but the availability of service providers in Australia [115]; Canada [116]; France [117,118]; and Germany [119,120] might serve as impetus for setting up other such facilities to cater to scientists in different parts of the world. The CGIAR centers and other centers of excellence for crop-related agricultural research could incorporate phenomics into their activities and hence serve as regional hubs. The Joint FAO/IAEA laboratory that traditionally provides services relating to induced mutations is also well positioned to take on this additional task.

5.2. Pre-Breeding

The extremely narrow genetic base of the world's important crops [2,121] is a major cause for concern as efforts are invested to crop productivities and enhance the resilience of cropping systems under climate change regimens.

It is ironic therefore that plant breeders largely continue to rely on the same "safe bet" parents in breeding programs. Many reasons, amongst which is the avoidance of the extra expense in time and resources required for breaking the linkage drags associated with the introgression of traits from crop wild relatives, landraces and other otherwise non-adapted materials, have been adduced for this predilection. The same mindset dampens the enthusiasm for generating induced mutants and incorporating them into crossing blocks. Unintended and undesirable induced mutations typically co-segregate with the desirable novelties, a "nuisance factor" that is expensive to get rid of. Pre-breeding, described by Nass and Paterniani [122] as "all activities designed to identify desirable characteristics and/or genes from unadapted (exotic or semi-exotic) materials, including those that, although adapted have [not] been subjected to any kind of selection for improvement" is a means to populate the gap in activities between curating germplasm collections and plant breeding with the generation of intermediate breeding materials. The Global Crop Diversity Trust elegantly describes this collaborative discipline between the germplasm curator and plant breeder as "... the art and science of identifying desired traits in otherwise unprepossessing and unpromising plants, often wild, and starting to incorporate them into modern varieties" [123].

For practical purposes, the raw putative induced mutants may be considered as those "otherwise unprepossessing and unpromising plants" that must be put through the paces of selections and hybridizations in order to elevate them to the stable and relatively "clean" materials that may be used as parents in breeding programs. The e-learning course [124] that was developed by the Global Partnership Initiative for Plant Breeding Capacity Building is a very useful tool for assisting germplasm curators (with in-depth knowledge and appreciation of the variations in germplasm collections) and the plant breeders (who work to incorporate the most desirable new traits into novel varieties) institutionalize this paradigm for crop improvement. It includes modules on germplasm management, classical plant breeding themes and molecular genetics techniques and therefore imparts skills that lend themselves to managing large mutant populations and integrating their novel traits into breeding lines. Pre-breeding, therefore, is indeed a most valuable appendage to all induced mutation programs as resources could be dedicated solely to transforming these "raw" induced mutants with all the deleterious alleles to "acceptable" parental materials for breeding programs. It is safe to assume that as pre-breeding becomes an established component of crop improvement, plant breeders will be under less pressure to account for the time invested in cleaning up induced mutants for use as parents in hybridization. Induced mutation projects should therefore include a pre-breeding component as means to forge the bridge to plant breeding.

6. CONCLUSIONS AND FUTURE PERSPECTIVES

Enhanced productivities on farmers' fields, with minimal ecological footprints, is a sustainable means for addressing the myriad constraints that threaten the 21st century global food security. Clearly, a diverse set of hardy, high yielding and input use-efficient crop varieties will, along with ecosystem-based agronomic practices, constitute the crop production packages needed to nourish an ever increasing human population, meet the demands of other industries for food-based substrates and yet leave the environment largely unharmed. The currently available cultivars of most staple crops do not fit into this envisaged highly efficient yet low-input crop production systems. This implies that a new portfolio of crop varieties will need to be bred.

Breeding the needed novel varieties will require a most judicious application of the most appropriate scientific and technological tools so that the inherent potentials of PGRFA are harnessed optimally [14,125]. Induced mutagenesis, albeit almost a century old technique, demonstrably can contribute to unleashing the potentials of PGRFA and thereby avail plant breeders and other scientists the raw materials needed to generate the envisaged "smart" crop varieties. Induced mutagenesis is a proven, safe, robust and cheap plant breeding strategy. Crop varieties generated through the exploitations of induced mutagenesis are contributing to global food and nutritional security and improved livelihoods. But, induced mutagenesis, as a crop improvement strategy, has for long been an inefficient means for accessing novel alleles of genes. This is largely on account of the requirement for generating and evaluating large populations of putative mutants, the incidence of chimeras and the recessive nature of mutations. This need not remain the *status quo* as the ever increasing access to publicly available genomic and genetic resources along with other advances in high throughput molecular genetics, cell biology and phenotyping techniques are proven means to mitigating these constraints and hence, enhancing efficiencies. These novel techniques facilitate the integration of induced mutations into seamless processes for crop improvement and classic genetics research.

Also, by inducing crops to mutate, the ability of scientists to understand and better exploit the underlying genic influences that modulate the expression of agronomic and crop quality traits is greatly enhanced. Induced mutagenesis is therefore now a strategy of choice in functional genomics as it greatly facilitates the identification of genes and the elucidations of their functions. Interestingly, the outputs of functional genomics, namely, elucidated genes, are, when used as molecular genetic markers, enhancing the efficiency of plant breeding (including mutation breeding). It is also noteworthy that phenomics will facilitate the generation of morphological and/or agronomic data or those of their contributory parameters. In like manner, pre-breeding will enhance the ease for integrating mutants into breeding programs. The model of the Joint FAO/IAEA Program, in being dedicated solely to the deployment of "atoms for peace" in food and agriculture, is innovative and unique. It is currently supporting R&D activities, providing training and niche services, facilitating technology transfer and fostering communities of practice. This model is deserving of emulation and scaling up.

In spite of the aforementioned advantages of, and benefits accruing from, induced mutations, it is hardly a mainstream endeavor in crop improvement. A majority of plant breeding programs have no active mutation induction activities while a number of irradiation facilities have been closed in the last two to three decades. It is plausible to infer that the decline in the applications of induced mutations—which parallels the general decline in capacities for crop improvement, especially, plant breeding—correlates inversely with the upsurge in the adoption and applications of biotechnologies, especially recombinant DNA technologies. But, it need not be one or the other being as the techniques are vastly different. Induced mutations, for instance, does not entail the introduction of extraneous hereditary materials which subsequently express in the recipient genome. This alone exempts induced mutants from the often times expensive and long regulatory regimes that genetically modified organisms are subject to. This relative simplicity of official processes couples with the earlier mentioned robustness and cheap start up and operational costs to make induced mutations particularly suitable for developing countries.

However, for induced mutagenesis to once again become a mainstream crop improvement technique for the 21st Century, significant efforts must be invested in a number of interventions. These include:

- Training of the new crop of plant breeders. Induced mutations, along with enabling biotechnologies, ought to be part of university curricula at both the honours and graduate levels. The annual training program of the Joint FAO/IAEA Division in induced crop mutations could provide the template for developing course curriculum that would be adaptable to other situations. In the same vein, students already undergoing formal plant breeding training programs in universities could also be deliberately targeted for inclusion in this training course. Equally, universities that have maintained strong programs in induced mutations may be supported to host trainees on ad-hoc basis.
- Research. Granted, biotechnologies, such as reverse genetics and cell biology, facilitate the induction, detection and transfer of desirable mutation events. But, the protocols, most of which are genotype-dependent, have been developed for the well-researched major crops. More investments are clearly required for adapting the protocols or

developing new ones for the "orphan" crops that are the mainstay of the cropping systems of many food insecure countries. Easily adaptable protocols for vegetatively propagated crops are particularly in short supply. Also, protocols that are less resource-intensive and are adaptable to the resource-poor research environments in developing countries are needed. For instance, for TILLING, neither the endonuclease cleavage of heteroduplexes nor the newer sequencing methods used in the identification of point mutations can be routinely carried out in the laboratories of developing countries that are technology challenged, resource-poor, lack skilled personnel and have weak infrastructures and utilities.

- Community of practice. FAO and the IAEA, through their ca. half a century partnership, is investing considerable resources in training, technology transfer, strengthening of infrastructure, provision of services and in research. To achieve the greatest impacts through engagements at scale, a critical mass of partners that are engaged in similar endeavors and whose activities complement those of this partnership need to be assembled. Such a community could arise around a theme, such as climate change, for instance. Induced mutations can unmask alleles of genes that can be introgressed into breeding materials. This would clearly complement several ongoing activities by a multitude of potential partners, even on global scales, but efforts need to be invested in coalescing these efforts. The TCP and CRP models of the Joint FAO/IAEA Division of Nuclear Techniques for Food and Agriculture can be further scaled up to achieve this purpose.

CONFLICT OF INTEREST

The author declares no conflict of interest. The views expressed in this publication are those of the author and do not necessarily reflect the views or policies of FAO.

REFERENCES

1. Food and Agriculture Organization of the United Nations. How to Feed the World in 2050; Food and Agriculture Organization of the United Nations: Rome, Italy, 2009.

2. Tester, M.; Langridge, P. Breeding technologies to increase crop production in a changing world. Science 2010, 327, 818–822.

3. Ejeta, G. Revitalizing agricultural research for global food security. Food Sec. 2009, 1, 391–401.

4. Nelson, G.C.; Rosegrant, M.W.; Koo, J.; Robertson, R.; Sulser, T.; Zhu, T.; Ringler, C.; Msangi, S.; Palazzo, A.; Batka, M.; Magalhaes, M.; Valmonte-Santos, R.; Ewing, M.; Lee, D. Climate Change: Impact on Agriculture and Costs of Adaptation; International Food Policy Research Institute: Washington, DC, USA, 2009; p. 19.

5. Hertel, T.W.; Burke, M.B.; Lobell, D.B. The poverty implications of climate-induced crop yield changes by 2030. Glob. Environ. Change 2010, 20, 577–585.

6. Rosegrant, M.W. Impacts of climate change on food security and livelihoods. In Food Security and Climate Change in Dry Areas, Proceedings of an International Conference, Amman, Jordan, 1–4 February 2010; Solh, M., Saxena, M.C., Eds.; International Center for Agricultural Research in the Dry Areas: Aleppo, Syria, 2011; pp. 24–26.

7. Food and Agriculture Organization of the United Nations. Global Survey of Agricultural Mitigation Projects; Food and Agriculture Organization of the United Nations: Rome, Italy, 2010.

8. Beddington, J.; Asaduzzaman, M.; Fernandez, A.; Clark, M.; Guillou, M.; Jahn, M.; Erda, L.; Mamo, T.; van Bo, N.; Nobre, C.A.; Scholes, R.; Sharma, R.; Wakhungu, J. Achieving Food Security in the Face of Climate Change: Summary for Policy Makers from the Commission on Sustainable Agriculture and Climate Change; CGIAR Research Program on Climate Change, Agriculture and Food Security (CCAFS): Copenhagen, Denmark, 2011.

9. Field, C.B., Barros, V., Stocker, T.F., Qin, D., Dokken, D.J., Ebi, K.L., Mastrandrea, M.D., Mach, K.J., Plattner, G.K., Allen, S.K., Tignor, M., Midgley, P.M., Eds.; Intergovernmental Panel on Climate Change. Summary for Policymakers.Managing the Risks of Extreme Events and Disasters to Advance Climate Change Adaptation, A Special Report of Working Groups I and II of the Intergovernmental Panel on Climate Change; Cambridge University Press: Cambridge, UK, 2012.

10. Chatham House. The Feeding of the Nine Billion: Global Food Security for the 21st Century; Chatham House: London, UK, 2009.

11. Food and Agriculture Organization of the United Nations. Save and Grow—A Policy Maker's Guide to the Sustainable Intensification of

Smallholder Crop Production; Food and Agriculture Organization of the United Nations: Rome, Italy, 2011.

12. McCouch, S. Diversifying selection in plant breeding. PLoS Biol. 2004, 2, 1507–1512.

13. Waines, G.J.; Ehdaie, B. Domestication and crop physiology: roots of green-revolution wheat. Ann. Bot. 2007, 100, 991–998.

14. Mba, C.; Guimaraes, P.; Ghosh, K. Re-orienting crop improvement for the changing climatic conditions of the 21st century. Agric. Food Secur. 2012, 1, 7.

15. All Nobel Prizes in Physics. Available online: http://nobelprize.org/nobel_prizes/physics/laureates (accessed on 2 November 2012).

16. Muller, H.J. Artificial transmutation of the gene. Science 1927, 66, 84–87.

17. Stadler, L.J. Genetic Effects of X-rays in maize. Proc. Nat. Acad. Sci. 1928, 14, 69–72.

18. Stadler, L.J. Mutations in barley induced by X-rays and radium. Science 1928, 68, 186–187.

19. Stadler, L.J. Some genetic effects of X-rays in plants. J. Heredity 1930, 21, 2–19.

20. Stadler, L.J. The experimental modification of heredity in crop plants 1. Induced chromosomal irregularities. 2. Induced mutations. Sci. Agr. 1931, 11, 557–572, 645–661.

21. Stadler, L.J. On the genetic nature of induced mutations in plants. Proc. 6th Int. Congr. Gen. 1932, 1, 274.

22. Mba, C.; Afza, R.; Shu, Q.Y. Mutagenic radiations: X-rays, ionizing particles and ultraviolet. In Plant Mutation Breeding and Biotechnology; Shu, Q., Forster, B.P., Nakagawa, H., Eds.; CABI: Oxfordshire, UK, 2012; pp. 83–90.

23. Mba, C.; Shu, Q.Y. Gamma Irradiation. In Plant Mutation Breeding and Biotechnology; Shu, Q., Forster, B.P., Nakagawa, H., Eds.; CABI: Oxfordshire, UK, 2012; pp. 91–98.

24. Mei, M.; Deng, H.; Lu, Y.; Zhuang, C.; Liu, Z.; Qiu, Q.; Qiu, Y.; Yang, T.C. Mutagenic effects of heavy ion radiation in plants. Adv. Space Res. 1994, 14, 363–372.

25. Wu, L.; Yu, Z. Radiobiological effects of a low-energy ion beam on wheat. Radiat. Environ. Biophys. 2001, 40, 53–57.

26. Mei, M.; Qiu, Y.; He, Y.; Bucker, H.; Yang, C.H. Mutational effects of space flight on Zea mays seeds. Adv. Space Res.1994, 14, 33–39.

27. Mei, M.; Qiu, Y.; Sun, Y.; Huang, R.; Yao, J.; Zhang, Q.; Hong, M.; Ye, J. Morphological and molecular changes of maize plants after seeds been flown on recoverable satellite. Adv. Space Res. 1998, 22, 1691–1697.

28. Ren, W.; Zhang, Y.; Deng, B.; Guo, H.; Cheng, L.; Liu, Y. Effect of space flight factors on alfalfa seeds. Afr. J. Biotechnol.2010, 9, 7273–7279.

29. Liu, L.; Zhao, L.; Guo, H. Current status and perspective outlook of space induced mutation breeding in crop plants.Rev. China Agric. Sci. Technol. 2007, 9, 26–29.

30. Auerbach, C. Tests of carcinogenic substances in relation to the production of mutationins Drosophila melanogaster.Proc. R. Soc. Edinb. 1940, B60, 164–173.

31. Auerbach, C.; Robson, J.M. Production of mutations by allyl isothiocyanate. Nature 1944, 154, 81.

32. Auerbach, C.; Robson, J.M. Chemical production of mutations. Nature 1946, 157, 302.

33. Auerbach, C. The induction by mustard gas of chromosomal instabilities in Drosophila melanogaster. Proc. R. Soc. Edinb.1947, B62, 307–320.

34. Auerbach, C.; Robson, J.M. The production of mutations by chemical substances. Proc. R. Soc. Edinb. 1947, B62, 27.

35. Auerbach, C.; Robson, J.M. Tests for chemical substances for mutagenic action. Proc. R. Soc. Edinb. 1947, B62, 284–291.

36. Auerbach, C. Chemical induction of mutations. In Hereditas 1949, Suppl. Vol. for Proc. 8th Intern. Congr. Genet. pp. 128–147.

37. Auerbach, C. Induction of changes in genes and chromosomes. Problems in chemical mutagenesis. Cold Spring Harbor Symp. Quant. Biol. 1951, 16, 199–213.

38. Auerbach, C. Genetical effects of radiations and chemicals. Experientia 1957, 13, 217–224.

39. Mba, C.; Afza, R.; Bado, S.; Jain, S.M. Induced mutagenesis in plants using physical and chemical agents. In Plant Cell Culture: Essential Methods; Davey, M.R., Anthony, P., Eds.; John Wiley & Sons, Ltd.: Chichester, UK, 2010.

40. Mba, C.; Afza, R.; Jankowicz-Cieslak, J.; Bado, S.; Matijevic, M.; Huynh, O.; Till, B.T. Enhancing Genetic Diversity Through Induced Mutagenesis in Vegetatively Propagated Plants. In Induced Plant Mutations in the Genomics Era; Shu, Q.Y., Ed.; Food and Agriculture Organization of the United Nations: Rome, Italy, 2009; pp. 293–296.

41. Van Harten, A.M. Mutation Breeding. Theory and practical Applications; Cambridge University Press: Cambridge, UK, 1998; p. 353.

42. Mutant Varieties Databse of the Joint FAO/IAEA Division of Nuclear Techniques in Food and Agriculture. Available online: http://mvgs.iaea.org/AboutMutantVarities.aspx (accessed on 2 November 2012).

43. Plant Breeding and Genetics Section of the Joint FAO/IAEA Division of Nuclear Techniques in Food and Agriculture. Available online: http://www-naweb.iaea.org/nafa/pbg/index.html (accessed on 2 November 2012).

44. Maluszynski, M.; Nichterlein, K.; van Zanten, L.; Ahloowalia, B.S. Officially released mutant varieties—The FAO/IAEA database. Mutat. Breed. Rev. 2000, 12, 1–88.

45. Maluszynski, M.; Szarejko, I. Induced mutations in the Green and Gene Revolutions. In the Wake of the Double Helix: From the Green Revolution to the Gene Revolution, Proceedings of the International Congress, Bologna, Italy, 27–31 May 2003; Tuberosa, R., Phillips, R.L., Gale, M., Eds.; 2005; pp. 403–425.

46. Ahloowalia, B.S.; Maluszynski, M.; Nichterlein, K. Global impact of mutation-derived varieties. Euphytica 2004, 135, 187–204.

47. Kharkwal, M.C.; Shu, Q.Y. The Role of Induced Mutations in World Food Security. In Induced Plant Mutations in the Genomics Era; Shu, Q.Y., Ed.; Food and Agriculture Organization of the United Nations: Rome, Italy, 2009; pp. 33–38.

48. Askew, M.F. Novel oil, fibre and protein crops in UK, A Future Perspective. Brighton Crop Protect. Conf. 1993, 2, 658.

49. Larkin, P.J. Introduction. In Somaclonal Variation and Induced Mutations in Crop Improvement; Jain, S.M., Brar, D.S., Ahloowalia, B.S., Eds.; Kluwer Academic Publishers: Dordrecht, Holland, 1998; pp. 3–13.

50. National Plant Germplasm System of the United States Department of Agriculture. Available online: http://www.ars-grin.gov/npgs/holdings.html (accessed on 2 November 2012).

51. The Maize Genetics Cooperation Stock Centre, University of Illinois, Urbana/Champaign, IL, USA. Available online: http://maizecoop.cropsci.uiuc.edu/mgc-info.php (accessed on 2 November 2012).

52. European Arabidopsis Stock Centre (also known as Nottingham Arabidopsis Stock Centre, NASC), University of Nottingham, Sutton Bonington Campus, UK. Available online: http://arabidopsis.info/ (accessed on 2 November 2012).

53. Arabidopsis Biological Resource Centre, (ABRC), Ohio State University, OH, USA. Available online: http://abrc.osu.edu/ (accessed on 2 November 2012).

54. CM Rick Tomato Genetics Resource Centre, University of California at Davis, CA, USA. Available online: http://tgrc.ucdavis.edu/ (accessed on 2 November 2012).

55. Cucurbit Genetics Cooperative (CGC), North Carolina State UniversityRaleigh, NC, USA. Available online: http://cuke.hort.ncsu.edu/cgc/index.html (accessed on 2 November 2012).

56. The Oryzabase of the National BioResource Project—Rice National Institute of Genetics, Japan. Available online: http://www.shigen.nig.ac.jp/rice/oryzabaseV4/ (accessed on 2 November 2012).

57. IR64 Rice Mutant Database of the International Rice Functional Genomics, International Rice Research Institute, Manila, Philippines. Available online: http://irfgc.irri.org/index.php?option=com_content&task=view&id=22&Itemid=44 (accessed on 2 November 2012).

58. Plant Functional Genomics Lab., POSTECH BIOTECH CENTER, San 31 Hyoja-dong, Nam-gu Pohang, Kyoungbuk, Korea. Available online: http://www.postech.ac.kr/life/pfg/ (accessed on 2 November 2012).

59. Barley mutants, Scottish Crop Research Institute, Dundee, Scotland. Available online: http://bioinf.scri.ac.uk/barley/?page_id=14 (accessed on 2 November 2012).

60. Barley and Wheat Genetic Stock of the USDA-ARS. Available online: http://www.ars.usda.gov/main/docs.htm?docid=2922 (accessed on 2 November 2012).

61. USDA-ARS Cereal Crops Research Unit, Fargo, ND, USA. Available online: http://wheat.pw.usda.gov/ggpages/GeneticStocks/Fargo_ARS_genetic_stocks.html (accessed on 2 November 2012).

62. Wheat Genetics Resource Center, Kansas State University, Manhattan, KS, USA. Available online: http://www.k-state.edu/wgrc/ (accessed on 2 November 2012).

63. Wheat Genetic Resources Database of the Japanese National BioResource Project. Available online: http://www.shigen.nig.ac.jp/wheat/komugi/strains/aboutNbrp.jsp (accessed on 2 November 2012).

64. Pea mutants, John Innes Centre, Norwich, UK. Available online: http://www.jic.ac.uk/GERMPLAS/pisum/index.htm(accessed on 02 November 2012).

65. McCallum, C.M.; Comai, L.; Greene, E.A.; Henikoff, S. Targeted screening for induced mutations. Nat. Biotechnol.2000, 18, 455–457.

66. Henikoff, S.; Till, B.J.; Comai, L. TILLING. Traditional Mutagenesis Meets Functional Genomics. Plant Physiol. 2004,135, 630–636.

67. McCallum, C.M.; Comai, L.; Greene, E.A.; Henikoff, S. Targeting induced local lesions IN genomes (TILLING) for plant functional genomics. Plant Physiol. 2000, 123, 439–442.

68. Colbert, T.; Till, B.J.; Tompa, R.; Reynolds, S.; Steine, M.N.; Yeung, A.T.; McCallum, C.M.; Comai, L.; Henikoff, S. High-throughput screening for induced point mutations. Plant Physiol. 2001, 126, 480–484.

69. Greene, E.A.; Codomo, C.A.; Taylor, N.E.; Henikoff, J.G.; Till, B.J.; Reynolds, S.H.; Enns, L.C.; Burtner, C.; Johnson, J.E.; Odden, A.R.; et al. Spectrum of chemically induced mutations from a large-scale reverse-genetic screen in Arabidopsis. Genetics 2003, 164, 731–740.

70. Till, B.J.; Reynolds, S.H.; Greene, E.A.; Codomo, C.A.; Enns, L.C.; Johnson, J.E.; Burtner, C.; Odden, A.R.; Young, K.; Taylor, N.E.; Henikoff, J.G.; Comai, L.; Henikoff, S. Large-scale discovery of induced point mutations with high-throughput TILLING. Genome Res. 2003, 13, 524–530.

71. Wu, J.L.; Wu, C.; Lei, C.; Baraoidan, M.; Bordeos, A.; Madamba, M.R.S.; Ramos-Pamplona, M.; Mauleon, R.; Portugal, A.; Ulat, V.J.; Bruskiewich, R.; Wang, G.; Leach, J.; Khush, G.; Leung, H. Chemical and irradiation-induced mutants of indica rice IR64 for forward and reverse genetics. Plant Mol. Biol. 2005, 59, 85–97.

72. Sato, Y.; Shirasawa, K.; Takahashi, Y.; Nishimura, M.; Nishio, T. Mutant Selection from Progeny of Gamma-ray-irradiated Rice by DNA Heteroduplex Cleavage using Brassica Petiole Extract. Breed. Sci. 2006, 56, 179–183.

73. Till, B.J.; Cooper, J.; Tai, T.H.; Colowit, P.; Greene, E.A.; Henikoff, S.; Comai, L. Discovery of chemically induced mutations in rice by TILLING. BMC Plant Biol. 2007, 7, 19.

74. Suzuki, T.; Eiguchi, M.; Kumamaru, T.; Satoh, H.; Matsusaka, H.; Moriguchi, K.; Nagato, Y. MNU-induced mutant pools and high performance TILLING enable finding of any gene mutation in rice. Mol. Genet. Genomics 2008, 279, 213–223.

75. Till, B.J.; Reynolds, S.H.; Weil, C.; Springer, N.; Burtner, C.; Young, K.; Bowers, E.; Codomo, C.A.; Enns, L.C.; Odden, A.R.; Greene, E.A.; Comai, L.; Henikoff, S. Discovery of induced point mutations in maize genes by TILLING. BMC Plant Biol. 2004, 4, 12.

76. Slade, A.J.; Fuerstenberg, S.I.; Loeffler, D.; Steine, M.N.; Facciotti, D. A reverse genetic, nontransgenic approach to wheat crop improvement by TILLING. Nat. Biotechnol. 2005, 23, 75–81.

77. Dong, C.; Dalton-Morgan, J.; Vincent, K.; Sharp, P. A Modified TILLING Method for Wheat Breeding. Plant Genome2009, 2, 39–47.

78. Hohmann, U.; Jacobs, G.; Jung, C. An EMS mutagenesis protocol for sugar beet and isolation of non-bolting mutants.Plant Breed. 2005, 124, 317–321.

79. Caldwell, D.G.; McCallum, N.; Shaw, P.; Muehlbauer, G.J.; Marshall, D.F.; Waugh, R. A structured mutant population for forward and reverse genetics in barley (Hordeum vulgare L.). Plant J. 2004, 40, 143–150.

80. Cooper, J.L.; Till, B.J.; Laport, R.G.; Darlow, M.C.; Kleffner, J.M.; Jamai, A.; El-Mellouki, T.; Liu, S.; Ritchie, R.; Nielsen, N.; Bilyeu, K.D.; Meksem, K.; Comai, L.; Henikoff, S. TILLING to detect induced mutations in soybean. BMC Plant Biol. 2008, 8, 9.

81. Triques, K.; Sturbois, B.; Gallais, S.; Dalmais, M.; Chauvin, S.; Clepet, C.; Aubourg, S.; Rameau, C.; Caboche, M.; Bendahmane, A. Characterization of Arabidopsis thaliana mismatch specific endonucleases: Application to mutation discovery by TILLING in pea. Plant J. 2007, 51, 1116–1125.

82. Galeano, C.H.; Gomez, M.; Rodriguez, L.M.; Blair, M.W. CEL I Nuclease digestion for SNP discovery and marker development in common bean (Phaseolus vulgaris L.). Crop Sci. 2009, 49, 381–394.

83. Porch, T.G.; Blair, M.W.; Lariguet, P.; Galeano, C.; Pankhurst, C.E.; Broughton, W.J. Generation of a mutant population for TILLING common bean genotype BAT 93. J. Am. Soc. Hort. Sci. 2009, 134, 348–355.

84. Minoia, S.; Petrozza, A.; D'Onofrio, O.; Piron, F.; Mosca, G.; Sozio, G.; Cellini, F.; Bendahmane, A.; Carriero, F. A new mutant genetic resource for tomato crop improvement by TILLING technology. BMC Res. Notes 2010, 3, 69.

85. Jankowicz-Cieslak, J.; Huynh, O.A.; Brozynska, M.; Nakitandwe, J.; Till, B.J. Induction, rapid fixation and retention of mutations in vegetatively propagated banana. Plant Biotechnol. J. 2012.

86. Till, B.J.; Jankowicz-Cieslak, J.; Sagi, L.; Huynh, O.A.; Utsushi, H.; Swennen, R.; Terauchi, R.; Mba, C. Discovery of nucleotide polymorphisms in the Musa gene pool by Ecotilling. Theor. Appl. Genet. 2010, 121, 1381–1389.

87. Tadele, Z.; Mba, C.; Till, B.J. TILLING for mutations in model plants and crops. In Molecular Techniques in Crop Improvement, 2nd; Jain, S.M., Brar, D.S., Eds.; Springer Publishing Inc.: Dordrecht, Holland, 2009; pp. 307–322.

88. Tsai, H.; Howell, T.; Nitcher, R.; Missirian, V.; Watson, B.; Ngo, K.J.; Lieberman, M.; Fass, J.; Uauy, C.; Tran, R.K.; Khan, A.I.; Filkov, V.; Tai, T.H.; Dubcovsky, J.; Comai, L. Discovery of Rare Mutations in Populations: TILLING by Sequencing. Plant Physiol. 2011, 156, 1257–1268.

89. Till, B.J.; Afza, R.; Bado, S.; Huynh, O.A.; Jankowicz-Cieslak, J.; Matijevic, M.; Mba, C. Global TILLING Projects. In Induced Plant Mutations in the Genomics Era, Proceedings of International Symposium on Induced Mutations in Plants, Vienna, Austria, 11–15 August 2008; Shu, Q.Y., Ed.; Food and Agriculture Organization of the United Nations: Rome, Italy, 2009; pp. 237–239.

90. The UC Davis TILLING Core, University of California, Davis, CA, USA. Available online: http://tilling.ucdavis.edu/index.php/Main_Page (accessed on 2 November 2012).

91. LycoTILL—Tomato Mutant Database, Metapontum Agrobios Centre, Metaponto di Bernalda, MT, Italy. Available online: http://www.agrobios.it/tilling/index.html (accessed on 14 January 2013).

92. The Maize TILLING Project, Purdue University, West Lafayette, IN, USA. Available online: http://genome.purdue.edu/maizetilling/ (accessed on 2 November 2012).

93. CAN-TILL, University of British Columbia. Available online: http://www3.botany.ubc.ca/can-till/Projects.html#napus(accessed on 2 November 2012).

94. RevGenUK, John Innes Centre, Norwich Research Park, Colney, Norwich, NR4 7UH, UK. Available online: http://revgenuk.jic.ac.uk/about.htm (accessed on 2 November 2012).

95. Seattle TILLING Project, Fred Hutchinson Cancer Research Center, North Seattle, WA, USA. Available online: http://tilling.fhcrc.org/ (accessed on 2 November 2012).

96. Plant Genomic Research, INRA/CNRS—URGV, Evry cedex, France. Available online: http://www.versailles.inra.fr/urgv/contact.htm (accessed on 2 November 2012).

97. Nickell, L.G. The continuous submerged cultivation of plant tissue as single cells. Proc. Nat. Acad. Sci. 1956, 42, 848–850.

98. Lopez-Torres, J. Personal Communication. Instituto de Investigaciones de Viandas Tropicales (INIVIT).

99. Forster, B.P.; Heberle-Bors, E.; Kasha, K.J.; Touraev, A. The resurgence of haploids in higher plants. Trends Plant Sci.2007, 12, 368–375.

100. Jauhar, P.P.; Xu, S.S.; Baenziger, P.S. Haploidy in cultivated wheats: Induction and utility in basic and applied research. Crop Sci. 2009, 49, 737–755.

101. Castillo, A.M.; Cistué, L.; Vallés, M.P.; Sanz, J.M.; Romagosa, I.; Molina-Cano, J.L. Efficient production of androgenic doubled-haploid mutants in barley by the application of sodium azide to anther and microspore cultures. Plant Cell Rep. 2001, 20, 105–111.

102. Forster, B.P.; Thomas, W.T.B. Doubled haploids in genetics and plant breeding. Plant Breed. Rev. 2005, 25, 57–88.

103. Mkuya, M.S.; Hua-min, S.; Wen-zhen, L.; Zong-xiu, S. Effect of 137Cs gamma rays to panicles on rice anther culture.Rice Science 2005, 12, 299–302.

104. Kasha, K.J. Chromosome Doubling and Recovery of Doubled Haploid Plants. In Biotechnology in Agriculture and Forestry; Palmer, C.E., Keller, W.A., Kasha, K.J., Eds.; Springer-Verlag: Berlin, Germany, 2005; Volume 56, pp. 123–152.

105. Takahata, Y.; Fukuoka, H.; Wakui, K. Utilization of Microspore-Derived Embryos. In Biotechnology in Agriculture and Forestry; Palmer, C.E., Keller, W.A., Kasha, K.J., Eds.; Springer-Verlag: Berlin, Germany, 2005; Volume 56, pp. 153–169.

106. Szarejko, I.; Forster, B.P. Doubled haploidy and induced mutation. Euphytica 2007, 158, 359–370.

107. Maluszynski, M.; Kasha, K.J.; Szarejko, I. Published doubled haploid protocols in plant species. In Doubled Hapolid Production in Crop Plants: A Manual; Maluszynski, M., Kasha, K.J., Forster, B.P., Szarejko,

I., Eds.; Kluwer Academic Publishers: Dordrecht, The Netherlands, 2003; pp. 309–335.

108. Joint FAO/IAEA Division of Nuclear Techniques in Food and Agriculture. Available online: http://www-naweb.iaea.org/nafa/about-nafa/index.html (accessed on 2 November 2012).

109. Coordinated Research Activities of the International Atomic Energy Agency. Available online: http://www-crp.iaea.org/default.asp (accessed on 2 November 2012).

110. Active crop improvement related Coordinated Research Projects managed by the Plant Breeding and Genetics Section of the Joint FAO/IAEA Division of Nuclear Techniques in Food and Agriculture. Available online: http://www-naweb.iaea.orp/rg/nafa/pbg/cactive-crps-pbg.html (accessed on 2 November 2012).

111. Plant Breeding and Genetics Laboratory of the Joint FAO/IAEA Division of Nuclear Techniques in Food and Agriculture. Available online: http://www-naweb.iaea.org/nafa/pbg/pbg-laboratory.html (accessed on 2 November 2012).

112. Houle, D.; Govindaraju, D.R.; Omholt, S. Phenomics: The next challenge. Nat. Rev. Genet. 2010, 11, 855–866.

113. Finkel, E. With 'Phenomics', plant scientists hope to shift breeding into overdrive. Science 2009, 325, 380–381.

114. Berger, B.; Parent, B.; Tester, M. High-throughput shoot imaging to study drought responses. J. Exp. Bot. 2010.

115. The Australian Plant Phenomics Facility. Available online: http://www.plantphenomics.org.au/ (accessed on 2 November 2012).

116. The Biotron Experimental Climate Change Research Centre. Available online: http://www.thebiotron.ca/ (accessed on 2 November 2012).

117. Ecotron. Available online: http://www.ecotron.cnrs.fr/ (accessed on 2 November 2012).

118. The Ecophysiology Laboratory of Plant under Environmental Stress. Available online: http://www1.montpellier.inra.fr/ibip/lepse/english/ (accessed on 2 November 2012).

119. Lemna Tec. Available online: http://www.lemnatec.com/ (accessed on 2 November 2012).

120. Jülich Plant Phenotyping Centre. Available online: http://www2.fz-juelich.de/icg/icg-3/jppc (accessed on 2 November 2012).

121. Martynov, S.P.; Dobrotvorskaya, T.V. Genealogical analysis of diversity of Russian winter wheat cultivars (Triticum aestivum L.). Genet. Resour. Crop Evol. 2006, 53, 379–386.

122. Nass, L.L.; Paterniani, E. Pre-breeding: A link between genetic resources and maize breeding. Sci. agric. 2000, 57, 581–587.

123. Pre-breeding theme by the Global Crop Diversity Trust. Available online: http://www.croptrust.org/content/pre-breeding (accessed on 2 November 2012).

124. The e-learning course on pre-breeding by the Global Partnership Initiative on Plant Breeding Capacity Building. Available online: http://km.fao.org/gipb/e-learning/gipb-pre-breeding-course/en/ (accessed on 2 November 2012).

125. Mba, C.; Guimaraes, E.P.; Guei, G.R.; Hershey, C.; Paganini, M.; Pick, B.; Ghosh, K. Mainstreaming the continuum approach to the management of plant genetic resources for food and agriculture through national strategy. Plant Genet. Resour. 2011.

Chapter 9

Mutagenesis and Functional Studies with Succinate Dehydrogenase Inhibitors in the Wheat Pathogen Mycosphaerella graminicola

Gabriel Scalliet[1*], Judith Bowler[2], Torsten Luksch[1], Lucy Kirchhofer-Allan[1], Diana Steinhauer1, Keith Ward2, Michael Niklaus1¤a, Andreas Verras1¤b, Michae' Csukai[2], Antoine Daina[1¤]c,Raymonde Fonne´ -Pfister[1]

[1] *Syngenta Crop Protection Mu¨ nchwilen AG, Stein, Switzerland,*

[2] *Syngenta Jealott's Hill International Research Center, Bracknell, United Kingdom*

ABSTRACT

A range of novel carboxamide fungicides, inhibitors of the succinate dehydrogenase enzyme (SDH, EC 1.3.5.1) is currently being introduced to the crop protection market. The aim of this study was to explore the impact of structurally distinct carboxamides on target site resistance development and to assess possible impact on fitness.

We used a UV mutagenesis approach in *Mycosphaerella graminicola*, a key pathogen of wheat to compare the nature, frequencies and impact of target mutations towards five subclasses of carboxamides. From this screen we identified 27 amino acid substitutions occurring at 18 different positions on the 3 subunits constituting the ubiquinone binding (Qp) site of the enzyme. The nature of substitutions and cross resistance profiles indicated significant differences in the binding interaction to the enzyme across the different inhibitors. Pharmacophore elucidation followed by docking studies in a tridimensional SDH model allowed us to propose rational hypotheses explaining some of the differential behaviors for the first time. Interestingly all the characterized substitutions had a negative impact on enzyme efficiency, however very low levels of enzyme activity appeared to be sufficient for cell survival. In order to explore the impact of mutations on pathogen fitness *in vivo* and *in planta*, homologous recombinants were generated for a selection of mutation types. *In vivo*, in contrast to previous studies performed in yeast and other organisms, SDH mutations did not result in a major increase of reactive oxygen species levels and did not display any significant fitness penalty. However, a number of Qp site mutations affecting enzyme efficiency were shown to have a biological impact *in planta*.

Using the combined approaches described here, we have significantly improved our understanding of possible resistance mechanisms to carboxamides and performed preliminary fitness penalty assessment in an economically important plant pathogen years ahead of possible resistance development in the field.

INTRODUCTION

Mycosphaerella graminicola (Fückel) J. Schröt. in Cohn (anamorph *Septoria tritici*) is the causal agent of Septoria leaf blotch (STB), a major threat to bread and durum wheat worldwide[1], [2], [3]. Breeding programs are ongoing to stack host resistance genes and create new varieties highly resistant to STB, but current control of this fungal disease relies heavily on fungicide use. Options for the chemical control of STB were decreased recently after the rapid development of resistance against the QoI fungicides in this pathogen [4], [5], [6]. For QoIs, the resistance mechanism was attributed to target site mutations in the *Cytb* gene of the target enzyme Cytochrome C reductase also known as the respiratory channel complex III [7]. Mainly, two amino acid substitutions namely F129L and G143A in the mitochondrial CytB protein were detected in field pathogens and these substitutions are responsible for the dramatic loss of efficacy observed for this whole class of fungicides. Currently, the great majority of the European *M. graminicola* population carries the G143A mutation, making its control highly reliant on C14-demethylase inhibitors (DMI) usage targeting the ergosterol biosynthesis pathway and on the multisite fungicide chlorothalonil (CTN). Gradual shifts in DMI sensitivity observed as an incremental reduction in sensitivity of pathogen population towards DMIs over time [8] further stress the importance of introducing novel modes of action for STB control. The availability of compounds with different modes of action is an essential component for effective anti-resistance strategies contributing to wheat yield security [9].

The introduction of novel carboxamide fungicides has delivered a major mode of action to almost all fungicide market segments including fruits, vegetables and cereals. These molecules display fungicidal activity by disrupting the mitochondrial tricarboxylic acid cycle (TCA) through inhibition of the succinate dehydrogenase (SDH) enzyme (also called succinate ubiquinone oxidoreductase (SQR), EC 1.3.5.1). The official term, as stated by the Fungicide Resistance Action Committee (www.FRAC.info) for this fungicidal class is "SDHIs" for succinate dehydrogenase inhibitors. At the molecular level, carboxamides inhibit ubiquinone reduction by binding to the ubiquinone binding site (Qp site) of the SDH enzyme [10]. The SDH enzyme is composed of four polypeptides which are nuclear encoded. SDHA and SDHB subunits assemble into the so called soluble catalytic dimer which faces the matrix whereas SDHC and SDHD subunits form the integral membrane component anchoring the heterotetrameric enzyme to the internal membrane of the mitochondria. Catalytic mechanisms by which electrons are transferred from succinate to ubiquinone involve: (i) oxidation of succinate at the level of SDHA which carries a covalent FAD (ii) transfer of electrons through the iron sulfur clusters [2Fe-2S], [4Fe-4S], and [3Fe-4S] carried by the SDHB subunit, (iii) two step reduction of the ubiquinone at the so called Qp site formed by the interface of SDHB SDHC and SDHD subunits. This later reaction involves transient formation of a semi quinone radical and the intervention of a heme which forms an integral part of the complex [11], [12],[13]. Crystal structures of the enzyme have been resolved for *Escherichia coli* [14], *Gallus gallus* (chicken) [15] and *Sus scrofa* (pig) [16].

Carboxin, was the first carboxamide to be developed for crop protection and was used as seed treatment displaying mainly a basidiomycete spectrum of control [17], [18]. Continuous research has led to the discovery of new chemical structures which modified and broadened this initial narrow biological spectrum and improved potency to the levels required from a modern fungal control agent. Newly discovered molecules include Penthiopyrad (Mitsui chemicals), Boscalid (BASF), Bixafen (Bayer), Fluopyram (Bayer), Sedaxane (Syngenta) and Isopyrazam (Syngenta), some of which display outstanding performance for STB control in the field. Even though SDHIs will be used in mixtures with a maximum of two applications per season in order to minimize the resistance development risk (FRAC), the almost simultaneous introduction of compounds displaying similar modes of action will impose a significant selection pressure on *M. graminicola* populations, in particular within European regions of high wheat production [19]. This further stresses the importance of understanding possible resistance mechanisms to better predict the emergence, spread and persistence of resistance to this class of fungicides in order to develop effective resistance monitoring and anti resistance strategies [20]. A number of target mutations have already been described both in the lab and in the field which can lead to carboxamide resistance. Artificial mutants with amino acid substitutions in the *SDH* genes encoding the

Qp site of SDH have previously been produced in various fungal species including *Ustilago maydis* (B_H257L) [21], [22], *M. graminicola* (SDHB_H267L/Y) [23], and more recently at various loci in *Aspergillus oryzae*(SDHB_H249Y/L/N, SDHC_T90I, SDHD_D124E) [24]. A spontaneous mutant conferring Flutolanil and Carboxin resistance was also reported and characterized in *Coprinus cinereus*(SDHC_N80K) [25]. In all these studies functional confirmation was obtained by expression of the mutated alleles in the WT background. In fact it has been suggested that these mutant genes may provide dominant selection markers that can be used in many commercially relevant fungal species [24], [26], [27]. Resistance towards Carboxin was claimed for barley field isolates of *Ustilago nuda* in France, Canada and Italy and the resistance mechanism although not elucidated was reported as monogenic [28], [29]. More recently, target site mutations which confer Boscalid resistance have been detected in various species in the field including *Botrytis cinerea* [30], [31], [32] and *Alternaria alternata* [33], [34]. Complex cross resistance patterns, including negative cross resistance were reported for Oxathiin Carboxamides in the late nineteen seventies [35]. Recent investigations performed with highly Boscalid resistant field isolates of *Corynespora cassiicola* and *Podosphaera xanthii* show that striking lack of cross resistance can be found across novel carboxamides [36]. A recent cross resistance study performed with a range of novel *M. graminicola* mutants which were selected on Carboxin only confirmed this is also true in *M. graminicola* [37]. This suggests that commercially introduced carboxamide SDHIs differ in their binding properties to the SDH enzyme.

The primary aim of this study was to understand possible target site resistance mechanisms to a range of newly introduced subclasses of carboxamides in *M. graminicola* by exploring the impact of target mutations on compound binding, enzyme efficiency and pathogen fitness. To this end *M. graminicola* was subjected to random UV mutagenesis and 5 structurally distinct carboxamides were used for selection. The characterization of more than 480 mutants enabled the identification of as many as 27 substitution types affecting in total 18 positions in the SDHB, SDHC and SDHD-Qp site forming proteins. Characterization of the mutants enabled the identification of substitution types that display selectivity to structurally distinct carboxamides. Docking studies, using a homology model of *M. graminicola* SDH, offer valuable insight to some of the experimental findings. Characterization of enzyme efficiencies showed that the ubiquinone reduction step was impaired in all mutants. Using transgenic strains expressing mutated copies of the enzyme SDH subunits we showed that very low levels of SDH activity is required for the establishment of resistance *in vivo*. Finally, homologous recombinant gene replacements for the most relevant substitutions types enabled preliminary fitness studies *in vitro* and *in planta* to be performed. Homologous recombinant strains developed in this haploid pathogen, correspond to the introduction of a single mutation in the whole genome enabling us to perform a very clean comparison of likely biochemical factors affecting fitness. Using these homologous recombinant strains we unexpectedly found that *M. graminicola* Qp site mutations did not significantly impact reactive oxygen species (ROS) production *in vivo*. However, *in planta* virulence was affected suggesting that these carboxamide selected Qp site mutations have an impact on the biology of this pathogen. Our study nicely complements recent results reported with *M. graminicola* Carboxin-selected mutants [37], whilst our modelling approach enables us to propose a more accurate model of the binding interaction which fits all our more extensive experimental findings for this class of inhibitors.

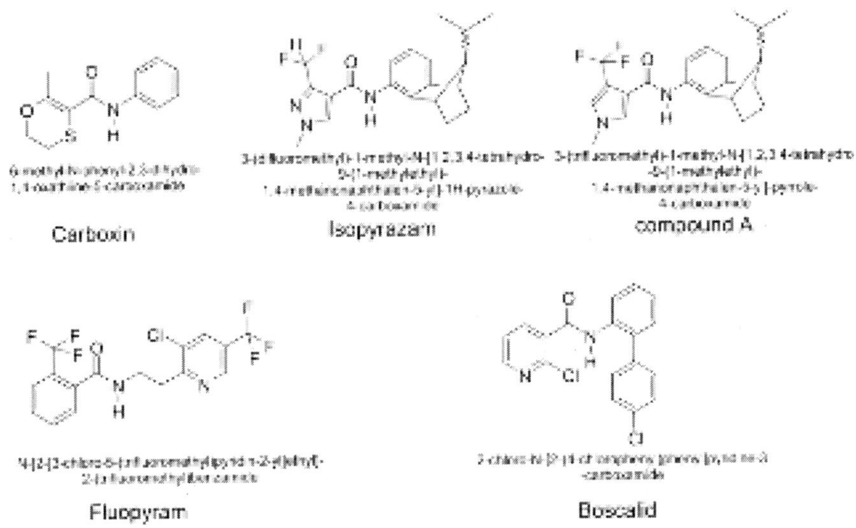

Figure 1. Chemical structure of carboxamides used in the study, trivial and IUPAC denominations. The generic structure of carboxamides can be divided into 3 main parts. (i) the "acid core" which differs from its ring structure: oxathiin (Carboxin), pyrazole (Isopyrazam), pyrrole (compound A), pyridine (Boscalid), substituted benzene (Fluopyram) (ii) the "linker" which is composed by a benzene ring or a 2-carbon aliphatic spacer in Fluopyram, (iii) the bulky hydrophobic rest, missing in Carboxin.The aim of the figure legend should be to describe the key messages of the figure, but the figure should also be discussed in the text. An enlarged version of the figure and its full legend will often be viewed in a separate window online, and it should be possible for a reader to understand the figure without switching back and forth between this window and the relevant parts of the text. Each legend should have a concise title of no more than 15 words. The legend itself should be succinct, while still explaining all symbols and abbreviations. Avoid lengthy descriptions of methods.

RESULTS

Generation of *M. graminicola* mutants resistant to carboxamides

Compounds used for selection following UV mutagenesis were chosen both for structural diversity and relevance to the crop protection market (Fig. 1).

Two different UV incidences were used for most selection compounds 75 mJ cm^{-2} or 50 mJ cm^{-2} leading to 80% and 50% lethality after UV treatment respectively. Frequency of resistance development ranged between 7.2 E^{-7} for Carboxin at the minimal selective concentration (MSC, see material and methods) to the absence of growing colonies for Fluopyram at 5×MSC (Table 1). The strength of UV treatment had a minor impact on resistance frequency suggesting that mutation rate, increasing with UV exposure and percentage of survival after UV treatment, decreasing with UV exposure, is finely balanced in this range of survival rates. For all compounds, except Boscalid and compound A, a pyrrole analogue of Isopyrazam, a decrease in the frequency of growing colonies was observed with 5×MSC selection which is consistent with higher concentrations of the active ingredient providing more stringent selection conditions. With Boscalid a higher frequency of resistance was observed with the 5×MSC selection. However, careful examination of the AE agar plates supplemented with 150 µM Boscalid (5×MSC) displayed white precipitate which developed over time. Taken together this suggests that the Boscalid was precipitating in the media and that the concentration of biologically available Boscalid at the time of selection was lower in the Boscalid 150 µM supplemented plates compared to the ones supplemented with Boscalid at 30 µM. No significant change in the frequency of resistant strain development could be observed between MSC and 2.5× MSC with the pyrrole carboxamide compound A.

Table 1. Mutagenesis experiments overview.

Selection	UV treatment	Plated cells (10^6)	Collected colonies	Frequency
Carboxin 85 µM (MSC)	50	90	56	6.2E-07
Carboxin 85 µM (MSC)	75	90	65	7.2E-07
Carboxin 425 µM (5× MSC)	50	90	1	1.1E-08
Carboxin 425 µM (5× MSC)	75	90	1	1.1E-08
Isopyrazam 3 µM (MSC)	50	90	29	3.2E-07
Isopyrazam 3 µM (MSC)	75	90	37	4.1E-07
Isopyrazam 15 µM (5× MSC)	50	90	13	1.4E-07
Isopyrazam 15 µM (5× MSC)	75	90	15	1.7E-07
compound A 100 µM (MSC)	50	90	13	1.4E-07
compound A 100 µM (MSC)	75	90	19	2.1E-07
compound A 250 µM (2.5× MSC)	50	90	19	2.1E-07
compound A 250 µM (2.5× MSC)	75	90	17	1.9E-07
Fluopyram 25 µM (MSC)	50	90	64	7.1E-07
Fluopyram 25 µM (MSC)	75	90	67	6.3E-07
Fluopyram 125 µM (5× MSC)	50	90	0	nd
Fluopyram 125 µM (5× MSC)	75	90	2	2.2E-08
Boscalid 30 µM (MSC)	50	50	64	1.3E-06
Boscalid 150 µM (5× MSC)	50	1.6	14	8.8E-06

Molecular characterization of carboxamide selected mutants

Following primary selection, 482 colonies were picked from the primary selection plates (Table 1) and re-isolated on the same selective media. Following this process, 30 strains unable to grow consistently were identified as false positives (Table 2). All other strains were cultured and sequence analysis of the *MgSDHB*, *MgSDHC* and *MgSDHD* genes was undertaken. Mutations leading to amino acid substitutions in the target proteins were detected in 444 out of 452 of the selected strains (98.2%). Target mutations were identified in all three Qp site encoding subunits and as many as 27 different substitution types were identified. Rare cases of double substitutions were observed in SDHB (Boscalid selection), SDHC (Isopyrazam selection) and SDHD (Carboxin selection). However, no mutants carrying substitutions in more than one subunit simultaneously were obtained. The absence of target mutation was only observed for 8 strains (1.8%). Additional controls performed on these strains including repeated isolation under selective conditions and re-sequencing of the four SDH encoding genes confirmed non target site resistance mechanisms can be selected under low compound selection pressure.

Table 2. Selected colonies and detailed quantitative and qualitative overview of the SDH substitutions selected with different carboxamides concentrations.

Compound		Carboxin		Isopyrazam		compound A		Fluopyram		Boscalid	
Concentration for selection		**MSC**	**5× MSC**	**MSC**	**5× MSC**	**MSC**	**2.5× MSC**	**MSC**	**5× MSC**	**MSC**	**5× MSC**
False positives		12	-	9	-	6	-	3	-	-	-
True positives		109	2	63	22	33	29	128	2	50	14
No SDH substitution		3	-	-	-	5	-	-	-	-	-
With SDH substitution		**106**	**2**	**63**	**22**	**28**	**29**	**128**	**2**	**50**	**14**
Codons	Substitution										
tct>ttt	SDHB_S218F	-	-	-	-	-	-	1	-	-	-
cca>aca	SDHB_P220T	-	-	-	-	-	-	1	-	-	-
cca>cta	SDHB_P220L	-	-	-	-	-	-	2	-	-	-
tcc>ccc	SDHB_S221P	2	-	-	-	-	-	-	-	-	-
aac>cac	SDHB_N225H	1	-	-	-	-	-	-	-	-	-
aac>atc	SDHB_N225I	-	-	-	-	-	-	1	-	-	-
cga>cca	SDHB_R265P	4	-	-	-	-	-	-	-	-	-
cac>cta/ctc/ctt	SDHB_H267L	7	2	16	3	2	-	7	-	6	-
cac>aac	SDHB_H267N	11	-	-	-	-	-	-	-	-	-
cac>cag	SDHB_H267Q	-	-	-	-	-	-	-	-	-	2
cac>tac	SDHB_H267Y	61	-	-	-	-	-	-	-	38	5
att>gtt	SDHB_I269V	11	-	-	-	-	-	-	-	-	-
aac>aag	SDHB_N271K	-	-	-	-	-	-	6	-	-	-
acc>atc	SDHC_T79I	-	-	-	-	-	-	-	-	-	4
tcg>ggg	SDHC_S83G	-	-	1	-	-	-	4	2	-	-
gcc>gtc/gtt	SDHC_A84V	-	-	-	-	-	-	23	-	-	-
gcc>atc	SDHC_A84I	-	-	-	-	-	-	1	-	-	-
ctc>ccc	SDHC_L85P	2	-	-	-	-	-	-	-	-	-
aac>aag/aaa	SDHC_N86K	-	-	8	3	4	9	6	-	-	1
cgc>tgc	SDHC_R87C	-	-	-	-	-	-	24	-	-	-
gac>gtc	SDHC_V88D	-	-	-	-	1	-	-	-	-	-
cac>cgt	SDHC_H145R	1	-	-	-	-	-	-	-	-	-
cat>cgt/cga	SDHC_H152R	-	-	37	16	21	20	-	-	-	1
gac>gag/gaa	SDHD_D129E	2	-	-	-	-	-	-	-	2	-
gac>ggc	SDHD_D129G	2	-	1	-	-	-	-	-	1	1
gac>tcc	SDHD_D129S	1	-	-	-	-	-	-	-	-	-
gac>acc	SDHD_D129T	-	-	-	-	-	-	1	-	-	-
	Multiple SDHB	-	-	-	-	-	-	1	-	-	-
	Multiple SDHC	-	-	-	-	-	-	-	-	3	-
	Multiple SDHD	1	-	-	-	-	-	-	-	-	-

Numbers correspond to the total number of colonies belonging to each sub-group (true/false positives and with/without mutation in the SDH encoding genes) for each selection condition (See material and methods). The second part details the number of colonies for the 27 SDH substitution type selected on each carboxamide concentration.

Selected substitution types appeared to be both compound and concentration dependent. Interestingly, clear decreases in resistance frequencies were observed for Fluopyram and Carboxin upon shifting selection from MSC to 5× MSC (Table 1). These were accompanied by a drastic reduction in the diversity of the different substitution types (Table 2). For Fluopyram the diversity of selected substitution types decreased from 13 at MSC to 1 at 5×MSC. The most frequent substitution at the MSC (SDHC_A84V, 57%) was not observed at 5×MSC. On the other hand SDHC_S83G substitution representing only 3.1% of the mutants at MSC was the only substitution type observed at 5× MSC. Similarly, Carboxin selected 13 substitutions at MSC but only the SDHB_H267L substitution could be found at 5×MSC. At the MSC, the most frequent substitution conferring Carboxin resistance was SDHB_H267Y accounting for 57.5% of the mutants, while SDHB_H267L represented only 6.6% of the mutants under these conditions. These results suggest that a diverse range of substitutions can lead to resistance to low levels of Fluopyram and Carboxin but at the more stringent 5× MSC selection substitution types leading to lower resistance factors are eliminated. Although less marked, diversity of substitution types was also reduced from 5 to 3 and from 5 to 2 with increased Isopyrazam and compound A selection respectively. When comparing the substitution types obtained across the different carboxamides, the substitution patterns were most similar for Isopyrazam and its pyrrole homologue. For these 2 molecules, the substitutions SDHB_H267L, SDHC_N86K and SDHC_H152R were the most frequently isolated. Given the similarity in the structures of these two carboxamides, similar substitution pattern might be expected, however the SDHB_H267L substitution was found much more frequently with Isopyrazam suggesting subtle differences in the molecular recognition of those two compounds by the enzyme. Seven of the isolated substitution types were found in multiple selection conditions, most notably the SDHB_H267L substitution which was selected with all compounds tested (Table 2). Interestingly, other substitutions were unique to the individual compounds used for selection, this was the case for many mutation types selected with Carboxin (8 out of 12) and with Fluopyram (9 out of 12). These different substitution patterns suggest subtle differences in the way the different carboxamide compounds interact with SDH.

Localization of amino acid substitutions in the primary structure of SDH proteins

Among the 3 subunits composing the Qp site, SDHB is by far the most conserved, displaying 57% identity across ascomycetes sequences used in the alignment (Fig. 2). By contrast SDHC and SDHD display rather poor primary sequence conservation with 22% and 21% identity respectively across the ascomycetes species used in the alignment.

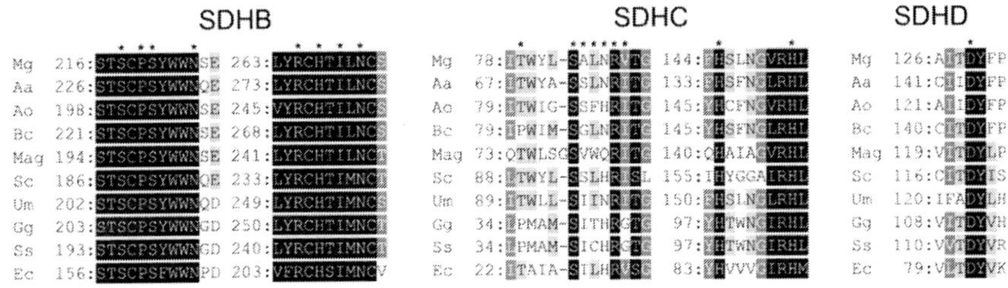

Figure 2. Localization of substituted amino acids in the SDH subunits of *M. graminicola* UV mutants and degree of conservation across species. Asterix indicate substituted residues, conserved residues are shaded in black, dark grey and light grey corresponding to 100%, 80% and 60% conservation respectively. Sequences are from *M. graminicola* (Mg), *Alternaria alternata* (Aa), *Alternaria oryzae*(Ao), *B. cinerea* (Bc), *Magnaporthe grisea* (Mag), *S. cerevisiae* (Sc), *Ustilago maydis*(Um), *G. gallus* (Gg), *S. scrofa* (Ss) and *E. coli* (Ec).

Our mutagenesis screen for carboxamide resistance led to the identification of 18 substituted amino acids, among which 13 were affecting highly conserved residues of the Qp site. A great proportion of these residues has been shown to be functionally involved in ubiquinone binding and reduction catalysis in other species [13], [38], [39], [40], [41].

Two regions prone to substitution could be identified on SDHB, one between amino acids S218 and P225 and the other between amino acids R265 and N271. All these SDHB amino acid positions are strictly conserved across species. On SDHC, one major region prone to amino acid substitution was identified between amino acid T79 and V88, with a continuous stretch of substituted positions between S83 and V88. This later contains the S83 and R87 highly conserved Qp site residues and four neighbouring positions which are not conserved across species (A84, L85, N86 and V88). The nature of the residues present at these positions may have an influence on Qp site structure and properties and thus, on the biological profile of carboxamides. Finally, the H145 and H152 histidines, which are also strictly conserved across species, were both found substituted with arginines. On SDHD, D129 was the only substituted position; together with its neighbouring Y130 these 2 positions represent the only conserved residues across the whole protein in the SDHD alignment.

Cellular and mitochondrial sensitivity of resistant mutants to carboxamides

To explore the sensitivity profiles conferred by amino acid substitutions at the cellular and mitochondrial level, we determined sensitivity towards 4 different compounds for 26 of the identified substitution types both *in vivo* using liquid

Table 3. *In vivo* and *in vitro* IC50s and resistance factors overview.

		IC50 (nM)							
		Carboxin		Isopyrazam		Fluopyram		Boscalid	
		in vivo	in vitro	in vivo	in vitro	in vivo	in vitro	in vivo	in vitro
WT (IPO323)		6470 ± 1085	1156 ± 102	55 ± 2	5 ± 0.3	584 ± 100	44 ± 2.2	455 ± 47	60 ± 1.8
		Resistance Factors (RFs)							
		Carboxin		Isopyrazam		Fluopyram		Boscalid	
Substitution	Strain	in vivo	in vitro	in vivo	in vitro	in vivo	in vitro	in vivo	in vitro
None	IPO323	1	1	1	1	1	1	1	1
B_S218F	Flu41	5.3	46.4	15.0	18.2	18.0	276.5	6.5	46.3
B_P220T	Flu110	0.1	0.7	3.8	1.8	11.5	84.9	6.1	28.3
B_P220L	Flu75	1.3	4.4	4.8	1.2	5.0	3.3	2.2	2.6
B_N225H	Cbx106	20.2	38.0	13.6	15.7	8.0	15.2	12.2	20.1
B_N225I	Flu7	10.2	6.9	15.0	7.0	17.5	25.0	12.9	18.5
B_R265P	Cbx56	20.1	271.1	7.6	88.3	2.8	63.0	13.3	256.0
B_H267L	Cbx29	14.5	1959.3	152.2	3248.5	22.4	766.1	258.0	9126.6
B_H267N	Cbx96	12.6	232.7	4.5	11.1	9.9	68.3	3.4	34.8
B_H267Q	Bos39	4.4	85.5	12.5	34.6	0.9	11.8	21.7	1034.3
B_H267Y	Cbx28	14.1	65.9	16.1	142.6	0.6	5.1	305.6	473.8
B_I269V	Cbx1	12.4	19.9	5.0	3.6	10.0	18.0	4.2	12.7
B_N271K	Flu33	2.6	1069.2	5.7	89.5	10.0	2481.0	3.4	223.7
C_T79N	Bos64	5.2	62.0	15.8	56.4	2.2	12.3	18.9	145.6
C_S83G	Flu1	10.7	11.7	61.2	38.6	126.8	201.0	564.3	1392.4
C_A84V	Flu4	0.8	1.2	3.5	3.1	19.0	102.7	2.6	3.7
C_A84I	Flu21	1.2	1.8	8.8	8.1	31.9	260.0	2.9	4.4
C_L85P	Cbx12	11.5	265.4	10.0	35.3	5.8	130.1	4.1	34.7
C_N86K	Izm7	23.8	135.9	464.4	849.9	49.9	245.8	78.1	160.2
C_R87C	Flu9	3.2	nd*	3.7	nd*	6.1	nd*	18.8	nd*
C_V88D	OI50	4.4	67.2	36.5	140.4	5.9	55.4	5.0	59.8
C_H145R	Cbx78	4.4	nd*	27.6	nd*	2.0	nd*	17.3	nd*
C_H152R	Izm1	13.2	45.4	133.9	95.0	11.5	14.3	30.2	58.1
D_D129E	Cbx22	35.3	631.6	6.8	145.3	1.3	10.5	42.8	1220.9
D_D129G	Cbx67	17.2	2030.3	22.9	583.2	5.6	1880.2	37.1	2052.4
D_D129S	Cbx101	39.7	377.7	10.8	82.5	1.9	12.3	15.9	311.2
D_D129T	Flu29	60.4	FR**	33.5	4095.5	11.5	3724.4	78.3	5515.8

Upper panel: *in vivo* and *in vitro* IC50 values obtained for the WT (IPO323) ± S.E. (triplicate). Lower panel: resistance factors (RFs) based on IC50 assessment for a selected subset of representative strains. Presented values are based on the ratio of the means of three individual determinations for the *in vivo* values and based on the ratio of a single determination for the *in vitro* values. Presented *in vitro* values were obtained by calibrating mitochondrial dilutions to obtain similar initial velocity (see material and methods). Cbx strains were originally obtained on Carboxin selection media, Flu on Fluopyram, Bos on Boscalid, Izm on Isopyrazam, OI on pyrrole compound A. nd*: IC_{50} could not be determined because fitted curves were not tending to 100% inhibition at infinite AI concentration. FR**: (full resistance), no sufficient inhibition detected at highest inhibitor concentration tested.

culture growth (AE media) and *in vitro*using purified mitochondria and the succinate: Q_0/DCPIP activity test (Table 3). To ensure that only target site resistance was studied, a series of comparative tests with strains carrying similar target mutations was performed and only strains that displayed consistent resistance profiles were selected for further study. One representative strain was selected for all substitution types apart from SDHB_S221P which seemed to result in a significant fitness penalty as the two strains carrying this substitution displayed very significant growth defects. All other strains displayed similar levels of growth in liquid AE media. Resistance factors (RF) were determined as the ratio IC_{50}mutant/IC_{50}WT. The values of *in vivo* resistance factors for strains selected on each compound ranged between the following values: 4.4 (SDHC_H145R) to 39.7 (SDHD_D129S) for Carboxin selection, 22.9 (SDHD_D129G) to 464.4 (SDHC_N86K) for Isopyrazam selection, 5 (SDHB_P220L) to 126.8 (SDHC_S83G) for Fluopyram and 21.7 (SDHB_H267Q) to 564.3 (SDHC_S83G) for Boscalid. Because of the very poor succinate: Q_0/DCPIP activity displayed by some mutants *in vitro*, we calibrated our test by adjusting enzyme concentrations at a similar initial velocity (see material and methods). As IC50s are dependent on amount of enzyme used in the test, the ranking of RFs obtained *in vivo*compared to the *in vitro* values can differ notably. For example the SDHD_D129T substitution which displayed an *in* *vitro* RF of 4095 towards Isopyrazam only ranks as the fifth strongest mutation *in vivo* (RF: 33). Similar discrepancies between the *in vivo* and *in vitro* rankings can be observed for all of the carboxamides tested. However, these data enable a comparison of the differential behaviours displayed towards the 4 carboxamides tested by a given mutated enzyme. Qp site mutations are well known to influence SDH enzyme activity and stability upon catalysis, [13], [38], [41]. Discrepancies observed between the *in vivo* and *in vitro* sensitivity data suggested that the impact of substitutions on enzymatic activity needed to be addressed to have a fuller understanding of the factors influencing the resistance displayed *in vivo*.

Carboxamide pharmacophore elucidation using available structural information

In order to obtain a clearer understanding of the structural basis for the differential resistance profiles observed across carboxamides in the SDH mutants, a prediction of the binding modes for the carboxamides used in this study was required. In order to generate putative binding modes, all publicly available crystal structures were analysed in a first step to identify key pharmacophoric features (see material and methods). All crystal structure ligands are equipped with a hydroxyl or carbonyl group making hydrogen bonds to B_Trp164 and D_Tyr83 (amino acid numbering according to *E. coli* SQR). In the crystal structure 1NEN the quinone ring is sandwiched by B_Pro160 and C_Ile28 [14]. The overlay of all structures onto each other shows that some ligands are not superimposing exactly on the quinone ring but all ligands display a scaffold that lie in the same plane (Fig. 3A). This was partly described already by Horsefield *et al.* who divided the binding pocket into a Q1 and Q2 sites [10]. The ligands of the structures with PDB codes 1NEK, 1NEN, and 2ACZ are equipped with lipophilic alkyl groups making hydrophobic interactions at the entrance of the quinone pocket. The alkyl groups are oriented perpendicular relative to the quinone plane (Fig. 3A). In carboxin SQR co-crystal structures, the orientation of the methyl group of the oxathiin ring was a matter of controversy. Initially it was proposed that the methyl group is oriented towards C_Ser27 (2FBW) [15] but more recent crystal structure elucidations with both chicken (2WQY) and *E. coli* SQR (2WDQ) enzymes demonstrated a 180° flipped oxathiin ring [40]. Moreover, recent structures elucidated with SQR-ubiquinone binding site inhibitors further confirm this trend. The structure 3AE4 shows that the Br atom, displaying a well defined electron density, has the same orientation as the methyl in the oxathiin ring of carboxin. In other structures the CF3, or methyl group were consistently built on the same side (e.g. PDB code 3ABV, and 3AE5). Mutagenesis and kinetic data have demonstrated the importance of the conserved SDHC Qp site serine (SDHC_S83 in*M.graminicola*, C_Ser27 in *E.coli*) in ubiquinone binding and catalysis [13]. In addition, there are two crystal structures with quinone

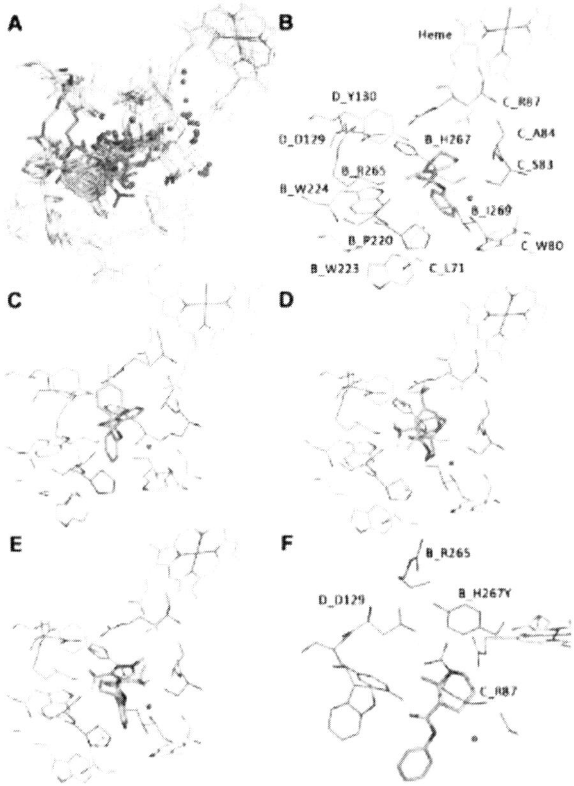

Figure 3. Tridimensional model of the *M. graminicola* SDH with Qp site docked carboxamides and interactions to substituted residues. A: Superposition of all complex II crystal structures with a resolution higher than 3 Å. Only amino acids that are in a 5 Å radius to the bound ligands in the quinone binding site are shown. Amino acids and the heme are represented in lines, the ligands in sticks and water molecules as non bonded spheres. The color code is presented according to atom types. Amino acid and heme carbons are colored in green, ligands carbons in salmon. B: putative binding mode of carboxin in a tridimensional model of *M.graminicola*SDH. The heme carbons are represented in cyan sticks, the carboxin carbon atoms in salmon. Amino acids that are involved in resistance after mutation are colored in dark blue, amino acids that make key interactions are shown in green and amino acids that are in close proximity to the ligand but which were not found substituted in this study are colored in grey. Hydrogen bonds are shown as yellow dotted lines. C: Putative binding mode of Boscalid in *M.graminicola* SDH. D: Putative binding mode of Fluopyram in *M. graminicola* SDH. E: Putative binding mode of Isopyrazam in *M. graminicola* SDH. F: Model of *M. graminicola* SDH where SDHB histidine 267 is mutated into a tyrosine. The putative binding mode of the mutated enzyme with Carboxin is shown.

scaffolds where a hydrogen bond between the ligands and C_Ser27 could be observed (1YQ3, 1YQ4) [42]. This residue is proposed to be a key residue interacting with ubiquinone during its proposed movement within the Qp site during catalysis [10]. Based on the high structural similarity to the carboxamides described in this study, the interaction observed in the chicken Carboxin-SQR co-crystal structure 2WQY gives a valuable indication of the binding mode of carboxamides. In this crystal structure a hydrogen bond mediated by a water molecule was observed between the ligand amide NH and the C_Ser27 hydroxyl [15].

Inhibitor docking studies in a *M. graminicola* SDH homology model

Utilizing observed protein ligand interactions, putative binding modes were generated for Boscalid, Isopyrazam, and Fluopyram (Fig. 3C–E). All three fungicides share similar key interactions displaying hydrogen bonds to SDHB_W224 and SDHD_Y130 through the amide oxygen and the hydrogen bond of the amide NH to SDHC_S83 mediated by a water molecule. The involvement of the SDHC_S83 in the establishment of a hydrogen bond upon carboxamide binding is supported experimentally by the high resistance factors (up to 564 fold *in vivo*) observed across all ligands with the SDHC_S83G mutant. The acid part of all carboxamides (oxathiin, pyrazole, phenyl and pyridine) is sandwiched between C_Arg87 and B_Ile269 and form Van-der-Waals interactions to SDHB_I269, and cation-π interaction to SDHC_R87. The SDHC_R87C mutation abolishes the cation-π interaction; *in vivo*, this mutation conferred only limited carboxamide sensitivity decrease (RFs between 3.2 and 18.8). *In vitro*, the effect of the SDHC_R87C substitution on enzyme inhibition looked more complex as only partial inhibition could be reached at maximal concentration of the inhibitor despite of an otherwise WT like response. The SDHB_I269V mutation also conferred limited resistance factors (ranging between 4 to 20 fold *in vitro*). A valine can still compensate for some of the lipophilic interactions carried out by SDHB_I269. The ortho substituents of the acid parts (methyl in Carboxin, CF2 in Isopyrazam, CF3 in Fluopyram, and Cl in Boscalid) are pointing towards SDHB_S221 which corresponds to the orientation suggested by Ruprecht *et al* [40].

SDHB_R265 and SDHD_D129 are not involved in direct interactions to the ligands according to the binding models but form a hydrogen bond network with the heme and the imidazole ring of SDHB_H267. SDHB_H267 is positioned at the bottom of the Qp site. Depending on the organism this residue either makes a direct hydrogen bond to SDHD_D129 (observed for all *G. gallus* and *S. scrofa* SQR crystal structures) or to one heme propionate (observed for all *E. coli*SQR structures). By analyzing the electron density for the histidine sidechain it was suggested that this amino acid can adopt multiple conformations [10]. In our proposed binding modes for*M. graminicola* SDH, SDHB_H267 makes a direct hydrogen bond to SDHD_D129 and edge to face stacking to the four different inhibitors. However, an alternative binding mode can be hypothesized, where a direct hydrogen bond of the histidine to the pyridine of Boscalid, to the pyrazole of Isopyrazam, and to the oxathiin of Carboxin occurs. In this case SDHB_H267 cannot make direct hydrogen bonds to SDHD_D129 or one heme propionate anymore. This alternative is supported by the recently solved crystal structure of the *E. coli* SQR enzyme containing the histidine to threonine substitution (PDB code 2WP9). In this study, the complex II was shown to remain stable and active in a case where neither a hydrogen bond to the heme propionate nor to SDHD_D129 are formed [43]. A mutation of B_H267Y leads to high *in vitro*resistance factors for Boscalid (RF=473.8), Isopyrazam (RF=142.6), and Carboxin (RF=65.9). In order to understand this in more detail a model for *M. graminicola* SQR with the SDHB_H267Y substitution was generated (Fig. 3F). The aromatic sidechain makes edge to face stacking to the inhibitors and the tyrosine hydroxyl group a hydrogen bond to SDHD_D129. In this configuration, a direct hydrogen bond of the tyrosine to the hydrogen bond accepting groups of Boscalid, Isopyrazam, and Carboxin is unlikely potentially impairing the binding of these molecules and thereby explaining the higher resistance factors observed for these molecules. Moreover, this model would predict the low resistance towards Fluopyram which is not equipped with an acceptor group and Carboxin with the second lowest resistance factor (*in vivo* RF=14.1) is equipped with the weakest accepting group followed by Isopyrazam and Boscalid.

SDHB_P220 is a fully conserved Qp site residue sandwiching the quinone ring together with a residue corresponding to SDHC_A84 in all ubiquinone-SDH co-crystal structures [14], [15],[16]. In the binding models, the proline displays hydrophobic interactions with the amine part of the carboxamides (aliphatic linker in Fluopyram; phenyl group in Carboxin, Boscalid, and Isopyrazam). *In vitro*, resistance factors are in a single digit range except for Fluopyram (RF=85) and Boscalid (RF=28) with the SDHB_P220T mutant. SDHC_A84 is predicted to interact via Van-der-Waals forces with the aliphatic linker of Fluopyram and the pyridine ring. This aliphatic linker is a specific feature of Fluopyram compared to the other carboxamides analysed in this study. The linker is in the z-dimension sterically more demanding and is getting closer to SDHC_A84 compared to phenyl ring systems of Carboxin, Boscalid, and Isopyrazam. In agreement with these predictions, Fluopyram binding is found to be specifically affected by the SDHC_A84V/I substitutions as shown by over 30 fold higher RFs *in vitro*compared to the phenyl ring carboxamides. Furthermore, higher resistance factors are observed with the bulkier amino acids substitution, SDHC_A84I which is in line with the proposed interaction. Interestingly, the amino acid at SDHC_A84 position is not conserved across fungal species and may contribute to some of the biological profile observed for these carboxamide subclasses.

The SDHB_W223, SDHC_L71, and SDHC_W80 residues, are all in Van-der-Waals distance to the amine part of all ligands. However, no substitutions at these positions were observed in this study. The substituted residues, SDHB_S218, SDHB_N225, SDHB_N271; SDHC_T79, SDHC_L85, SDHC_N86, SDHC_V88, SDHC_H145, and SDHC_H152 are not predicted to directly interact with carboxamides but amino acid substitutions more distant from the binding pocket may influence the shape and electronics of the Qp site and therefore have a negative impact on ligand binding. A putative long-range effect and/or a rearrangement of a large zone affecting part of the Qp site is particularly likely for the SDHC_H145 position which was substituted by an arginine upon Carboxin selection as this is the furthest substituted residue from the Qp site. Despite a distance of ~10 Å to the closest atom of docked Carboxin, the SDHC_H145 is a ligand of the heme and therefore may influence the Qp site structure and/or properties. According to available SDH crystal structures, the propionate side chains of the heme can be considered part of the Qp site. Furthermore, it was shown that the heme itself directly participates in electron transfer during catalysis [11]. In yeast and in *E. coli*, modification of the coordinated histidine impairs ubiquinone binding and catalysis efficiency of the SDH enzyme [39], [42], [44]. Therefore, the SDHC_H145R substitution in *M. graminicola*may influence heme properties and/or positioning which in turn might have a negative impact on inhibitor binding.

SDHC_N86K and SDHC_H152R were the most frequent substitutions upon Isopyrazam and compound A selection. Both were also selected by Boscalid but less frequently and the SDHC_N86K substitution was also found upon Fluopyram selection. In the *M. graminicola*tridimensional SDH model, these residues are not predicted to directly interact with the inhibitor yet are in the close vicinity to the heme, again making a direct influence on heme positioning possible.

Enzyme efficiency determination

To compare specific activities conferred by the different amino acid substitutions we first determined the level of SDH enzyme in mitochondrial preparations for each of the resistant strains (Table 4). SDH enzyme carries a covalently bound FAD at the succinate oxidation site located within the SDHA subunit [45], this feature enables the accurate quantification of the SDH enzyme even in complex samples [46]. Overall, covalent FAD values differed at most by a factor 4 between samples, suggesting that the strains carrying less active SDHs are not compensating for this effect by overexpression of the enzyme. This result was further validated by western blot using anti *M. graminicola* SDHB antibodies (data not shown).

Table 4. Enzymatic efficiencies determinations for selected subset of strains representative of each mutation type.

Mutation	Strain	Covalent FAD[a]	MTT/PMS activity[b]			Q_o/DCPIP activity[c]		
none	IPO323	330.3 ± 23.6	474	±62	(100%)	484.4154407	±42	(100%)
B_S218F	Flu41	344.4 ± 14	416	±67	(88%)	92.87292759	±18	(19%)
B_P220T	Flu110	378.0 ± 4.7	720	±110	(152%)	107.7682646	±9	(22%)
B_P220L	Flu75	304.5 ± 5.8	717	±84	(151%)	231.8025324	±16	(48%)
B_N225H	Cbx106	299.5 ± 3.8	850	±175	(179%)	334.8801055	±33	(69%)
B_N225I	Flu7	281.0 ± 14.2	651	±98	(137%)	188.7997691	±19	(39%)
B_R265P	Cbx56	363.1 ± 19.7	217	±53	(46%)	41.7782216	±10	(9%)
B_H267L	Cbx29	519.5 ± 10.4	381	±37	(80%)	64.57824113	±7	(13%)
B_H267N	Cbx96	246.7 ± 4.3	553	±77	(117%)	94.85037408	±15	(20%)
B_H267Q	Bos59	453.8 ± 8.5	334	±40	(70%)	65.25108944	±6	(13%)
B_H267Y	Cbx28	438.4 ± 25.5	190	±44	(40%)	45.65897328	±18	(9%)
B_I269V	Cbx1	387.2 ± 21.0	545	±147	(115%)	416.7898614	±73	(86%)
B_N271K	Flu33	286.9 ± 7.8	218	±26	(46%)	25.98331188	±5	(5%)
C_T79I	Bos64	241.9 ± 3.0	501	±70	(106%)	124.887427	±16	(26%)
C_S83G	Flu1	368.6 ± 12.4	714	±97	(151%)	286.9379225	±34	(59%)
C_A84V	Flu4	229.6 ± 9.1	850	±119	(179%)	419.0858506	±59	(87%)
C_A84I	Flu21	631.5 ± 15.4	596	±95	(126%)	245.6437942	±32	(51%)
C_L85P	Cbx12	320.8 ± 24.9	334	±56	(70%)	143.5533934	±28	(30%)
C_N86K	Izm7	284.7 ± 6.6	407	±86	(86%)	200.6388817	±35	(41%)
C_R87C	Flu9	333.5 ± 9.3	389	±46	(82%)	117.638248	±12	(24%)
C_V88D	OI50	200.5 ± 7.8	838	±170	(177%)	139.7192089	±21	(29%)
C_H145R	Cbx76	377.8 ± 27.2	555	±98	(117%)	97.84219544	±12	(20%)
C_H152R	Izm1	156.7 ± 8.8	227	±34	(48%)	106.7682223	±22	(22%)
D_D129E	Cbx22	211.4 ± 7.6	491	±123	(104%)	202.5551707	±30	(42%)
D_D129G	Cbx67	180.9 ± 3.8	808	±80	(171%)	91.9077442	±9	(19%)
D_D129S	Cbx101	453.8 ± 8.5	534	±84	(113%)	111.2427749	±17	(23%)
D_D129T	Flu29	253.9 ± 10.5	568	±69	(120%)	29.33687605	±6	(6%)

[a]Covalent FAD is expressed in pmol covalend FAD mg of protein^{-1} and is an estimation of the total amount of succinate dehydrogenase enzyme present in the mitochondrial preparation, presented values are the average of four individual determinations ± S.D.
[b]MTT/PMS activity expressed as µmol PMS reduced MTT min^{-1} µmol covalent FAD^{-1}. Values represent the means of at least 10 determinations ± S.D.
[c]Qo mediated DCPIP reduction expressed in µmol reduced DCPIP min^{-1} µmol covalent FAD^{-1} in the presence of 1 mM Qo. Values represent the means of at least 10 determinations ± S.D.

The malonate sensitive succinate: PMS/MTT activity test is classically considered as a measurement of the SDHA-B dimer [47]. This activity does not require the functional reduction of the ubiquinone at the Qp site and was accordingly not affected upon carboxamide addition in*M. graminicola* [48]. Activity levels varied greatly among mutants, ranging between 46% and 179% of the WT enzyme activity. A wide range of effects could be observed even when substitions affected similar position as observed with the SDHB_H267Y and SDHB_H267N which displayed succinate: PMS/MTT activities of 40% and 117% of the WT activity respectively. Interestingly, mutations on SDHC and SDHD also have a major impact on this activity. As the PMS electron donor site has not been identified yet, we suspect that electron distribution within the enzyme might be affected in our SDH mutants, which could in turn favour or disfavour reduction of this substrate at its reduction site(s) [43], [47], [49]. Similar variations have also been reported for other Qp site mutations in other studies [41], [50].

In vivo, the electrons derived from succinate have to be transferred to its acceptor (ubiquinone) to enable the enzymatic oxidizing of novel molecules of succinate. The succinate: Qo/DCPIP activity is a measure of the succinate: ubiquinone reductase activity, which is the most relevant one biologically. *In vitro*, full inhibition of the WT enzyme can be reached using all four carboxamides compared in this test.

All mutants displayed weaker ubiquinone reductase activity compared to the wild type. The weakest effect was detected for the SDHC_A84V mutant which was 87% as active as the WT. The strongest impairment was displayed by the SDHB_N271K mutant with only 5% residual activity. As might be expected, different substitutions at the same residue can result in differential impact on enzyme efficiency. This effect seems to be linked to the degree of steric or physico chemical conservation displayed by the substitutive amino acid. For example, the SDHD_D129E conservative substitution maintains 42% of WT activity whiles the non conservative SDHD_D129G/S/T substitutions impact enzyme activity much more strongly (from 6 to 23% remaining activity). The same observation can be made for the SDHC_A84V variant which is more active (87%) than the SDHC_A84I (51%) counterpart which carries a larger substituent.

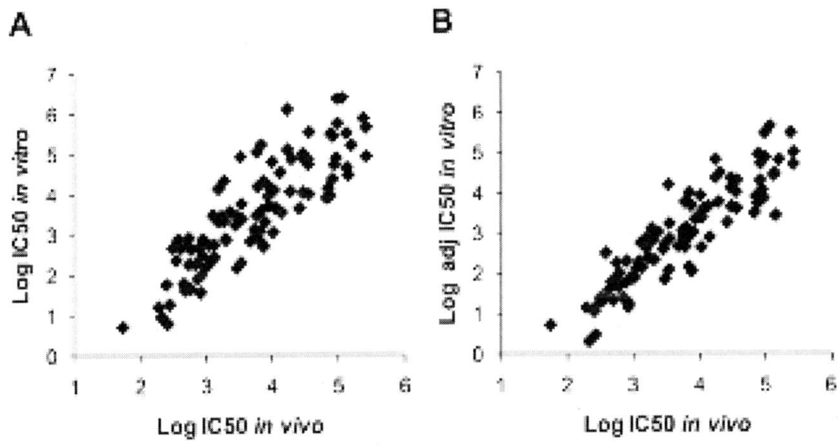

Figure 4. Scatter plot presenting *in vivo* LogIC$_{50}$ (nM) versus *in vitro* LogIC$_{50}$ (nM) adjusted (B) or not (A) for enzyme efficiency. All data extracted from table 3 and adjusted for amount of enzyme used in the sensitivity test in panel B. Adjustment was performed using enzyme efficiency data extracted from table 4 following a simple equation: if percent efficiency is denoted by E then the efficiency adjusted IC$_{50}$=observed IC$_{50}$×E/100. Following this adjustment the correlation was improved for all compounds.

The straight comparison of the *in vivo* log IC_{50} estimates and *in vitro* log IC_{50} estimates across the different strains for any given compound displayed reasonable correlation for each of the four compounds considered here (Fig. 4A). We attempted to correct IC_{50} values using enzyme efficiency as a correction factor for total amount of enzyme used in the tests. Interestingly, using this simplified adjustment (see Figure 4 legend); the correlations betweenin *in vitro* and *in vivo* log IC_{50} were improved for all compounds (Fig. 4B). However depending on the Ki and Km values displayed by each mutant, the relationship between IC_{50} and amount of enzyme can be complex and a determination of these values through classical enzymology would be required to further improve the correlation factor applied here [51]. In addition to enzyme efficiency, stability upon catalysis [42], [50] and modification of preferred substrate[43] are factors which might also have an influence on the establishment of resistance *in vivo*.

In vivo and *in vitro* sensitivity comparison between ectopic transformants and homologous recombinants carrying resistant alleles

To explore the influence of SDH mutations *in vivo* in more detail, we introduced ectopic expression cassettes in the *M. graminicola* IPO323 WT strain creating a strain which is artificially heterozygous for a particular *SDH* gene (Materials and Methods, Fig. 5B). Corresponding homologous recombinant (HR) strains where the wild type SDH gene was replaced with the mutant form were also generated for comparison. These HR events were fully characterized molecularly, confirming the presence of a unique mutation in the whole genome differentiating the WT strain (IPO323) from its corresponding homologous recombinants (cfMaterials and Methods). The ectopic transformant Tr_SDHB_H267L expresses a mixture of both sensitive (SDHB_WT) and resistant (SDHB_H267L) SDHB containing enzymes. As a result, the mitochondrial Boscalid inhibition curve is biphasic reflecting the two populations of enzymes present (Fig. 5A). The WT enzyme is inhibited at low doses but mitochondrial activity is maintained at higher doses indicating the contribution of mutated enzyme to the mitochondrial activity in this transformant. Interestingly, in the ectopic transformant expressing SDHB_H267L (Tr B_H267L), the total SDH activity which can be attributed to the mutated enzyme (see Fig. 5A, remaining activity at WT inhibiting concentrations) is much lower than the one attributable to the WT SDH. Despite its lower contribution to the total SDH activity in the mitochondria, clear dominance of the mutated allele can be seen *in vivo* using liquid culture tests (Fig. 5C). However, resistance levels reached by TrH267L strains remained significantly lower (unpaired *t* test, $p<0.01$) compared to the one reached by the corresponding homologous recombinant HR_H267L strains. This effect was not restricted to SDHB, as similar results were obtained with the SDHC_A84V mutation towards Fluopyram (Fig. 5D, also p<0.01). These results may indicate competition between the WT and mutated subunit for assembly into the functional enzyme in the mitochondria of the ectopic transformants. This assumption is further supported by the observed carboxamides oversensitivity displayed by ectopic transformants expressing non functional subunits (data not shown, [48]). This would explain the weaker phenotype seen with the ectopics carrying less mutated enzyme compared to the pure mutant.

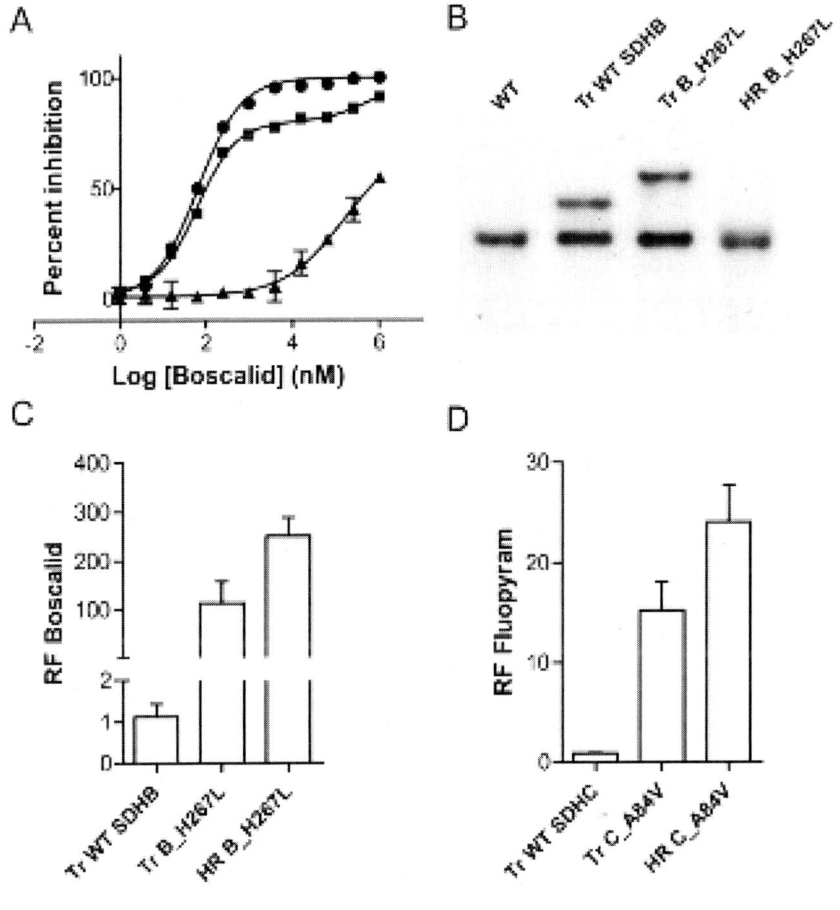

Taken together, our results confirmed that total SDH activity level remaining upon carboxamide inhibition is the driver for cell survival. Despite the major loss in SDH enzyme efficiency observed in many of the mutants compared to the WT (Table 4), here we show that very low levels of activity are sufficient to confer survival upon carboxamide treatment (compare Fig. 5A and Fig. 5C). This suggests that *in vivo*, carboxamide inhibitors have to almost completely block SDH enzymatic activity in order to deliver their fungicidal effect.

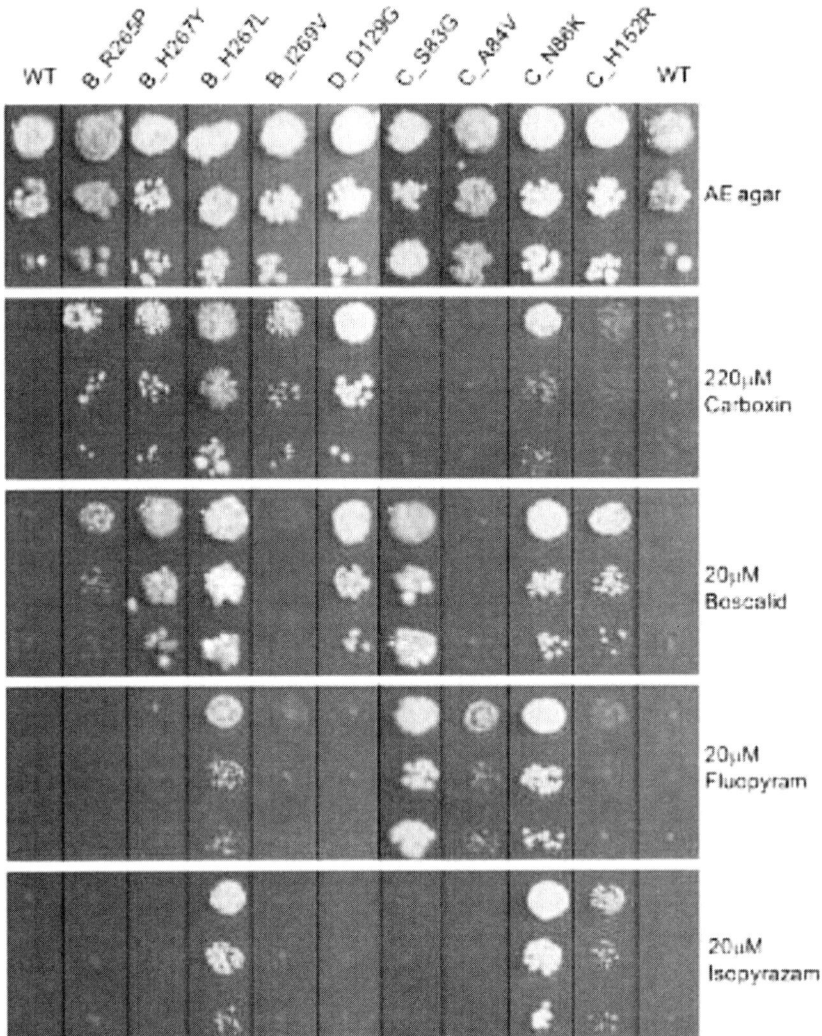

Figure 6. Comparative growth of homologous recombinant strains on various carboxamide amended media. All strains were grown on AE agar supplemented (or not) with the indicated active ingredient (final concentration indicated). Cell suspensions obtained from 4 days old V8 agar plates were serially diluted by 10 fold factor steps in liquid AE and 2 ul were spotted onto the plates. Pictures were taken 10 days post inoculation for carboxamides amended plates and 6 days post inoculation for AE plate. Pictures were taken from same plates and reordered for visualization.

Generation of homologous recombinants for most relevant mutation types

To fully explore a potential fitness cost conferred by SDH mutations we generated homologous recombinant strains for some of the strongest and most frequent substitutions types. The following substitutions were selected for further analysis: SDHB_R265P and SDHB_I269V uniquely obtained upon Carboxin selection at MSC, SDHB_H267Y most frequent substitution occurring mutant following selection at the MSC for Carboxin and Boscalid, SDHB_H267L leading to high resistance levels towards all classes *in vitro*, SDHC_S83G leading to the strongest Fluopyram and Boscalid resistance *in vivo*, SDHC_A84V being the most frequent mutation occurring upon Fluopyram selection at MSC and conferring specific resistance to this compound and SDHC_N86K and SDHC_H152R, strongest and most frequent mutations found with Isopyrazam selection. As shown in figure 6, the 9 HR strains displayed similar growth on non selective AE media even though some pigmentation difference can be noticed. Similar growth behaviour was maintained on a minimal media (data not shown). The strength of growth obtained on SDHIs amended media correlated well with the *in vivo* resistance factors obtained using their corresponding UV mutants in liquid culture test (Table 3) validating these target mutations fully conferred observed cross resistance profiles.

SDH mutants are not hypersensitive towards oxidative stresses

During catalysis, the Qp site of SDH is responsible for the 2 step electron transfer from the [3Fe-4S] iron sulphur cluster of SDHB to the ubiquinone substrate. The reaction involves the formation of a transient ubisemiquinone radical which is stabilized by electron transfer to the heme [13]. SDH Qp site mutants of *Saccharomyces cerevisae* and *E. coli* have been reported to display increased superoxide radical formation [38], [41], [52]. SDH Qp site mutations can also affect lifespan of organisms by causing a continuous oxidative stress which has been associated to premature aging and tumor formation [52], [53], [54], [55]. Mutations of the Qp site residues corresponding to SDHB_P220, SDHC_A84 and SDHD_D129 in *M. graminicola*have been shown to confer increased ROS production and oxidative stress hypersensitivity in other organisms including *E. coli* [38], *Caenorhabditis elegans* [52], [53], and *S. cerevisae* [41]. To assess whether carboxamide-selected Qp site mutations isolated in this study display a similar impact on oxidative stress sensitivity in *M. graminicola* we compared sensitivity of a subset of the generated homologous recombinant strains to various oxidative stresses. Two superoxide inducing conditions were used, treatment with 10 mM Paraquat (methyl viologen), a herbicide reported to induce oxidative stress through the generation of superoxide anions and a continuous 100% oxygen atmosphere simulating hyperoxia. Paraquat supplemented media had a similar impact on

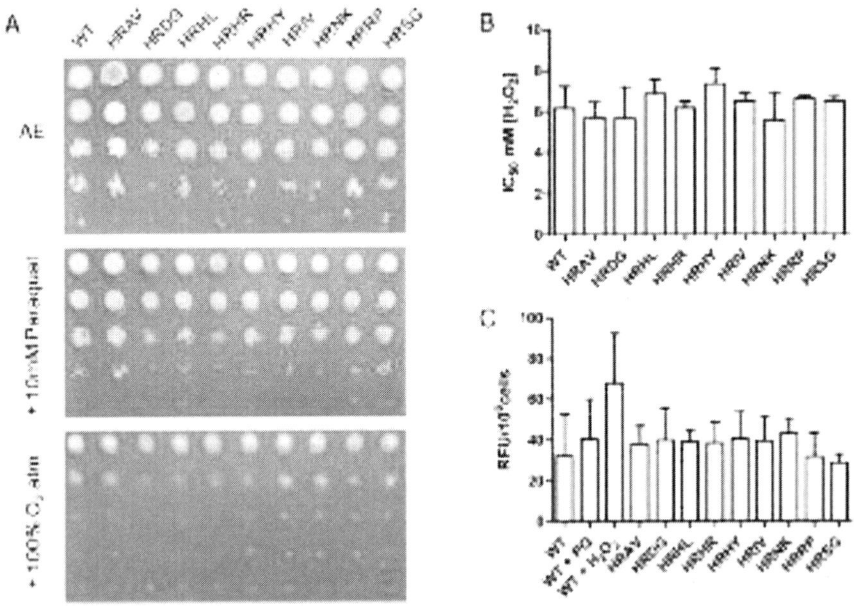

Figure 7. Absence of a major oxidative stress related fitness penalty in *M. graminicola* homologous recombinant strains carrying Qp site mutations. A: Respiratory growth of wild type and mutants in the presence of oxidative stresses. Upper panel, AE agar plate growth, middle panel: AE agar supplemented with 10 mM Paraquat, lower panel: AE agar plates were placed in a constant 100% atmosphere. All plates were incubated in same room at room temperature for 6 days. B: Sensitivity towards hydrogen peroxide in liquid AE media. Values presented are the averages from replicates within 3 individual experiments. C: Evaluation of mitochondrial ROS produced*in vivo* as determined with MitoSOX™ Red fluorescent indicator. Evaluations based on values from over 3 replicates within 3 individual experiments. +PQ corresponds to 1 mM Paraquat supplementation during the incubation period, +H_2O_2 correspond to 10 mM H_2O_2 supplementation during the incubation period. WT (IPO323), and homologous recombinant strains HRAV: SDHC_A84V, HRDG: SDHD_D129G, HRHL: SDHB_H267L, HRHR: SDHC_H152R, HRHY: SDHB_H267Y, HRIV: SDHB_I269V, HRNK: SDHC_N86K, HRRP: SDHB_R265P, HRSG: SDHC_S83G.

growth of both the homologous recombinant strains and the WT (Fig. 7A, middle panel). UV strains representative of all mutations types (excluding SDHB_P221 mutants, strongly impaired in growth) were tested using same conditions but did not display any clear difference in growth (data not shown). Similarly, hyperoxia simulated by a continuous 100% oxygen atmosphere showed a very minor impairment in growth for the SDHB_R265P and SDHD_D129G mutants (Fig. 7A, lower panel). This suggests that all carboxamide selected SDH mutations have no or a minor influence on ROS production in *M. graminicola in vivo*. When hydrogen peroxide was used as a further oxidative stress inducer and the data subjected to analysis of variance, again no significant shift in the sensitivity towards this oxidizing agent could be detected across WT and HR strains ($p>0.1$) (Fig. 7B). This result is further supporting these carboxamide-selected *M. graminicola* SDH mutations have a minor or no oxidative stress related fitness penalty *in vivo*. To evaluate ROS production in the mitochondria of SDH Qp mutants *in vivo*, we used the intracellular ROS indicator MitoSOX™ Red (Molecular Probes). As could be expected from the lack of hypersensitivity to oxidative stresses in previous *in vivo* tests, comparison of our subset of homologous recombinant strains showed no clear evidence for a difference across the WT and the target mutants ($p>0.1$) (Fig. 7C). However, in all conditions tested, fluorescence intensity remained very low, even hydrogen peroxide and Paraquat driven changes in fluorescence signal were not significantly higher than WT. Poor signals were also obtained with the cytosolic ROS marker dihydroxyethidium bromide (DHE, Molecular probes) (data not shown). These results may be caused by a poor uptake of these small molecules by the fungal cells or highlight a very good defence against oxidative agents in this pathogen.

Impact of SDH mutations on necrosis symptom development *in planta*

In order to evaluate the impact of SDH mutations on *in planta* growth, a smaller subset of the recombinant strains was compared for necrosis symptoms displayed *in planta*. For comparing the diverse SDH mutants we used various homologous recombinant strains carrying SDHB_H267L, SDHC_H152R, SDHC_N86K and SDHC_A84V mutations. Homologous recombinant strains were similarly inoculated on the sensitive WT plant Riband and *in planta*symptoms assessed visually 14 days after inoculation (Fig. 8). The data were subjected to analysis of variance which detected clear evidence of differences amongst strains ($p<0.01$). The statistical significance of differences between the WT and each of the HR strains was determined using an LSD test which showed that the HR strains displayed significantly higher symptomology compared to the WT ($p<0.05$). Furthermore, no difference was seen between strains carrying a similar SDH mutation ($p>0.1$ in all cases) confirming these *in planta*phenotypes are fully SDH driven.

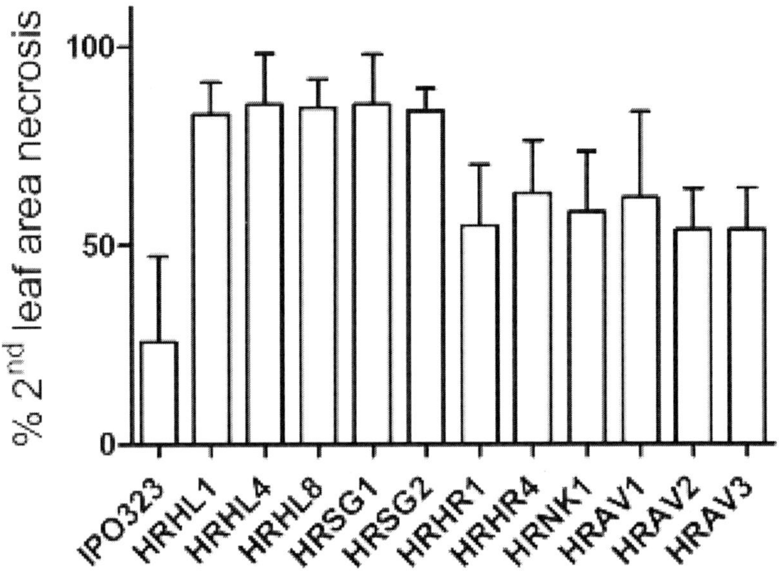

Figure 8. Comparison of *in planta* necrosis symptoms displayed by individual homologous recombinant strains carrying mutations conferring resistance to carboxamides. Data presented correspond to the individual means (average of 12 leaves each) of 3 individual experiments ± S.D. Numbering corresponds to individual homologous recombinants strains carrying: HRHL1, 4 and 8: SDHB_H267L substitution, HRSG1 and 2: SDHC_S83G substitution, HRHR1 and 4: SDHC_H152R substitution, HRNK1: SDHC_N86K substitution, HRAV1, 2 and 3: SDHC_A84V substitution.

Interestingly, comparison of these results with those in table 4 shows that the magnitude of necrosis symptoms are not correlating to SDH activity impairment as quantified by the succinate: Qo/DCPIP activity test. For example the SDHC_S83G mutants displayed a similar increase in symptomology compared to the SDHB_H267L mutants in spite of their different impact on enzyme efficiency (mutations conferring 59% and 13% residual activity respectively). Similarly the SDHC_H152R mutants (22% residual activity) displayed similar necrosis size compared to SDHC_A84V mutants (84% residual activity), and to the SDHC_N86K mutant (41% residual activity). This result suggests mutations in the SDH enzyme have an impact on the biology of the pathogen.

DISCUSSION

In this study, we developed a better understanding of the binding properties and resistance mechanisms for a range of new carboxamides recently introduced as crop protection fungicides. The different biological spectrum displayed by the new carboxamides demonstrates that an incredibly broad range of biological specificities can be developed from a single core structure. By comparing enzyme inhibition and biological profiles, we have previously found that biological activity is primarily driven by the affinity of a molecule to the SDH enzyme in targeted organisms [48]. Poor conservation in residues belonging to subunits SDHC or SDHD surrounding the Qp site of SDH is observed across fungal species. One of the challenges in delivering good agrochemical solutions from carboxamide chemistry has been to overcome this variation in order to deliver an effective balance between binding efficacy and fungal spectrum. Partly because of this wide structural variation in the target enzyme, a unique solution enabling the control of all fungal pathogens could not be found. Therefore, further SDHIs that display additional fungicide spectrum might be introduced in the coming years.

Our mutagenesis study led us to identify 27 different substitution types affecting 18 positions in 3 of the 4 subunits encoding the Qp site of the target SDH enzyme. The pattern and frequency of mutations selected was found to be highly dependent on the compound used for selection. Accordingly, sensitivity profiles are substitution dependent, as a result of specific interaction of different classes of inhibitors to specific structural features of the enzyme. The large majority of the mutations lead to a sensitivity decrease across all carboxamides *in vivo*, but the level of reduced sensitivity shows a high degree of variation across the carboxamide/substitution pairs studied. More practically, this suggests that the use of carboxamides of different structures to control the same pathogens will strongly influence the nature and composition of the mutant population in the field as was found in *A. alternata* field trials [56].The nature of carboxamide-selected *M. graminicola* target mutations found in the laboratory display striking similarities with the mutations found in *B. cinerea* field populations following several years of Boscalid usage [30], [32]. Among the various substitutions detected in *B. cinerea* field isolates the SDHB_H272Y/L, SDHB_P225L/T and SDHB_N230I substitutions correspond to SDHB_H267Y/L, SDHB_P220L/T and SDHB_N225I carboxamide-selected substitutions respectively in *M. graminicola*. This shows that carboxamide resistance can be conferred by similar substitutions at similar positions within the structure of the SDH enzyme across species. Contrasting with the SDHB_P225L/T substitutions conferring high levels of resistance towards Boscalid and Fluopyram in *B. cinerea* ([30], Syngenta, unpublished result). In our screen, the *M. graminicola* SDHB_P220L/T substitutions were obtained upon Fluopyram selection only and conferred limited resistance towards this active ingredient as well as weak Boscalid resistance (Table 3).

Therefore, subtle differences in the structure of the Qp site of SDH within targeted organisms are likely to influence the nature of substitutions conferring resistance to a given carboxamide and this highlights the problems in extrapolating resistance prediction from one pathogen to another. This is further exemplified by the different substitution patterns and associated resistance factors displayed by *A. alternata*pistachio field isolates after a few years of Boscalid usage [33], [34]. Despite the difficulties in extrapolating between species some key conserved interactions are beginning to emerge. For example, Fluopyram hypersensitivity is observed in SDHB histidine to tyrosine Qp site mutants in a range of species including *M. graminicola* (SDHB_H267Y), *B. cinerea*(SDHB_H272Y) and *A. alternata* (SDHB_H277Y) ([30], [56], Syngenta, unpublished results). A similar substitution may also explain the similar negative cross resistance behavior observed in some Boscalid resistant isolates of *C. cassiicola* and *P. xanthii* [36]. Using the homology model developed in this study, a possible explanation for this conserved negative cross resistance was proposed. In the WT enzyme, a key H-bond interaction may occur between the rotated histidine of the Qp site and the acceptor group of Boscalid. This crucial interaction for binding is removed by the tyrosine substitution which consequently impairs Boscalid binding in the mutant. Contrastingly, Fluopyram which has no H-bond acceptor group does not rely on this specific interaction for binding and is then unaffected by the histidine to tyrosine substitution. Further confirming this assumption, compound A which also lacks a H-bond acceptor group gives also better control of the *M. graminicola* SDHB_H267Y mutant compared to the WT (data not shown). Given the degree of conservation for cross resistance profiles seen with this particular mutant it seems that the depicted interaction is consistently conserved across species.

Using transformation we evidenced that the remaining SDH activity present in the cells at a given inhibitor concentration is responsible for survival. Interestingly, very low levels of SDH activity were sufficient for the establishment of resistance, as confirmed by the selection of substitutions leading to over 90% loss in activity. This suggests that for each mutant, *in vivo*survival upon carboxamide treatment is a balance between a negative impact brought by decreased enzyme activity/stability caused by substitutions affecting the Qp site and a positive one brought by poorer binding of carboxamide inhibitors resulting in weaker inhibition of the enzyme. From a cellular perspective and considering the central role of SDH for energy production, it seems logical that the remaining SDH activity, which is necessary to maintain an active TCA cycle, is the driver for survival.

A balance between substrate and inhibitor binding would explain why some highly conserved residues of the Qp site which are predicted to be essential for carboxamide inhibitor binding in the tridimensional model were neither found substituted in our screen nor reported yet in field populations. Notably the fully conserved Qp site residues SDHB_W224 and SDHD_Y130 which are predicted to hydrogen-bond to the amide oxygen of carboxamides [15], [40]. In agreement with the essential involvement of the conserved SDHD tyrosine in the establishment of a crucial hydrogen bond to one quinone oxygen, *E. coli*, SDHD_Y83F and *S. cerevisae*SDHD_Y89F substitutions impair 85% and 95% of the ubiquinone reductase activity respectively [13], [57]. We introduced the SDHD_Y130F substitution in the *M. graminicola MgSDHD* gene using site directed mutagenesis and found that ectopic transformants expressing SDHD_Y130F are more sensitive to carboxamides compared to the WT (data not shown). The absence of any mutation at this residue for all carboxamides tested might indicate that substitutions at this position could not confer selective advantage in the balance between catalysis and inhibition.

Because SDH enzyme activity was impaired in all mutants we expected to find some degree of fitness penalty *in vivo*. Furthermore, similar Qp site substitutions have been shown to have biological influence on the lifespan of organisms through the increased production of ROS by the mutated SDH enzyme [38], [41], [52], [53], [54], [55]. To primarily address this and to prevent the likely interference caused by mutations in other genes in UV mutants, we generated homologous recombinants to some of the most relevant substitution types. The absence of any significant growth defect in our carboxamide-selected Qp site mutants and homologous recombinant strains suggested that the likely increased ROS production by the mutated enzyme was not exceeding the capacity of the antioxidant defense system in *M. graminicola*. One explanation for this result may be that our initial selection for carboxamide resistance is strongly biased against the selection of mutants which display high level of oxidative stress. This could occur if the presence of ROS producing mutations impacted growth as shown with many yeast Qp site mutants on non-fermentable media [38], [41]. Impaired growth could have resulted in the absence of identifiable colonies at the point of colony selection. Carboxamide-selected Qp site substitutions were all leading to decreased ubiquinone reductase activity suggesting that, at least in some cases, the stability of the transient ubiquinone semiradical generated during the course of the reaction might be affected and thus lead to locally increased ROS production [38]. Therefore, the absence of strong oxygen hypersensitivity phenotypes is surprising and may be explained by a combination of factors. First the specific activity displayed by *M. graminicola* SDH is 10 times lower compared to that of *S. cerevisiae* SDH [41]. More generally, *M. graminicola* is a slow growing pathogen and might have a much lower mitochondrial ROS production rate compared to fast growing fungi like *S. cerevisiae*. Secondly, *M. graminicola* is a hemibiotrophic pathogen and therefore has to survive plant induced oxidative burst in colonized plant tissue. The pathogen requires and induces an exceptionally efficient ROS detoxification enzyme toolset in order to survive the necrotrophic stage of its lifecycle [58]. In fact it has been suggested that ROS production by *M. graminicola* itself and ROS signalling may also contribute to pathogenicity[58], [59]. Therefore, the combination of a slower turnover with the presence of an extraordinary large enzymatic toolset capable of defend against ROS in this species might lead to the very poor impact of oxidative stresses towards carboxamide-selected Qp site mutants when compared to other species. These hypotheses are further supported by the poor fluorescence signals obtained with intracellular ROS indicators showing that no significant ROS accumulation occurs in both WT and mutants. *In vitro* tests with a purified mitochondrial fraction will be required to measure the potentially increased production of ROS by the mutated*M. graminicola* SDH enzyme in more detail [38], [41], [52]. This and further work would be required to understand the marked differences in ROS generation and sensitivity between *M. graminicola* and other species.

In our test conditions, *in planta* virulence tests performed with some of the HR strains showed that some of the SDH mutations can lead to increased symptoms in the absence of carboxamide treatment. The reason for this remains unclear, but the increased symptomology observed in the climatic room is unlikely to result in increased fitness in the field as nature would have selected such SDH variants during the course of evolution if it was the case. One explanation for the observed phenotype is that the presence of less efficient SDH enzyme might lead to increased intracellular levels of succinate as observed in yeast mutants [41] and in *M. graminicola* WT upon treatment with sub-lethal doses of carboxamides (data not shown). In yeast, perturbations of the succinate dehydrogenase function have a large impact on metabolism. Moreover, it was shown that succinate dehydrogenase mutations displaying similar impairment in quinone reductase activity can lead to very distinct metabotypes [60]. The metabolic impact of impairments of the *M. graminicola* SDH enzyme may lead to developmental effects on *in planta* growth and symptoms development as complex regulation of metabolic fluxes have been shown to play a major role in the infection process [58], [61],[62]. Necrotrophic growth involves massive expression of transporters and of enzymes involved in primary metabolite uptake and degradation and it is very likely to involve secondary metabolites (small molecules toxins) production [58], [61], [62]. Modifications in primary metabolism might have an influence on this metabolic switch and potentially increase secretion of small toxin molecules thus explaining the observed increased symptomology. Even though the reasons for this increased symptomology *in planta* remain elusive, our data indicate that SDH mutations can have an influence on the biology of the pathogen. Aside additional fitness penalty and resistance studies to be performed *in planta*, the question as to know whether this phenomenon is influenced by ROS or TCA metabolites or both will require further evaluations now rendered possible by the homologous recombinants generated during the course of this study.

Finally, our data clearly shows that a combination of factors might have to be considered for a diagnosis of mutations likely to occur in the field. In our mutagenesis screen, high resistance factors, frequency of occurrence and maintained *in planta* growth point towards some mutations for which sensitive molecular tests will be designed and applied in the monitoring of field populations. These results combined with the resistance situation in other pathogens further stress the importance of a proper anti-resistance strategy for the SDHIs fungicides. At this point in time and in order to prolong the efficacy of this class of fungicide in wheat, recommendations include restrictions in the number and timing of applications as well as the mandatory usage of mixtures (see www.FRAC.info).

MATERIALS AND METHODS

Strains, media and culture conditions

M. graminicola isolate IPO323 was used for all experiments [63]. The isolate was inoculated from stocks stored in liquid nitrogen onto solid V8 agar [64]. Yeast like cells were harvested from these plates and used immediately as an inoculum for all experiments. The following media were used throughout: Aspergillus minimal medium (AMM) [64]; TSM40 (4 g glucose, 10 g malt extract, 4 g yeast extract, pH 7.0); AE medium [23]; induction medium (IM) [64]; potato dextrose agar (PDA) [64]. DH5α or ccdB survival cells (Invitrogen) were used for the maintenance of plasmids in *E. coli*. *Agrobacterium. tumefaciens* strain EHA105 [65], [66], was used for maintenance of constructs and for *A. tumefaciens* mediated transformation.

Carboxamides, mutagenesis and UV mutant isolation

Minimal selective concentration (MSC) is the minimal concentration of carboxamide necessary to fully inhibit growth of the wild type IPO323 *M. graminicola* strain after inoculation of AE agar plates with 10^7 cells followed by incubation for 2 weeks at 20°C in the dark. Selection plates were produced after adding 1% of a 100× DMSO solution of the active ingredients to pre-melted AE-agar. A higher concentration corresponding to 5× MSC was applied with Carboxin, Isopyrazam, Fluopyram and Boscalid and 2.5× MSC with the Isopyrazam pyrrole analogue compound A.

Cells were harvested from 4 days old V8 agar plates and conidial cells were suspended in liquid AE. 10^7 cells were spread onto carboxamide amended AE agar plates and mutagenized using UV Stratalinker 2400 (Stratagene) with 75 mJ cm^{-2} or 50 mJ cm^{-2}. These UV treatments produce lethality of 80% and 50% respectively. Selection plates were incubated in the dark at 20°C for 14 days. Visible colonies were picked from the primary selection plate and re-grown on fresh selection media for 2 additional weeks under the same conditions.

Molecular characterization of UV mutants

Strains were grown in 5 ml liquid TSM40 on a reciprocal shaker (220 rpm, 20°C) for 5 days prior to genomic DNA extraction using a Qiagen mini plant kit (Qiagen). PCR primer design was based on the *M. graminicola* IPO323 genome sequence [63]. The four SDH-encoding genes were PCR amplified and sequenced using primers listed in Table S1. For comparison purposes, SDHB, SDHC and SDHD proteins sequences were retrieved for a number of different species as listed in supplemental data section (Table S2).

Production of *M. graminicola* transformants

For ectopic integration of SDH subunit expression cassettes, the multisite binary pNOV2114_gateway vector was used with the *A. nidulans* gpda (glyceraldehyde-3-phosphate dehydrogenase) promoter to drive the expression [64]. Wild type or mutated *MgSDH* gene were amplified from genomic DNA using PfuTurbo-DNA polymerase (Stratagene) and two rounds of amplification with primers listed in table S1. The attB PCR fragments were cloned into pDONRzeo plasmid (Invitrogen). Finally, multiple gateway reaction was performed using multisite Gateway cloning (Invitrogen) to produce pNOV_gpda_*MgSDHB*_TrpCter, pNOV_gpda_*MgSDHB-H267L*_TrpCter, pNOV_gpda_*MgSDHC*_TrpCter, and pNOV_gpda_*MgSDHC-A84V*_TrpCter plasmids.

For homologous recombination, genomic fragments containing part of the *SDH* gene without the start codon but containing the mutation(s) leading to the specified amino acid substitution plus ~800 bp down-stream of the gene were PCR amplified using primers described in Table S1 and cloned into the multiple cloning site of pNOV2114 vector [67], [68]. Oligonucleotides were designed to avoid in-frame start codons upstream of the cloned fragment.

M. graminicola transformation procedure has been described previously [64]. For isolation of homologous recombinants, selection of transformants was performed on AE agar supplemented with an appropriate carboxamide. Three rounds of selection were applied. Absence of the start codon in the homologous recombination constructs ensured low frequency of functional expression from ectopic insertions.

Molecular characterization of transformants

For ectopic transformants, primer pairs were designed to screen for the presence and completeness of the gpda promoter, *MgSDH* gene and hygromycin resistance gene. For homologous recombinants, primer pairs were generated to verify the unique presence of the mutated copy using differential restriction profiles displayed by the mutated allele versus the non mutated allele. In addition primer pairs were designed to screen for the absence of T-DNA downstream of the integrated fragment. All strains which passed PCR verification were sequence verified. All PCR primers and details mentioned above are listed in Table S1.

The copy number of the SDH gene in the transformed strains was assessed by Southern blot. Hybridization probes were produced by PCR amplification of the SDH gene of interest and labeled using the AlkPhos Direct™ system (GE Healthcare). Genomic DNA (~10 µg) were digested using BamH1 for *SDHB* and *SDHD* transformants; BamH1/EcoR1 for *SDHC*transformants. Digested DNA was electrophoresed and subsequently blotted and UV cross-linked to Hibond™ Nylon membranes (GE-Healthcare) [69]. Hybridization was performed using AlkPhos Direct™ reagents (GE-Healthcare). A hybridization temperature of 55°C was used for labeling all three *SDH* genes, followed by washes at 55°C (*SDHC* and *SDHD*) or 60°C (*SDHB*). Luminescence signal was generated with CDP-Star (GE-Healthcare) and visualized on Kodak Biomax light films (Sigma).

Mitochondria isolation

Strains were grown in 1 liter TSM40 units. Cultures were inoculated at 0.05 $OD_{600\ nm}$ and grown on a reciprocal shaker (110 rpm, 22°C) for 5 days. Cells were harvested by vacuum filtration and disrupted in liquid nitrogen using a mortar and pestle. Resultant powder was re-suspended to 10% w/v in mitochondrial extraction buffer (10 mM $KHPO_4$ pH7.2; 10 mM KCl, 10 mM $MgCl_2$, 0.5 M sucrose, 0.2 mM EDTA, 2 mM PMSF). Extract was clarified by centrifugation (5,000 g, 4°C for 10 minutes, 2 times), intact mitochondria were then pelleted at 10,000 g for 20 minutes at 4°C. The pink-reddish pellet was re-suspended in a small volume (1 ml for 10 grams initial fresh weight) of extraction buffer. This volume was transferred to conical 1.5 ml tubes and re-centrifuged, 10,000 g for 10 minutes at 4°C. The pellet separates into two layers; the upper reddish colored being easily re-suspended by pipetting is composed of nearly pure mitochondria as assessed by microscopy. Subsequent sucrose gradient centrifugation did not improve specific activity of the mitochondrial fractions; therefore the purification procedure was stopped at this stage. Mitochondrial suspensions were brought to a final concentration of 10 mg ml^{-1} in extraction buffer and stored at −80°C until used. SDH activity was found to remain stable for months.

Enzyme assays

Mitochondrial suspensions were incubated at 30°C for 30 minutes in extraction buffer supplemented with 10 mM succinate prior to enzymatic measurements.

Succinate: ubiquinone/DCPIP measurements were performed by adding 5 µl of pre-activated mitochondria to 100 µl assay buffer (50 mM Phosphate-Na pH 7.2, 250 mM sucrose, 3 mM NaN_3, 10 mM succinate) supplemented with 140 µM Dichlorophenolindophenol (DCIP) and 1 mM of 2,3-dimethoxy-5-methyl-1,4-benzoquinone (Q_0), from a 0.1 M DMSO stock solution. A reaction plate was equilibrated for 10 minutes at 30°C before starting the reaction. DCPIP reduction was conducted at 30°C and monitored at 595 nm using a HTS7000 microtiterplate reader (Perkin Elmer). For calculations a molecular extinction coefficient of $\varepsilon = 17\ mM^{-1}\ cm^{-1}$was used [70].

The succinate: PMS (Phenazine methosulfate) coupled reduction of MTT ((4,5-dimethylthiazol-2-yl)-2,5-diphenyltetrazolium bromide) measurements were performed using final concentrations of 400 µM PMS and 150 µM MTT freshly diluted in assay buffer (50 mM Phosphate-Na pH 7.2, 250 mM sucrose, 3 mM NaN_3, 10 mM succinate). Reactions were started with the addition of 5 µl of preactivated mitochondrial suspension to 100 µl of the reaction mixture at 30°C using an experimental setup similar to DCPIP measurements. The MTT reduction was monitored by reading the absorbance increase at 570 nm using a HTS7000 microtiterplate reader (Perkin Elmer). For calculations a molecular extinction coefficient of $\varepsilon = 21\ mM^{-1}.cm^{-1}$ was used [71].

In the succinate: ubiquinone/DCPIP test, activity displayed by some mutants was so low that an adjustment for enzyme activity was necessary for accurate IC50 determination. For this reason, all mitochondrial suspensions were adjusted to a similar initial velocity (1 $OD_{595\ nm}hour^{-1}$) following conditions described above. Inhibitor concentrations ranged between 1 mM and 0.95 nM with uniform 4× dilution factor steps (11 inhibitor concentrations+DMSO control). 96 wells plates were pre equilibrated at reaction temperature (30°C) for 10 minutes before the reactions were started by the addition of 5 µl preactivated *M. graminicola* mitochondrial suspension. Calculated absorbance slopes (OD/hour) were used for IC50 calculations using Graphpad Prism 5.0 software non linear curve fitting against log inhibitor concentrations.

Flavin Quantification

Fluorometric quantification of the covalent FAD of SDH was carried out as described in [72]using 1 mg of protein as starting material.

Comparative homology modeling.

The four *M. graminicola* SQR subunits were modelled by using SQR crystal structures from three different species as templates: *E. coli* SQR co-crystallized with the natural substrate ubiquinone-2 at a 2.60 Å resolution (PDB code 1NEK) [14]; *G. gallus* (chicken) SQR in complex with the inhibitor carboxin resolved at 2.10 Å (PDB code 2FBW) [15]; *S. scrofa* (pig) SQR cocrystallized with ubiquinone-2 at 2.40 Å (PDB code 1ZOY) [16].

ClustalW [73] was used to align the individual *M. graminicola* SDH protein sequences with the bacterial (1NEK), the avian (2FBW) and the porcine (1ZOY) protein subunits. The alignments were analyzed and edited in the Prime 1.5 workflow within the Maestro environment (Schrödinger, LLC, Portland, OR) [74]. In contrast to SDHA and SDHB, the alignments of the SDHC and SDHD anchoring subunits of *M. graminicola* displayed limited identity with the sequences of the three crystallographic templates. Therefore, to further improve the alignments, the multiple template strategy together with a secondary structure prediction (SSPro4.0 [75]) was applied. In the next step, the actual three-dimensional models of each four subunits were built according to the standard superimposition and side-chain orientation of Prime protocols. The individual sub-models were then assembled by superimposition on the 2FBW structure keeping carboxin inside the Qp site as well as the heme, the FAD and malate-like intermediate inside the so-called "substrate binding site" to help in the definition of pockets and cavities. The refinement module of Prime (PLOP, Protein Local Optimization Program [76]) was used to optimize the sidechain conformations with particular attention given to the SDHB, SDHC and SDHD interfaces and other regions in the vicinity of Qp-site. Moreover default loop refinement method was performed on sections coating the Qp-site, 256–271, 226–234 on SdhB and 62–78 on SdhC subunits.

The final *M. graminicola* SQR model was optimized within the OPLS2005 force field in Prime[74], [77].

Pharmacophore elucidation. All SDH crystal structures were extracted from the Protein Data Bank (PDB) and superimposed onto each other. The Qp binding sites of the superimposed structures with a resolution higher than 3 Å were visually analyzed in more detail for common pharmacophoric features (PDB codes 3AE4, 3AEF, 2WQY, 2H88, 2H89, 1YQ3, 1YQ4, 2FBW, 1ZOY, 1NEK, 1NEN, 2WDQ, 2WP9, 2WU5).

Molecular Docking. Carboxin was manually placed in the Qp binding site of *Mg*SDH with Moloc [78] in an analogous way as observed in the carboxin *G.gallus* SQR complex crystal structure 2WQY. In a second step the protein ligand complex was minimized using Moloc MAB force field [78], allowing full flexibility for the ligand and keeping the *M. graminicola* SQR protein rigid.

Boscalid, Fluopyram, and Isopyrazam were manually docked into the *M. graminicola* SQR Qp binding site. Interactions to key residues which were determined through the pharmacophore elucidation (see results) were formed. In a second step the protein ligand complexes were minimized using Moloc MAB force field [78], allowing full flexibility for the ligands and keeping the *M. graminicola* SQR protein rigid.

Mutated Qp-site modeling. Mutations were generated within the *M. graminicola* SQR homology model with the "mutate residue" procedure included in Maestro (Schrödinger, LLC, Portland, OR), followed by a contact analysis involving multiple rotamers and a geometry sanitation within the OPLS2005 force field.

In vivo tests for strain sensitivity determination. Conidial cells were collected from 4 day old V8 agar plates (cultured at 18°C in the dark) using liquid AE and a cell spreader. Cell density was determined and cells re-suspended to a concentration of 5×10^5 cells ml^{-1} in AE media. Sterile 96 wells microtiter plates with lids (Costar) were inoculated with 198 µl of this cell suspension (10^5 cells per well) and 2 µl of inhibitor (100× in DMSO). Final inhibitor concentrations were between 0.95 nM and 1 mM with uniform 4× dilution factor steps (11 inhibitor concentrations+1 DMSO control). Plates were incubated for 6 days at 19°C in the dark before cells were re-suspended and measured for growth ($OD_{600\ nm}$). IC_{50} values were determined by non linear curve fitting on the basis of measured OD values versus log inhibitor concentrations (GraphPad Prism 5.0 software).

In vivo mitochondrial ROS production assay Conidial cells were collected from 4 day old V8 agar plates (cultured at 18°C in the dark) using 2 ml of assay buffer (PIPES 10 mM pH 7.2, 2% w/v glycerol) and a cell spreader. Cell density was determined and cells re-suspended to a concentration of 3×10^7 cells ml^{-1} in assay prior to their incubation for 2.5 hours in the dark in the presence of 5 µM MitoSOX red (Invitrogen). After the incubation period, 200 µl aliquots of each cell suspension were distributed (3 replicates per strain) into 96well plates (AcroWell™ 96 filter plates) equipped with 0.45 µm filters (Pall life sciences). First step of centrifugation 500 g for 3 minutes eliminated non incorporated fluorescent dye. Cells were resuspended in 200 µl assay buffer and same centrifugation procedure was applied. Finally, cells were resuspended in 110 µl assay buffer and 100 µl were transferred to clear bottom/black framed 96 wells Optilux™ plates (Becton Dickinson) for measurement. Fluorescence was measured with an Enspire 2300 multimode reader (Perkin Elmer) using λ_{exc} of 510 nm and λ_{em} of 580 nm (well scan mode, 100 flashes/read, 25 reads/well).

In planta pathogenicity test

Conidial cells were collected from 4 day old V8 agar plates (cultured at 18°C in the dark) using 2 ml of de-ionized water. Cell density was determined and cells re-suspended to a concentration of 10^6 cells ml^{-1} in de-ionized water supplemented with 0.05% Tween 20. Each spore suspension was used for the inoculation of three pots of two weeks old wheat (var. Riband) plantlets (3 to 4 plants/pot). The plants were homogenously sprayed and covered with a Plexiglas chamber for incubation at 100% humidity for 48 hours at 24°C in the dark. After this incubation period, pots were placed in a climatic room with alternation periods of 8 hours darkness (18°C), and 16 hours light (24°C), with 60% constant humidity. Evaluation of symptoms was performed 16 days after inoculation by visually estimating the percentage of 2nd leaf area covered with pathogen–related necrosis symptoms. Negative controls (uninfected plants) confirmed the absence of necrosis at the evaluation time.

ACKNOWLEDGMENTS

We would like to thank Daniel Stierli and Harald Walter for chemical support, and Nikola Kokosar and Laura Quaranta for experimental support with the oxygen test. Thanks also to Tyler Harp, Dietrich Hermann, Helge Sierotzki and Andy Leadbeater for reviewing this manuscript. Thanks are also due to Gert Kema for supplying us with the IPO323 strain used in this study.

AUTHOR CONTRIBUTIONS

Conceived and designed the experiments: GS JB TL AV AD RFP. Performed the experiments: GS JB LKA DS MN AV AD. Analyzed the data: GS KW. Contributed reagents/materials/analysis tools: GS JB. Wrote the paper: GS MC.

REFERENCES

1. Eyal Z, Scharen AL, Prescott JM, Van Ginkel M (1987) *Septoria* diseases of wheat: concepts and methods of disease management. CIMMYT, Mexico city.
2. Mehrabi R, Zwiers LH, de Waard MA, Kema GH (2006) MgHog1 regulates dimorphism and pathogenicity in the fungal wheat pathogen *Mycosphaerella graminicola*. Molecular Plant-Microbe Interactions 19: 1262–1269.
3. Polley RW, Thomas PL (1991) Surveys of diseases of winter wheat in England and Whales, 1976–1988. Annals of Applied Biology 119: 1–20.
4. Amand O, Calay F, Coquillart L, Legat T, Bodson B, et al. (2003) First detection of resistance to QoI fungicides in *Mycosphaerella graminicola* on winter wheat in Belgium. Communications in Agricultural & Applied Biological Sciences 68: 519–531.
5. Fraaije BA, Cools J, Fountaine DJ, Lovell J, Motteram JS, et al. (2005) Role of Ascospores in Further Spread of QoI-Resistant Cytochrome b Alleles (G143A) in Field Populations of M*ycosphaerella graminicola*. Phytopathology 95: 933–941.

6. Grasso V, Palermo S, Sierotzki H, Garibaldi A, Gisi U (2006) Cytochrome b gene structure and consequences for resistance to Qo inhibitor fungicides in plant pathogens. Pest Management Science 62: 465–472.

7. Gisi U, Sierotzki H, Cook A, McCaffery A (2002) Mechanisms influencing the evolution of resistance to Qo inhibitor fungicides. Pest Management Science 58: 859–867.

8. Leroux P, Albertini C, Gautier A, Gredt M, Walker AS (2007) Mutations in the CYP51 gene correlated with changes in sensitivity to sterol 14 alpha-demethylation inhibitors in field isolates of *Mycosphaerella graminicola*. Pest Management Science 63: 688–698.

9. Brent KJ (1995) Fungicide resistance in crop pathogens: How can it be managed? Brussels: GIFAP.

10. Horsefield R, Yankovskaya V, Sexton G, Whittingham W, Shiomi K, et al. (2006) Structural and computational analysis of the quinone-binding site of complex II (succinate-ubiquinone oxidoreductase): a mechanism of electron transfer and proton conduction during ubiquinone reduction. Journal of Biological Chemistry 281: 7309–7316.

11. Anderson RF, Hille R, Shinde SS, Cecchini G (2005) Electron transfer within complex II. Succinate:ubiquinone oxidoreductase of *Escherichia coli*. Journal of Biological Chemistry 280: 33331–33337.

12. Oyedotun KS, Lemire BD (1999) The *Saccharomyces cerevisiae* succinate-ubiquinone oxidoreductase. Identification of Sdh3p amino acid residues involved in ubiquinone binding. Journal of Biological Chemistry 274: 23956–23962.

13. Tran QM, Rothery RA, Maklashina E, Cecchini G, Weiner JH (2006) The quinone binding site in *Escherichia coli* succinate dehydrogenase is required for electron transfer to the heme b. Journal of Biological Chemistry 281: 32310–32317.

14. Yankovskaya V, Horsefield R, Tornroth S, Luna-Chavez C, Miyoshi H, et al. (2003) Architecture of succinate dehydrogenase and reactive oxygen species generation. Science 299: 700–704.

15. Huang LS, Sun G, Cobessi D, Wang AC, Shen JT, et al. (2006) 3-nitropropionic acid is a suicide inhibitor of mitochondrial respiration that, upon oxidation by complex II, forms a covalent adduct with a catalytic base arginine in the active site of the enzyme. Journal of Biological Chemistry 281: 5965–5972.

16. Sun F, Huo X, Zhai Y, Wang A, Xu J, et al. (2005) Crystal structure of mitochondrial respiratory membrane protein complex II. Cell 121: 1043–1057.

17. Snel M, von Schmeling B, Edgington LV (1970) Fungitoxicity and structure-activity relationships of some oxathiin and thiazole derivatives. Phytopathology 60: 1164–1169.

18. von Schmeling B, Kulka M (1966) Systemic fungicidal activity of 1,4-oxathiin derivatives. Science 152: 659.

19. Parnell S, Gilligan CA, Lucas JA, Bock CH, Van Den Bosch F (2008) Changes in fungicide sensitivity and relative species abundance in *Oculimacula yallundae* and *O. acuformis* populations (eyespot disease of cereals) in Western Europe. Plant Pathology 57: 509–517.

20. Hobbelen PHF, Paveley ND, Fraaije BA, Lucas JA, van den Bosch F (2011) Derivation and testing of a model to predict selection for fungicide resistance. Plant Pathology 60: 304–313.

21. Broomfield PL, Hargreaves JA (1992) A single amino-acid change in the iron-sulphur protein subunit of succinate dehydrogenase confers resistance to carboxin in *Ustilago maydis*. Current Genetics 22: 117–121.

22. Keon JP, White GA, Hargreaves JA (1991) Isolation, characterization and sequence of a gene conferring resistance to the systemic fungicide carboxin from the maize smut pathogen, *Ustilago maydis*. Current Genetics 19: 475–481.

23. Skinner W, Bailey A, Renwick A, Keon J, Gurr S, et al. (1998) A single amino-acid substitution in the iron-sulphur protein subunit of succinate dehydrogenase determines resistance to carboxin in *Mycosphaerella graminicola*. Current Genetics 34: 393–398.

24. Shima Y, Ito Y, Kaneko S, Hatabayashi H, Watanabe Y, et al. (2009) Identification of three mutant loci conferring carboxin-resistance and development of a novel transformation system in *Aspergillus oryzae*. Fungal Genetics & Biology 46: 67–76.

25. Ito Y, Muraguchi H, Seshime Y, Oita S, Yanagi SO (2004) Flutolanil and carboxin resistance in *Coprinus cinereus* conferred by a mutation in the cytochrome b560 subunit of succinate dehydrogenase complex (Complex II). Molecular Genetics & Genomics: MGG 272: 328–335.

26. Honda Y, Matsuyama T, Irie T, Watanabe T, Kuwahara M (2000) Carboxin resistance transformation of the homobasidiomycete fungus *Pleurotus ostreatus*. Current Genetics 37: 209–212.

27. Kilaru S, Collins CM, Hartley AJ, Burns C, Foster GD, et al. (2009) Investigating dominant selection markers for *Coprinopsis cinerea*: a carboxin resistance system and re-evaluation of hygromycin and phleomycin resistance vectors. Current Genetics 55: 543–550.

28. Menzies J, McLeod R, Tosi L, Cappelli C (2005) Occurence of carboxin resistant strain of *Ustilago nuda* in Italy. Phytopathologia mediterranea 44: 216–219.

29. Newcombe G, Thomas PL (2000) Inheritance of Carboxin Resistance in a European Field Isolate of *Ustilago nuda*. Phytopathology 90: 179–182.

30. Leroux P, Gredt M, Leroch M, Walker AS (2010) Exploring mechanisms of resistance to respiratory inhibitors in field strains of *Botrytis cinerea*, the causal agent of gray mold. Applied & Environmental Microbiology 76: 6615–6630.

31. Stammler G (2008) Mode of action, biological performance and latest monitoring results of boscalid sensitivity. 18th Symposium of the Research Committee on Fungicide Resistance. Shimane, Japan: The Phytopathological Society of Japan. pp. 30–43.

32. Yin Y, Kim YK, Xiao CL (2011) Molecular Characterization of Boscalid Resistance in Field Isolates of *Botrytis cinerea* from Apple. Phytopathology 101: 986–995.

33. Avenot HF, Sellam A, Karaoglanidis G, Michailides TJ (2008) Characterization of mutations in the iron-sulphur subunit of succinate dehydrogenase correlating with Boscalid resistance in *Alternaria alternata* from California pistachio. Phytopathology 98: 736–742.

34. Avenot HF, Sellam A, Michailides TJ (2009) Characterization of mutations in the membrane-anchored subunits AaSDHC and AaSDHD of succinate dehydrogenase from*Alternaria alternata* isolates conferring field resistance to the fungicide boscalid. Plant Pathology 58: 1134–1143.

35. White GA, Thorn GD, Georgopoulos SG (1978) Oxathiin derivatives highly active against carboxin-resistant Succinic Dehydrogenase complexes from Carboxin-selected mutants of *Ustilago maydis* and *Aspergillus nidulans*. Pesticide Biochemistry and Physiology 9: 165–182.

36. Ishii H, Miyamoto T, Ushio S, Kakishima M (2011) Lack of cross-resistance to a novel succinate dehydrogenase inhibitor, fluopyram, in highly boscalid-resistant isolates of*Corynespora cassiicola* and *Podosphaera xanthii*. Pest Management Science 67: 474–482.

37. Fraaije BA, Bayon C, Atkins S, Cools HJ, Lucas JA, et al. (2011) Risk assessment studies on succinate dehydrogenase inhibitors, the new weapons in the battle to control *Septoria* leaf blotch in wheat. Molecular Plant Pathology 13: 263–275.

38. Guo J, Lemire BD (2003) The ubiquinone-binding site of the Saccharomyces cerevisiae succinate-ubiquinone oxidoreductase is a source of superoxide. Journal of Biological Chemistry 278: 47629–47635.

39. Oyedotun KS, Yau PF, Lemire BD (2004) Identification of the heme axial ligands in the cytochrome b562 of the *Saccharomyces cerevisiae* succinate dehydrogenase. Journal of Biological Chemistry 279: 9432–9439.

40. Ruprecht J, Yankovskaya V, Maklashina E, Iwata S, Cecchini G (2009) Structure of*Escherichia coli* succinate:quinone oxidoreductase with an occupied and empty quinone-binding site. Journal of Biological Chemistry 284: 29836–29846.

41. .Szeto SS, Reinke SN, Sykes BD, Lemire BD (2007) Ubiquinone-binding site mutations in the *Saccharomyces cerevisiae* succinate dehydrogenase generate superoxide and lead to the accumulation of succinate. Journal of Biological Chemistry 282: 27518–27526.

42. Maklashina E, Rothery RA, Weiner JH, Cecchini G (2001) Retention of heme in axial ligand mutants of succinate-ubiquinone xxidoreductase (complex II) from *Escherichia coli*. Journal of Biological Chemistry 276: 18968–18976.

43. Ruprecht J, Iwata S, Rothery RA, Weiner JH, Maklashina E, et al. (2011) Perturbation of the quinone-binding site of complex II alters the electronic properties of the proximal [3Fe-4S] iron-sulfur cluster. Journal of Biological Chemistry 286: 12756–12765.

44. Vibat CR, Cecchini G, Nakamura K, Kita K, Gennis RB (1998) Localization of histidine residues responsible for heme axial ligation in cytochrome b556 of complex II (succinate:ubiquinone oxidoreductase) in *Escherichia coli*. Biochemistry 37: 4148–4159.

45. Kearney EB (1960) Studies on succinic dehydrogenase. XII. Flavin component of the mammalian enzyme. Journal of Biological Chemistry 235: 865–877.

46. Singer TP, Salach J, Hemmerich P, Ehrenberg A (1971) Flavin peptides. Methods in Enzymology 18: 416–427.

47. Hagerhall C (1997) Succinate: quinone oxidoreductases. Variations on a conserved theme. Biochimica et Biophysica Acta 1320: 107–141.

48. .Scalliet G, Boehler M, Green PS, Kilby PM, et al. (2011) SDHIs and the Fungal Succinate Dehydrogenase. In: Dehne HW, Deising HB, Gisi U, Kuck KH, Russell PE et al, editors. Modern Fungicides and Antifungal Compounds VI. DPG-Verlag, Braunschweig, Germany. pp. 171–178.

49. Cheng VW, Ma E, Zhao Z, Rothery RA, Weiner JH (2006) The iron-sulfur clusters in Escherichia coli succinate dehydrogenase direct electron flow. Journal of Biological Chemistry 281: 27662–27668.

50. Oyedotun KS, Lemire BD (1997) The carboxyl terminus of the *Saccharomyces cerevisiae* succinate dehydrogenase membrane subunit, SDH4p, is necessary for ubiquinone reduction and enzyme stability. Journal of Biological Chemistry 272: 31382–31388.

51. Henderson PJF (1972) A linear equation that describes the steady-state kinetics of enzymes and subcellular particles interacting with tightly bound inhibitors. Biochemical Journal 127: 321–333.

52. .Huang J, Lemire BD (2009) Mutations in the *C. elegans* succinate dehydrogenase iron-sulfur subunit promote superoxide generation and premature aging. Journal of Molecular Biology 387: 559–569.

53. Ishii N, Fujii M, Hartman PS, Tsuda M, Yasuda K, et al. (1998) A mutation in succinate dehydrogenase cytochrome b causes oxidative stress and ageing in nematodes.[see comment]. Nature 394: 694–697.

54. Ishii T, Yasuda K, Akatsuka A, Hino O, Hartman PS, et al. (2005) A mutation in the SDHC gene of complex II increases oxidative stress, resulting in apoptosis and tumorigenesis. Cancer Research 65: 203–209.

55. Walker DW, Hajek P, Muffat J, Knoepfle D, Cornelison S, et al. (2006) Hypersensitivity to oxygen and shortened lifespan in a *Drosophila* mitochondrial complex II mutant. Proceedings of the National Academy of Sciences of the United States of America 103: 16382–16387.

56. Sierotzki H, Frey R, Morchoisne M, Olaya G, Mösch M, et al. (2011) Sensitivity of fungal pathogens to SDHI fungicides. In: Dehne HW, Deising HB, Gisi U, Kuck KH, Russell PE et al, editors. Modern Fungicides and Antifungal Compounds VI. DPG-Verlag, Braunschweig, Germany. pp. 179–186.

57. Silkin Y, Oyedotun KS, Lemire BD (2007) The role of Sdh4p Tyr-89 in ubiquinone reduction by the *Saccharomyces cerevisiae* succinate dehydrogenase. Biochimica et Biophysica Acta 1767: 143–150.

58. Keon J, Antoniw J, Carzaniga R, Deller S, Ward JL, et al. (2007) Transcriptional adaptation of *Mycosphaerella graminicola* to programmed cell death (PCD) of its susceptible wheat host. Molecular Plant-Microbe Interactions 20: 178–193.

59. Heller J, Tudzynski P (2011) Reactive Oxygen Species in Phytopathogenic Fungi: Signaling, Development, and Disease. Annual Reviews in Phytopathology 49: 369–390.

60. Szeto SS, Reinke SN, Sykes BD, Lemire BD (2010) Mutations in the *Saccharomyces cerevisiae* succinate dehydrogenase result in distinct metabolic phenotypes revealed through (1)H NMR-based metabolic footprinting. Journal of Proteome Research 9: 6729–6739.

61. Keon J, Antoniw J, Rudd J, Skinner W, Hargreaves J, et al. (2005) Analysis of expressed sequence tags from the wheat leaf blotch pathogen *Mycosphaerella graminicola* (anamorph *Septoria tritici*). Fungal Genetics & Biology 42: 376–389.

62. Stergiopoulos I, Zwiers LH, De Waard MA (2003) The ABC transporter MgAtr4 is a virulence factor of *Mycosphaerella graminicola* that affects colonization of substomatal cavities in wheat leaves. Molecular Plant-Microbe Interactions 16: 689–698.

63. Goodwin SB, Ben M'barek S, Dhillon B, Wittenberg AH, Crane CF, et al. (2011) Finished Genome of the Fungal Wheat Pathogen *Mycosphaerella graminicola* Reveals Dispensome Structure, Chromosome Plasticity, and Stealth Pathogenesis. PLoS Genetics 7: e1002070.

64. Bowler J, Scott E, Tailor R, Scalliet G, Ray J, et al. (2010) New capabilities for*Mycosphaerella graminicola* research. Molecular Plant Pathology 11: 691–704.

65. Hellens R, Mullineaux P, Klee H (2000) Technical Focus:a guide to *Agrobacterium*binary Ti vectors. Trends in Plant Science 5: 446–451.

66. .Hood EE, Gelven SB, Melchers LS, Hoekema A (1993) New *Agrobacterium* helper plasmids for gene transfer to plants. Trangenic Research 2: 208–218.

67. Pazour GJ, Ta CN, Das A (1992) Constitutive mutations of *Agrobacterium tumefaciens*transcriptional activator virG. Journal of Bacteriology 174: 4169–4174.

68. Scheeren-Groot EP, Rodenburg KW, den Dulk-Ras A, Turk SC, Hooykaas PJ (1994) Mutational analysis of the transcriptional activator VirG of *Agrobacterium tumefaciens*. Journal of Bacteriology 176: 6418–6426.

69. Sambrook J, Fritsch EF, Maniatis T (1989) Molecular Cloning: A Laboratory Manual. NY: Cold Spring Harbor Laboratory Press.

70. Kistler WS, Lin EC (1971) Anaerobic L- -glycerophosphate dehydrogenase of*Escherichia coli*: its genetic locus and its physiological role. Journal of Bacteriology 108: 1224–1234.

71. Baginsky ML, Hatefi Y (1969) Reconstitution of succinate-coenzyme Q reductase (complex II) and succinate oxidase activities by a highly

purified, reactivated succinate dehydrogenase. Journal of Biological Chemistry 244: 5313–5319.

72. Singer TP, Edmondson DE (1980) Structure, Properties, and Determination of Covalently Bound Flavins. Methods Enzymol 66: 354–361.

73. Chenna R, Sugawara H, Koike T, Lopez R, Gibson TJ, et al. (2003) Multiple sequence alignment with the Clustal series of programs. Nucleic Acids Research 31: 3497–3500.

74. Jacobson MP, Friesner RA, Xiang Z, Honig B (2002) On the Role of Crystal Packing Forces in Determining Protein Sidechain Conformations. Journal of Molecular Biology 320: 597–608.

75. Pollastri G, Przybylski D, Rost B, Baldi P (2002) Improving the prediction of protein secondary structure in three and eight classes using recurrent neural networks and profiles. Proteins 47: 228–235.

76. Li X, Jacobson MP, Friesner RA (2004) High-resolution prediction of protein helix positions and orientations. Proteins 55: 368–382.

77. Jacobson MP, Pincus DL, Rapp CS, Day TJ, Honig B, et al. (2004) A hierarchical approach to all-atom protein loop prediction. Proteins 55: 351–367.

78. Gerber PR, Muller K (1995) MAB, a generally applicable molecular force field for structure modelling in medicinal chemistry. Journal of Computer-Aided Molecular Design 9: 251–268.

Chapter 10

Application of Genomics-Assisted Breeding for Generation of Climate Resilient Crops: Progress and Prospects

Chittaranjan Kole at el.

[1]*Bidhan Chandra Krishi Viswavidyalaya, Mohanpur, India*

ABSTRACT

Climate change affects agricultural productivity worldwide. Increased prices of food commodities are the initial indication of drastic edible yield loss, which is expected to increase further due to global warming. This situation has compelled plant scientists to develop climate change-resilient crops, which can withstand broad-spectrum stresses such as drought, heat, cold, salinity, flood, submergence and pests, thus helping to deliver increased productivity. Genomics appears to be a promising tool for deciphering the stress responsiveness of crop species with adaptation traits or in wild relatives toward identifying underlying genes, alleles or quantitative trait loci. Molecular breeding approaches have proven helpful in enhancing the stress adaptation of crop plants, and recent advances in high-throughput sequencing and phenotyping platforms have transformed molecular breeding to genomics-assisted breeding (GAB). In view of this, the present review elaborates the progress and prospects of GAB for improving climate change resilience in crops, which is likely to play an ever increasing role in the effort to ensure global food security.

INTRODUCTION

Three major events in agricultural history, namely domestication, displacement of native crops by major crops, and intensification of agricultural production through the Green Revolution, have contributed significantly toward reduced genetic and trait diversity within major crop species (Pingali, 2012). Despite this decrease in crop diversity, global production of the major staple crops increased in the last century (Fischer et al., 2009). This increase in productivity has largely been driven by conventional plant breeding coupled with intensification and simplification of production systems. This includes selection for edible yield and adaptation, and against yield reducing factors such as susceptibility to pathogens to pests, as well as optimization of crop husbandry practices (through high inputs such as the use of fertilizers, herbicides, pesticides, and mechanization) to minimize the impact of environmental flux. However, selection under such "ideal," high-input environments has led to the loss of certain genes which are responsible for efficiency or adaptation to stress(es) (Brown, 2003). This situation presents three potential challenges: (i) to modify the selection criteria to focus on efficiency or adaptation to stress(es) rather than total edible yield, (ii) to ensure the presence and efficiency of stress-tolerance genes and their exploitability in elite material and wider breeding germplasm, and (iii) to expand the use of minor crops, which may possess better nutritional qualities, environmental sustainability or resilience and require lower inputs than major crops.

At present, global agriculture is facing a serious threat from climate change, which is predicted to result in reduced productivity. Increasing food prices and greater global food insecurity are the outcomes of decreased productivity (FAO, 2014) and this scenario, if it persists, would lead to further increase in food prices, and could lead to social unrest and famine in certain instances. Climate change will affect food supply unless actions are taken to increase the resilience of crops: projections have shown a drastic decrease in the production of major cereals by 2020, including 9% for maize, 11% for rice, and 14% for wheat (Hisas,2011). Global warming, changes in rainfall patterns and other extreme weather events may mostly contribute to this disaster, and the changing pattern of climate would result in increased attack of pathogens and pests. Moreover, the elevated CO_2 levels are predicted to reduce the nutritional quality of many crops, while some crops may become toxic due to changes in the chemical composition of their tissues (Dwivedi et al., 2013).

Therefore, increasing the resilience of crops to climate change represents a critical component towards ensuring food and nutritional security, which could be achieved through genetic engineering-based approaches or molecular breeding strategies. Genetic engineering allows direct transfer of beneficial gene(s) or manipulation of existing gene(s) in the crop of interest for generating expected phenotype(s), whereas breeding approaches involve the improvement of germplasm through introduction of novel alleles into target crops by breeding. Since genetic modification remains controversial in a number of countries (though it serves as an invaluable tool in tailoring modifications to produce alleles and phenotypes beyond the range available through exploitation of existing genetic variation), molecular breeding could offer an easy-to-accept approach for crop improvement.

Potential of genomics-assisted breeding in producing climate resilient crops

Genomics offers tools to address the challenge of increasing food yield, quality and stability of production through advanced breeding techniques. Application of DNA markers to facilitate marker-aided selection (MAS) for crop improvement have proved successful in crossbreeding. Advances in plant genomics provide further means to improve the understandings of crop diversity at species and gene levels, and offer DNA markers to accelerate the pace of genetic improvement (Muthamilarasan et al., 2013, 2014). A genomics-led breeding strategy for new cultivars for the development of new cultivars that are "climate change ready" (Varshney et al., 2005) commences by defining the stress(es) that will likely affect crop production and productivity under certain climate change scenarios. Data from multi-environment testing provide an opportunity for modeling "stress-impacts" on crops and target populations of environments. Plant breeders and genebank curators will search for morphological and physiological traits in available germplasm that could enhance crop adaptation under such climate variability. In this regard, crop physiology may help define the ideotypes to be pursued for enhancing such adaptation. Moreover, the use of geographic information systems and passport data can allow identification of accessions from stress-prone environments, whereas the available characterization, including DNA fingerprinting, and evaluation data as well as mapping of desired genes or quantitative trait loci (QTL) will assist in selecting promising accessions for further screening against specific stress(es). Similarly, precise phenotypic assessments and appropriate biometric analysis will assist in identifying unique responses of a set of genotypes in a given physiological stage influenced by variation of weather patterns. This information will be further used in genomics-aided breeding approaches such as genome-wide selection of promising germplasm for further use in crop breeding aiming at both population improvement and cultivar release.

Genetic mapping and QTL analysis, via bi-parental or association mapping (AM) populations, have accelerated the dissection of genetic control of agricultural traits, potentially allowing MAS, QTL, and AM studies or direct calculation and genomic selection (GS) of high value genotypes to be made in the context of breeding programs (Kulwal et al., 2011). Until recently, AM and GS were hampered by the need for very high marker density coverage of the genome. Advancement of next-generation sequencing (NGS) methods has facilitated the development of large numbers of genetic markers, such as single nucleotide polymorphisms (SNP), insertion-deletions (InDels), etc. even in relatively research-neglected crop species. Discovery of novel genes/alleles for any given trait could be then performed through genotyping-by-sequencing (GBS) approaches. Similarly, genome-wide association studies (GWAS) could be used to identify the genomic regions governing traits of interest by performing statistical associations between DNA polymorphisms and trait variations in diverse collection of germplasms that are genotyped and phenotyped for traits of interest. NGS coupled with GWAS increases the mapping resolution for precise location of genes/alleles/QTL (Ma et al., 2012; Liu et al., 2013; Varshney et al., 2014).

During the course of evolution, nature has evolved new genes, and shuffled and selected these genes in a wide range of environments to produce the diversity evidenced in wild species. In contrast, the selection and domestication of crops by humans is relatively recent, having occurred over the last 10,000 years. During the domestication and breeding process, there has been a significant reduction of genetic diversity in major crops, alongside a selection for yield under highly managed agricultural environments. Currently, breeders are shuffling relatively few alleles to produce enhanced combinations that provide increased yield and other attractive agronomic characteristics. In many large-genome crop species, even this reshuffling process is limited by restricted recombination patterns within the species, leading to the consistent inheritance of blocks of genes, raising issues of linkage drag and fixed linkage blocks, which may not contain the best possible combination of alleles. Breaking down these linkage constraints will allow breeders to access novel allele/gene combinations from within their current elite parents. The need to evaluate the genetics of the processes that allow genes to be recombined between parental genotypes in crops is a critical requisite. Genomics possesses the potential to increase the diversity of alleles available to breeders through mining the gene pools of crop wild relatives (CWRs). Genomics tools also enable rapid identification and selection of novel beneficial genes and their controlled incorporation into novel germplasm. In the genomics era, this technology will be used to safeguard the future through improved food security. Taken together, the application of genomics for crop germplasm enhancement offers the greatest potential to increase food production in the coming decades. With continued rapid advances in genome technologies, the application of genomics to identify and transfer valuable agronomic genes from allied genepools and crop relatives to elite crops will increase in pace and assist in meeting the challenge of global food security.

Genomics of climate resilience in major crops

The following section summarizes the state of knowledge of the genetic blueprints of many leading crops, together with information about breeding needs and priorities related to climate resilience. Genomic tools and resources are widely available and being employed in most of these species and will soon be ubiquitous, aiding "MAS" strategies that can be successful even based only on phenotypic information. Knowledge of gene functions is less consistent, leveraging to varying degrees the accumulated information from plant model species. However, even in model crops, the exact functions of most genes remain unknown, and exploring the variation conferred during angiosperm diversification represents an opportunity to identify a host of solutions to agricultural challenges.

CEREALS

Cereals are a staple to billions and their production is increasingly threatened by the recent changes in weather patterns due to global warming, particularly in less-developed countries where the consequences of changing climate have devastating socio-economic impact. Reaching a level of cereal production sufficient to sustain an adequate level of global food security will require the effective integration of crossbreeding with "omics" approaches that allow dissecting and more effectively manipulating the genetic make-up of adaptation to abiotic stresses (Langridge and Fleury, 2011). In the past decade, genomics-based approaches have been extensively deployed to dissect the genetic make-up of abiotic stress adaptation and given the quantitative nature of abiotic stress tolerance, QTL have been the main target of research to identify the genetic loci regulating the adaptive response of cereal crops to unfavorable environmental conditions. This includes drought-adaptive traits (Serraj et al., 2009; Tuberosa, 2012), root architecture (Wasson et al., 2012; Uga et al., 2013; Lynch et al., 2014), accumulation of water-soluble carbohydrates and their partitioning to storage organs (Landi et al., 2005; Salem et al., 2007; Snape et al., 2007; Rebetzke et al., 2008), abscisic acid concentration (Rebetzke et al., 2008; Rehman et al., 2011), stay-green (Yang et al., 2007; Borrell et al.,2014), canopy temperature (Lopes et al., 2014), and carbon isotope discrimination ($\Delta^{13}C$) (Pinto et al.,2010).

Global warming is intimately associated with an increase in temperature that accelerates leaf senescence, disrupts starch accumulation and curtails yield, particularly when combined with drought. In wheat, a major QTL located on chromosome 4A explained 27 and 17% of phenotypic variance for reduction in yield under drought and heat stress, respectively (Pinto et al., 2010). The same study also identified common QTL for drought and heat stress adaptation on chromosomes 1B, 2B, 3B, 4B, and 7A. Yield QTL were shown to be associated with components of other traits, supporting the prospects for dissecting crop performance under abiotic stress conditions into physiological and genetic components in order to facilitate a strategic approach to breeding (Reynolds and Tuberosa, 2008). Additional QTL with concurrent effects under both heat and drought conditions have been described by Wang et al. (2012) and Paliwal et al. (2012).

In rice, the result of a study based on 227 intensively managed irrigated farms forecast a net negative impact on yield from the warming expected in the coming decades, and clearly show that diurnal temperature variation must be considered when investigating the impact of climate change (Welch et al., 2010). Higher temperatures are speculated to reduce rice grain yields through two main pathways: (i) high maximum temperatures that in combination with high humidity cause spikelet sterility, and (ii) increased nighttime temperatures, which may reduce assimilate accumulation (Wassmann and Jagadish, 2009).

Flooding is one of the abiotic stresses, whose frequency and intensity is increasing due to global warming and changes in rainfall patterns. Therefore, it is important to produce cereal crops with the ability to withstand the anoxic conditions associated with waterlogging and/or extended submergence. Among cereals, rice is more prone to submergence stress, which periodically affects approximately 15 million hectares of rain-fed lowland areas in Asia to cause annual losses in excess of US $1 billion (Mackill et al., 2012). In rice, the*Sub1* QTL accounts for a major portion of variability for survival under prolonged submergence. Positional cloning of *Sub1* has revealed a cluster of three putative ethylene response factors (ERFs), namely *Sub1A, Sub1B,* and *Sub1C*. Further work unequivocally assigned the functional polymorphism to *Sub1A* (Xu et al.,2006). Following the identification of *Sub1A* QTL, marker-aided backcrossing (MABC) was used to efficiently convert submergence-susceptible rice cultivars into tolerant cultivars in only three backcross generations. Accordingly, DNA markers were developed for introgressing *Sub1* into six popular cultivars to meet the needs of farmers in flood-prone regions (Bailey-Serres et al., 2010). This clearly demonstrates the effectiveness of MAS for introgressing agronomically beneficial QTL alleles into elite material. The success of this work is largely due to the major effect of *Sub1* QTL and the stability of its effect in different genetic backgrounds under submergence conditions. In maize, Mano et al. (2005a) identified QTL for adventitious root formation at the soil surface, one of the most important adaptations to soil waterlogging, which can severely impair root growth at an early stage, thus reducing the capacity of the plant to extract soil moisture at a later stage when water shortage is more likely to occur. Several QTL for adventitious root formation have been mapped, and a major QTL was mapped on chromosome 8 (Mano et al., 2005b).

Salinity is also an impact of global climate change, which affects over 20% of the world's agricultural soils and thereby affecting cultivation. In durum wheat (genome AABB), two major QTL have been shown to control Na^+ accumulation in shoot via Na^+ exclusion (James et al., 2006). Both exclusion genes represent introgressions from an accession of *Triticum monococcum* (genome AA). Remarkably, under standard conditions, durum wheat containing the salinity tolerant allele at *TmHKT1;5-A*, which is one of the two salt-tolerance loci showed the phenotype similar to durum wheat that lacked the beneficial allele at this locus. But under saline conditions, it outperformed its durum wheat parent, with increased yields of up to 25% (Munns et al., 2012). In barley, evaluation of a mapping population derived from a cross between a wild barley (*Hordeum vulgare* ssp. *spontaneum*) accession and cultivated barley (*H. vulgare*) allowed the identification of a major QTL capable of limiting Na^+ accumulation in the shoots under saline conditions (Shavrukov et al.,2010). In rice, several QTL for salinity tolerance have been identified (Wang et al., 2012) indicating that pyramiding by marker-assisted selection (MAS) of QTL can be applied to enhance salt tolerance of rice.

Oilseeds and pulses

Oilseeds and pulses are major food crops, known for their unique protein and oil rich characteristics. Major biotic and abiotic stresses are the most serious production constraint for global oilseed and pulse production, and are predicted to worsen with anticipated climate change. Among the oilseeds, soybean has the highest protein content (40%) and the second highest oil content (20%). In spite of this importance, efforts are yet to be invested toward improving stress tolerance and other traits in soybean. *Phaseolus* beans are an essential part of the human diet and are a source of proteins, vitamins, and minerals (Gepts and Aragão, 2008). Of the five domesticated *Phaseolus* species, common bean (*P. vulgaris* L.) is the most economically important bean. Genetic studies and cultivar breeding in *P. vulgaris* have shown that heat and drought tolerance are under complex genetic control, although a single instance of a major gene has also been observed (Schneider et al.,1997; Asfaw et al., 2012). Selection of lines with improved drought adaptation has also been successful (Singh, 2007; Beebe et al., 2008; Urrea et al., 2009). Development of MAS methodology for drought adaptation has been initiated (Schneider et al., 1997; Asfaw et al., 2012) with the assistance of genomic resources developed through whole-genome sequencing of Andean (accession: G19833) (Schmutz et al.,2014) and Mesoamerican (accessions: BAT93, OAC Rex) bean genomes, and a bean breeder's genome toolbox and database (http://phaseolusgenes.bioinformatics.ucdavis.edu/).

In case of chickpea and pigeonpea, several abiotic and biotic stresses pose a threat to high and stable grain yields. To overcome these production constraints and meet the growing demand for these crops, efforts at national and international levels have led to the development of large-scale genetic and genomic resources (Varshney et al., 2013a). These resources have been used to understand the existing genetic diversity and exploit it in breeding programs. In chickpea, several intra- and inter-specific genetic maps have been developed (Gaur et al., 2011; Gujaria et al., 2011; Thudi et al., 2011; Hiremath et al., 2012) and genomic regions responsible for different biotic stresses (Anbessa et al., 2009; Kottapalli et al., 2009; Anuradha et al.,2011), abiotic stress (Rehman et al., 2011; Vadez et al., 2012) and agronomic traits (Cobos et al., 2009; Rehman et al., 2011; Bajaj et al., 2014, 2015; Das et al., 2015; Kujur et al., 2015a,b) have been reported. In pigeonpea, more than 3000 SSR markers (Saxena et al., 2010; Bohra et al., 2011; Dutta et al., 2011), ESTs (Raju et al., 2010), 454/FLX transcript reads (Dubey et al., 2011; Dutta et al., 2011), transcriptome assemblies (Dubey et al., 2011; Kudapa et al., 2012), and SNPs (Saxena, 2008) have been developed for their use in genomics-assisted breeding for crop improvement.

The draft genome sequence of both Kabuli (http://www.icrisat.org/gt-bt/ICGGC/GenomeSequencing.htm) and Desi (http://www.nipgr.res.in /CGWR/home.php) chickpeas have recently been published (Jain et al.,2013; Varshney et al., 2013b). Similarly, International Initiative on Pigeonpea Genomics (IIPG,http://www.icrisat.org/gt-bt/iipg/Home.html) released the draft genome of pigeonpea (Varshney et al., 2012). These sequence data of chickpea and pigeonpea will assist in enhancing their crop productivity and lead to conserving food security in arid and semi-arid environments. Further, attempts have been made toward improvement of oilseed crops such as peanut (or groundnut) using genomics-assisted breeding. Large-scale genomic resources were developed during recent years to facilitate molecular breeding in peanut and QTL have been identified for stress adaptation related traits (Varshney et al., 2009; Gautami et al., 2012), rust and late leaf spot resistance (Khedikar et al., 2010; Sujay et al., 2012), and oil quality (Sarvamangala et al., 2011).

Millets

Millets are small-grained graminaceous crops, well-known for their water-use efficiency, excellent nutrient content, adaptation to a range of ecological conditions and ability to flourish in nutrient-poor soils. Foxtail millet, proso millet, pearl millet, barnyard millet, finger millet, and kodo millet are few notable millet crops cultivated worldwide and of these, foxtail millet is considered as a C_4 crop model for studying the biology of other millets and biofuel grasses (Lata et al., 2013; Muthamilarasan and Prasad, 2015). Therefore, the Beijing Genomics Institute, China and the US Department of Energy–Joint Genome Institute have sequenced the foxtail millet genome

(Bennetzen et al., 2012; Zhang et al., 2012). As foxtail millet serves as a rich source of genes, alleles, or QTL for genetic improvement of major cereals and bioenergy grasses, large-scale genomic resources were developed including simple sequence repeats (SSRs) (Pandey et al., 2013; Zhang et al.,2014b), intron length polymorphisms (Muthamilarasan et al., 2014), eSSRs (Kumari et al., 2013), miRNA-based markers (Yadav et al., 2014), and transposable-elements based markers (Yadav et al., 2015). Moreover, open access online databases such as foxtail millet marker database (FmMDb) (Suresh et al.,2013), foxtail millet miRNA database (FmMiRNADb) (Khan et al., 2014) and foxtail millet transposable elements-based marker database (FmTEMDb) (Yadav et al., 2015) have been constructed. In addition to development of these markers, their utility in population genetics, association mapping, comparative genomics and genomics-assisted breeding have also been demonstrated (Muthamilarasan and Prasad, 2015). An allele-specific marker developed from an SNP in *SiDREB2* gene linked to drought tolerance in foxtail millet (Lata et al., 2011) is being used in allele mining and MAS for drought tolerance (Lata and Prasad,2012, 2013).

In pearl millet, three major QTL for grain yield with low QTL × environment interactions were identified across a range of post-flowering moisture environments (Bidinger et al., 2007). One of these major QTL accounted for up to 32% of the phenotypic variance of grain yield under drought. The effects of this QTL were validated in two independent MABC programs in which 30% improvement in general combining ability for grain yield expected from this QTL under terminal drought stress was recovered in introgression lines, based on the information provided by the markers flanking the QTL (Yadav et al., 2011). Compared to other crops, research on millets is at initial stage. Being predominantly climate resilient crops, millets could serve as valuable source of novel genes, alleles and QTL for stress tolerance, which needs to be identified and characterized. The close phylogenetic relationships between millets and other cereals could enable the introgression of novel alleles, genes or QTL identified in millets for better agronomic traits into other cereals toward ensuring food security under changing climate.

Forest and fruit tree crops

Clones of trees, namely populus, pinus, abies, and eucalyptus are used in afforestation, as they are dedicated to the production of wood and other wood-derived products. Therefore, it is imperative to develop climate-change resilient clones or populations of these forest trees. Several procedures have been developed for high-throughput DNA genotyping and genome-wide marker identification in forest trees. The genome complexity reduction DArT (Alves-Freitas et al., 2011) and whole-exome capture using in-solution target enrichment (Neves et al., 2011) have been tested successfully for genome-wide marker identification needed for GS in*Pinus taeda*. Considering the importance of genome sequence for development of genetic markers, Conifer Genome Project (http://www.pinegenome.org/cgp/) has been launched with an aim of promoting advance genome research in loblolly pine (*P. taeda*; 21.7 Gbp/1C; *n* = 12), white pines (*Pinus* subgenus *strobus*; 25.1 Gbp/1C; *n* = 12), as well as *Sequoia sempervirens* (31.4 Gbp/1C; *n* = 3x = 33) and Douglas-fir (*Pseudotsuga menziesii*; 18.6 Gbp/1C; *n* = 13). An extensive genetic resources and gene catalog was developed for *P. taeda* and *Picea glauca* (white spruce; 19.7 Gbp/1C; *n* = 12) (http://www.pinegenome.org/cgp/). The GENOAK project (http://urgi.versailles.inra.fr/Projects/GenOak) aims to establish a high quality reference genome sequence for pedunculate oak (*Quercus robur*; 905 Mbp/1C; *n* = 12). The *Eucalyptus grandis* (640 Mbp/1C, *n* = 11) genome has been deciphered (http://phytozome.jgi.doe.gov/pz/portal. html#!info?alias=Org_Egrandis) and will benefit agro-foresters utilizing this fast-growing hardwood tree to support industries based on Eucalypt fiber and hardwood products, and the production of Eucalypt feedstock for cellulosic biofuels. Importantly, this would assist in accelerating forest tree breeding for fast response to the need of adapted populations facing environmental modifications induced by climate change.

Fruit trees are also as important as the pulse and cereal crops and climate-change resilient clones or populations of fruit tree crops are necessary to maintain the source of nutrients that help the daily intake of healthy food ingredients. Genomics-based breeding approaches, along with bioinformatics capability and other omics resources will be the essential components of perennial fruit crop breeding and particularly, to adapt their cropping to combat or mitigate climate change effects. Genome sequencing and annotation projects include perennial fruit crops such as apple (Velasco et al., 2010), grape (Velasco et al., 2007), banana (D'Hont et al., 2012), cocoa (Argout et al., 2011), peach (Ahmad et al., 2011), and sweet orange (Xu et al., 2013). The advances in genome sequencing, along with high-resolution genetic mapping and precise phenotyping will accelerate the discovery of functional alleles and allelic variations that are associated with traits of interest for perennial fruit crop breeding. However, very less progress has been made in this aspect and particularly, enhancing climate resilience needs more attention. For achieving this, genetics and genomics methodologies could provide the toolbox for identifying genomic regions associated with the desired phenotype, and assist the selection from the wild genetic resources of the parental plants that will be intercrossed to provide the progenies for commencing breeding procedures for recurrent selection.

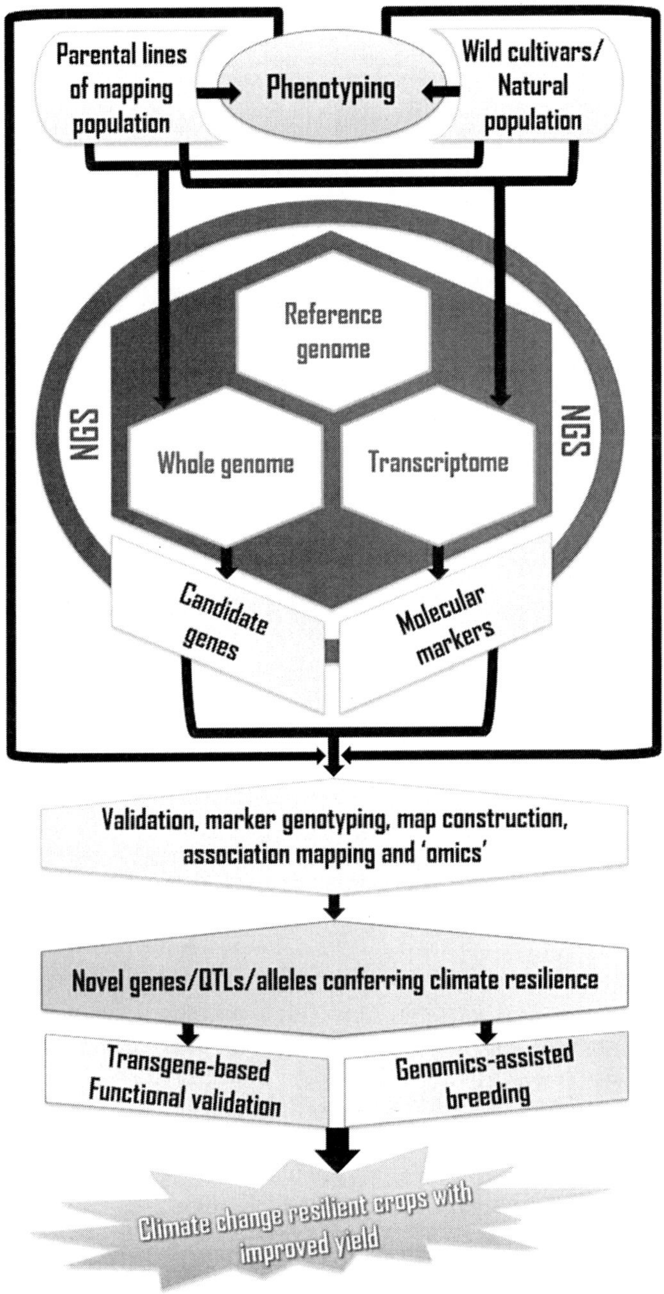

Figure 1. Flow-chart demonstrating the steps involved in generating climate resilient crops using genomics and next-generation sequencing technology.

Genomics-assisted breeding strategies for climate resilient traits

Genomics-based approaches and NGS have ushered in sequence-based breeding strategies that will expedite the dissection and cloning of the loci controlling abiotic stress tolerance, while providing unparalleled opportunities to tap into wild relatives of crops, hence expanding the reservoir of genetic diversity available to breeders (Tuberosa et al., 2011; Edwards et al., 2013) (Figure (Figure1).1). In view of the genetic complexity of yield, particularly under drought and other abiotic stresses, GS will provide the most powerful approach to raise the yield potential to the levels required to meet the fast-increasing global demand in cereal grain. However, MAS will remain a valid option for major loci or QTL, while QTL cloning will become a more routine activity facilitated through increased utilization of high-throughput, accurate phenotyping (Araus and Cairns, 2014), sequencing (Imelfort et al., 2009; Edwards and Wang, 2012), and identification of suitable candidate genes through "omics" profiling (Gupta et al., 2013). Cloned QTL will provide novel opportunities for genetic engineering for abiotic stress tolerance and for a more targeted search for novel alleles in wild germplasm (Salvi et al., 2007). Even with the application of advanced genomics technologies, mitigating the negative effects of climate change on crop productivity will remain a daunting undertaking. This requires a multidisciplinary and integrated approach, which will eventually allow plant breeders to more effectively select crops that are more resilient to climate change and ensure a sufficient level of food security for the decades to come.

Flowering time and drought adaptation

Temperature influences crop development in concert with additional floral pathways such as day-length (photoperiod), which collectively control floral transition through interconnected genetic pathways. Global warming will result in increased ambient temperature with unchanged photoperiods at given latitudes. Annual plants generally respond to increased temperatures with accelerated growth and development, having shortened lifecycles, less opportunity for photosynthesis (Reynolds et al., 2010), a shorter reproductive phase and lower yield potential (Ainsworth and Ort, 2010). There is also an increased risk of damage to reproductive tissue caused by the coincidence of high temperatures and sensitive developmental stages. Therefore, detailed knowledge of the interplay between genetic control of flowering, allelic variants, epistatic interactions, and phenotypic variations in varied growth conditions is necessary in order to identify breeding targets for climate change scenarios.

There are increasing number of germplasm resources including precise near isogenic lines (NILs) (Bentley et al., 2011, 2013) as well as next-generation populations such as multi-founder populations (e.g., multi-parent advanced generation intercross populations), which have been developed in wheat (Mackay et al., 2014) and other crops to facilitate further research and validation of climate-smart crops. New variation incorporated into elite backgrounds from landraces, ancestral or wild crop relatives (e.g., www.wheatisp.org) also offers potential for discovery of functional variation for manipulating flowering time to suit future climate permutations. However, initial work should focus on understanding the effect of flowering time on yield potential across environments and environmental stresses. Identifying the potential utility of loci of minor effect and/or which affect various stages of reproductive development could offer the ability to shorten or lengthen various phases of the flowering process, thereby enabling fine-tuning of flowering to suit particular regional climatic conditions, and to adapt to any changes in these conditions.

In case of drought tolerance, multi-disciplinary research is underway to improve plants' response to drought and water-use efficiency. With the advent of molecular breeding, QTL identification and QTL use in breeding programs assist in developing new cultivars with improved drought tolerance. In maize, extensive work has been carried out to investigate the role of root in mitigating the negative effects of drought. QTL for root traits have been described in a number of maize populations (Ruta et al., 2010; Hund et al., 2011; Tuberosa et al., 2011) in which some QTL showed a concurrent effect on grain yield performance under drought (Landi et al., 2005). Recently, Syngenta and Pioneer-DuPont deployed proprietary genomics-assisted approaches to select drought-tolerant maize hybrids (Agrisure Artesian™ and AQUAmax™, respectively) (Cooper et al., 2014). The superior performance of these maize hybrids in the severe drought that plagued the US Corn Belt in summer 2012 underlines their validity under dry soil conditions (Cooper et al., 2014). In wheat, yield QTL were shown to be associated with components of other traits, supporting the prospects for dissecting crop performance under abiotic stress conditions into its physiological and genetic components in order to facilitate a strategic approach to breeding (Reynolds and Tuberosa, 2008). At least 15 different populations have been used to map drought adaptation in rice and four regions were identified as key for yield or yield components under stress, and drought-tolerant component traits were identified across populations with interval lengths of 35–64 cm (Kamoshita et al., 2008). The first region (on chromosome 1) was associated with grain yield drought-resistance traits, plant type traits (Zhang et al., 2001a), and QTL for cell-membrane stability (Tripathy et al., 2000) and osmotic adjustment (Lilley et al., 1996), and root traits (Robin et al., 2003). Second genomic region on chromosome 4 was rich in root trait QTL (Hemamalini et al.,2000; Zheng et al., 2000; Zhang et al., 2001b; Kamoshita et al., 2002; Nguyen et al., 2004; Boopathi et al.,2005) under well-watered and drought conditions. The third region located on chromosome 8 contained QTL for plant water status, grain yield, cell membrane stability, osmotic adjustment, rate of non-stomatal water loss and deep and thick root traits (Hemamalini et al., 2000; Zheng et al., 2000; Zhang et al., 2001b; Nguyen et al., 2004; Boopathi et al., 2005). The fourth important region for drought was located in chromosome 9, which was characterized by QTL for root traits, cell membrane stability, plant water status, leaf rolling and leaf drying, biomass, number of grains per panicle, relative spikelet fertility and delay in flowering time (Hemamalini et al., 2000; Tripathy et al., 2000; Zhang et al., 2001a; Kamoshita et al., 2002; Price et al., 2002a,b; Babu et al., 2003; Courtois et al., 2003; Robin et al., 2003; Zheng et al., 2003; Lafitte et al., 2004; Lanceras et al., 2004; Nguyen et al., 2004; Boopathi et al., 2005; Gomez et al., 2005; Li et al.,2005; Xu et al., 2005; Yue et al., 2006; Jearakongman, 2005). Since these four regions are consistently reported to be associated with drought response and stood above the average, these regions should be part of marker assisted breeding program for drought tolerance in rice.

Cold and heat stress tolerance

Tolerance to freezing temperatures is the most important component for winter survival, but also of considerable importance is the capability to withstand combinations of stresses due to desiccation, wind, ice-encasement, heaving, low light, snow cover, winter pathogens, and fluctuating temperatures. Resistance to desiccation through the maintenance of cell membrane integrity and retention of cellular water is essential, and it is unsurprising that the same genetic response to the onset of freezing temperatures is often observed with drought or salinity stress (Yue et al., 2006). Indeed, cold acclimation (CA) can frequently improve adaptation to a mild drought stress and *vice versa* (Seki et al., 2002). Major genes or gene clusters involved in the control of frost and drought adaptation are located on a region of the long arm of Triticeae group 5 chromosomes. Traits such as winter hardiness (Thomas and James, 1993), vernalization response and frost tolerance (Galiba et al., 1995; Hayes et al., 1998), cold- and drought-induced ABA production (Laurie et al.,1995), and osmotic stress-tolerance (Galiba et al., 1993), have all been mapped to this region. Across the grasses and cereals, this chromosome region has been a major focus for genome research and for plant breeding. It may well be as consequence of climate change from the perspective of future crop design that in many locations where winter temperatures are on the increase and favoring continued plant growth, and where this is accompanied by a decrease in winter rainfall, that unseasonal winter droughts will ensue, which will require a new breeding strategy for common stress tolerance to both stress factors.

The C-repeat binding factor (CBF) genes are key regulators of the expression of COR (cold regulated genes), which are conserved among diverse plant lineages such as eudicots and monocots. The CBF transcription factors recognize the cis-acting CRT/DRE (C-repeat/dehydration responsive element) element in the regulatory regions of COR genes (Stockinger et al., 1997). Twenty CBF genes have been identified in barley (*H. vulgare*), of which 11 are found in two tight tandem clusters on the long arm of chromosome 5H in the same region as the *Fr-H2* frost tolerance locus (Skinner et al., 2006; Francia et al., 2007). An orthologous genomic region in *T. monococcum* contains similar CBF gene clusters located at the *Fr-A^m2*frost tolerance QTL (Vagujfalvi et al., 2003; Miller et al., 2006). Studies of the organization of CBF cluster in barley and wheat have shown that the number of CBF genes at *Fr-H2/Fr-A1* locus may vary among cultivars with winter forms having a higher copy number of some CBFs (Francia et al., 2007; Knox et al.,2010). The co-segregation of CBF gene clusters with barley *Fr-H2* and wheat *Fr-A^m2* frost tolerance loci, their role in cold acclimation (Stockinger et al., 1997), and the association of transcript levels of CBF genes with frost tolerance loci (Vagujfalvi et al., 2003) make them obvious candidates for one of the two major frost tolerance QTL on Triticeae group 5 chromosomes. The locations of two frost tolerance/winter survival QTL on the chromosome 5F of forage grass *Festuca pratensis* correspond most likely to the *Fr-A1* and *Fr-A2* loci on wheat homoeologous group 5A chromosomes. One of these QTL (*QFt5F-2/QWs5F-1*) has*FpCBF6* as a candidate gene shown to be rapidly up-regulated during CA (Alm et al., 2011).

Conversely, many crops are currently grown in places, where high temperature prevails and field studies have indicated that increase in temperature reduces grain yield of cereals and legumes by 4–14% per 1°C increase (Quarrie et al., 1997). Current projections indicate that both day and night temperatures are likely to increase during this century (Hall and Ziska, 2000) and ideally, heat-resistant cultivars should not only have higher grain yields in hot environments but also similar grain yields as current cultivars in cool atmosphere. Public plant breeding programs have developed heat-resistant cultivars of cowpea, common bean, tomato and Pima cotton that are more productive in hot environments than standard cultivars. Commercial plant breeding companies rarely divulge their methods, but from the available heat-resistant commercial cultivars, it is clear that they have had some success in breeding for heat tolerance during reproductive development in tomato and upland cotton. In the past, very few public or commercial plant-breeding programs gave any emphasis to breeding heat-resistant cultivars. For crops that are sensitive to high temperatures during reproductive development the way forward is to give great emphasis to breeding and finding DNA markers for heat adaptation during flowering.

Submergence and salinity tolerance

Waterlogging is a major problem for cereal production worldwide, as in sodic environments, soils are affected by seepage from irrigation canals, and excess wetting due to rainfall or floods, especially if it rains after irrigation. Genetic diversity in waterlogging tolerance was reported in various crops, including wheat, barley, maize, and oats (Kerr, 1986), and diverse mechanisms have been associated with tolerance. They are associated with phenology and morphology, nutrition balances, metabolism, including anaerobic catabolism and anoxia tolerance, and post-anoxia damage and recovery (Setter and Waters, 2003). Tolerance of flooding during germination and early seedling growth is essential for direct seeding of rice, both in rainfed and irrigated areas, where even waterlogging is sufficient to cause considerable reduction in crop stand because of their high sensitivity to hypoxia at this stage (Ismail et al., 2009). Substantial genetic variation was recently observed in the ability to germinate and establish in flooded soil. Tolerant genotypes are capable of catabolizing starch reserves in seeds germinating under hypoxia into simple sugars, and use them as substrates to generate energy via anaerobic pathways for the growing embryos (Miro and Ismail, 2013; Septiningsih et al., 2013). Several QTL originating from a few rice landraces were identified, two of them with large effects; on chromosome 9 (*qAG-9-2*) (Angaji et al., 2010) and chromosome 7 (*qAG-7-1*) (Septiningsih et al., 2013). These QTL are being targeted for cloning and for use through MAB, which could eventually result in high yielding rice cultivars for direct seeding in flood-borne areas. Recently, tolerant rice varieties carrying *SUB1* locus became available. *SUB1* is a major QTL on chromosome 9 that has been cloned and the gene responsible for tolerance identified as *SUB1A-1*. This gene encodes an ERF that suppresses ethylene-mediated responses under submergence, and subsequently limits excessive elongation and halts chlorophyll degradation. Both processes are essential to prevent carbohydrate starvation of the submerged plants. These varieties can survive 4–18 days of complete submergence, with yield benefits of 1 to over 3.5 t ha^{-1} (depending on flood duration and floodwater condition), compared to current farmers' varieties, and without any undesirable effects on the features of the original varieties (Singh et al., 2009; Bailey-Serres et al., 2010; Mackill et al., 2012; Ismail et al., 2013). Additional genes are being targeted for submergence tolerance, and once identified they could be combined with *SUB1* for higher tolerance during complete submergence. Further, the progress made in rice could potentially be exploited to improve flood tolerance of other crop species and provide more resilient varieties for current and future flood-affected areas.

Progressive salt accumulation due to excessive irrigation with poor water quality coupled with poor or improper drainage results in high salt levels (Ismail et al., 2007). Numerous studies have characterized responses mediated by salt stress in different plant species and highlighted the complexity of the mechanisms involved (Munns and Tester, 2008). Studies have shown that few major loci and many minor ones were associated with various aspects of salinity tolerance. The best known for rice is *Saltol* on chromosome 1 (Thomson et al., 2010), which possesses a major gene, *OsHKT1;5* (Ren et al., 2005). In wheat, two members of *HKT* gene family (including the wheat *HKT1;5* ortholog) have also been shown to co-localize with major QTL (Byrt et al., 2007). Apparently, many other QTL have been identified in rice and other cereals, and several of them are common across mapping populations. In addition, numerous genes have been identified through functional genomics studies of salt-stress responses, and many of them lead to improved tolerance when they are over- or under-expressed. Some even co-localize with QTL regions, but there has been no further success in using them for breeding tolerant cereal crops or in cloning additional QTL.

Current approaches in this aspect involves using NGS to target major QTL for cloning, and to develop efficient SNP and InDel marker systems to manipulate these loci during MAB. The substantial genetic diversity in the tolerance of salt stress and mechanisms used by various crops to cope with increasing salt concentrations in soil and water provides opportunities to enhance salt-stress tolerance in cereals. However, this will require large investments to dissect and combine the genetic determinants of various traits. Developing such cultivars that are highly tolerant of salt stress is a requisite to cope with the current worsening climatic conditions and to meet the urgent need of producing more food from marginal land and limited water resources.

Host plant resistance to pathogens and pests

The climatic variables including changes in temperature, rainfall, and atmospheric chemical composition along with predominantly elevated CO_2 levels would accelerate the reproduction time of many plant pathogens and pests, thereby increasing their infection pressure on crop plants (Boonekamp, 2012). Climate change may also affect the ability of plants to express resistance to pathogens and insects. Experiments conducted by Huang et al. (2009) indicated a 45% increase in leaf lesions in oilseed rape, when the surrounding temperature was increased by 5°C. This finding suggests that the expression and efficacy of R-genes in host plants may be affected where both crop and associated pathogen or pest are affected by climatic variation. This may be influenced by different combinations of selective pressures, and each may respond to these pressures at different rates. Improved understanding on the host-pest/pathogen interactions and knowledge on different effects of climate change is a requisite for the development of climate-resilient crops. To date, research on the impact of climate change on plant diseases has been limited, with many studies focusing on the effects of a single atmospheric constituent or meteorological variable on the host, pathogen, or the interaction of the two, under controlled conditions. Whilst this work is a valuable base to start from, the combined effects of biotic and abiotic stresses must be studied (Ramegowda and Senthil-Kumar, 2015).

Recent advances in genome sequencing and genotyping assays allow for many strategies at the genomics level, which can be developed to understand the impact of climate change on plant diseases. The newly available genome sequences for plants, pathogens and pests provide the resources to study their co-evolution in response to climate change. An understanding of the co-evolution of genes responsible for virulence and resistance will lead to improved plant protection strategies and provide a model to understand plant-pathogen and plant-insect interactions in diverse species. Though it is important to understand the genomics of disease resistance in crop species, and how allelic differences are altering resistance, combining this with studies of CWR or germplasm collections further allows the identification of novel variants. These variants can be used for the introgression of novel resistance genes into cultivars, utilizing the germplasm for breeding and developing new cultivars, or genetic engineering with the advantageous genes. Taken together, it is obvious that the impact of climate change on disease resistance is difficult to predict and is likely to be variable depending on the crop and local environment. However, crop disease is an important factor when considering the impact of climate change on food production and intensive studies applying advanced genomics tools will be required to help ameliorate the impact of climate change on future cropping scenarios in relation to plant disease.

Genomic engineering tools for targeted mutagenesis by editing genes for adaptation

Plant breeders have been applying mutagenesis to induce genetic variation for increasing crop yield and later, the strategy has been used for improving the adaptability of crop plants. Initially, X-ray radiation was used as a mutagen since it was readily available to researchers (Muller, 1927). Subsequently, gamma-ray radiation has been used to induce point mutations, although chromosomal mutations were also produced (Devreux and Scarascia Mugnozza, 1964). From recent times, chemical mutagenesis is being practiced since they are easy to use, do not require any specialized equipment, and can provide a very high mutation frequency. Compared to radiation methods, chemical mutagens tend to induce SNPs rather than chromosomal mutations. Currently, chemical mutagens, such as Ethyl methanesulfonate (EMS) are being used to induce random mutations into the genome and have become a useful complement to the isolation of nuclear DNA from mutated lines by TILLING (Targeting Induced Local Lesions in Genomes) technology and screening of the M_2 population at the DNA level using advanced molecular techniques. Single mutations in specific genes for adaptation could be identified by cleavage of mismatched bases formed as a result of repeated melting and reannealing of PCR products amplified from a pool of alleles for the specific gene in a pool of DNA from a set (usually 8) of M_2plants (McCallum et al., 2000; Caldwell et al., 2004). NGS can efficiently accelerate the identification of mutations at the whole-genome level. Promotor mutations and mutation in other regulatory elements responsible for the downstream effect can be identified by qPCR, microarray and RNA-seq. Once a mutant allele is identified within gene(s) of interest, those mutations may be linked to a specific phenotype for stress resistance by backcrossing the mutant to the parental line. This TILLING approach is a reverse genetics procedure to associate a mutant allele to its phenotype. Of note, TILLING can also be used for a forward genetic approach by screening phenotypes for adaptation to stresses and then characterize the phenotype using a combination of whole-genome resequencing, linkage maps and microarrays, to gain a broad picture of gene expression changes due to the newly introduced SNPs compared to the original line.

Other molecular tools and resources are now available for genome engineering and reverse genetics experiments in crop plants in order to implement precise manipulation of genetic building blocks and regulatory machinery that underlie yield improvement under stress condition and directly correct harmful mutations by genome editing (Hsu et al., 2014). Targeted genome engineering has emerged as an alternative to classical plant breeding and transgenic methods to improve crop plants and ensure sustainable food production (Belhaj et al., 2013; Osakabe and Osakabe, 2015). Currently, four types of engineered nucleases are used for genome editing: engineered homing endonucleases/meganucleases (EMNs) (Silva et al., 2011), zinc finger nucleases (ZFNs) (Townsend et al., 2009), transcription activator-like effector nucleases (TALENs) (Cermak et al., 2011), and CRISPR (clustered regularly interspaced short palindromic repeats)/Cas (CRISPR-associated)9 (Cong et al., 2013; Mali et al., 2013).

Sequence-specific nucleases (SSN) enable precise genome engineering by introducing DNA double-strand breaks (DSB) that subsequently trigger endogenous DNA repair by non-homologous end joining (NHEJ) or homology directed repair (HDR) recombination mechanisms in different species. Site-directed mutagenesis mediated via NHEJ can be achieved while HDR cause directed gene knock-in/correction at specific locations in the genome (HDR uses a DNA template to replace the DNA sequence at the break point). NHEJ functions throughout the entire cell cycle whereas HR is restricted to late S/G2 phases in the cell cycle. Therefore, NHEJ is the major DSB repair pathway in eukaryotes. Belhaj et al. (2013) and Osakabe and Osakabe (2015) display a clear illustration of genome editing assays in model (*Arabidopsis thaliana* and*Nicotiana benthamiana*) and crop (*Oryza sativa, Triticum aestivum* and *Sorghum bicolor*) plant species. These SSN effects generate targeted genome modifications including mutations, insertions, replacements and chromosome rearrangements and have been induced in a variety of important crops, such as rice, maize, wheat, barley, and soybean. Each technology has advantages and disadvantages with regard to cost, ease of construction, efficiency of targeting, and specificity (Chen and Gao, 2014; Gao, 2015). Major advantages of ZFNs are related to the acceptance of the technology as no transgenic is produced because viral vectors have been used for expressing transiently the nuclease, which do not integrate into the genome. However, it has disadvantages such as difficulties to design the experiments, limited number of target sites, and the regeneration of juvenile and chimeric mutated plants when custom-designed nucleases have been delivered in tree explants.

CRISPR was first discovered as an immune system of prokaryotes, which subsequently became a powerful tool for genome editing in eukaryotes (Gaj et al., 2013). It has emerged as an alternative to classical plant breeding and transgenic methods to improve crop plants. Plant transformation and co-expression of the Cas9 with a chimeric guide-RNA (gRNA) targeting a GN19NGG motif in the gene of interest, results in a double-strand non-self DNA cleavage on both strands at a specific site near the protospacer adjacent motif (PAM) (Gasiunas et al., 2012; Jinek et al., 2012). The Cas9 from *Streptococcus pyogenes* (SpCas9) recognizes 5′-NGG-3′ as the PAM sequence (Hsu et al., 2013). PAM plays an important role in target binding and cleavage by the Cas9–gRNA complex (Lei et al., 2014). CRISPR/Cas9 has greater number of advantages, including the straightforward construct design and assembly and the achievement of high mutation rates, matching or exceeding those obtained with ZFNs and TALENs. Only 20 nucleotides in the gRNA need to be modified to recognize a different target making unnecessary the sophisticated protein engineering for each target that is crucial for the other SSN approaches.

So far, the CRISPR-Cas9 technology has been applied in *A. thaliana* (Feng et al., 2013, 2014; Jiang et al.,2013; Li et al., 2013; Mao et al., 2013) *Nicotiana benthamiana* (Jiang et al., 2013; Li et al., 2013; Nekrasov et al., 2013), *Oryza sativa* (Feng et al., 2013; Jiang et al., 2013; Mao et al., 2013; Miao et al., 2013; Shan et al., 2013; Xie and Yang, 2013; Zhang et al., 2014a), *Solanum lycopersicum* (Brooks et al., 2014), *Sorghum bicolor* (Jiang et al., 2013), *Triticum aestivum* (Wang et al., 2014), *Citrus sinensis* (Jia and Wang, 2014), and*Populus tremula* x *alba* (Zhou et al., 2015). Genes controlling traits of importance for adaptation have also been edited by CRISPR-Cas9 technology. gRNAs were designed to target three specific sites of the rice*OsMPK5* gene which encodes a stress-responsive rice mitogen-activated protein kinase and the targeted mutation of *OsMPK5* enhanced rice disease resistance (Xie and Yang, 2013). Transgenic wheat plants carrying mutations in *TaMLO-A1* allele were susceptible to powdery mildew diseases (Wang et al., 2014). The bacterial blight susceptibility genes, *OsSWEET14* and *OsSWEET11*, were targeted for mutation at the promoter region in Arabidopsis, tobacco, sorghum and rice (Jiang et al., 2013). High CRISPR-Cas9 mutational efficiency was achieved for three 4-coumarate:CoA ligase (4CL) genes, *4CL1*, *4CL2* and *4CL5*, associated with lignin and flavonoid biosynthesis in *Populus tremula* x *alba* (Zhou et al., 2015).

Moreover, accelerated breeding of crop plants carrying targeted gene mutation(s) without foreign DNA is possible using CRISPR genome editing. In fact, although transgene Cas9 and selectable marker integration is hemizygous,

CRISPR editing at the target loci is biallelic. Therefore, in autogamous plants, self-fertilization of T_1 plants will provide 25% of the T_2 plants without the transgene but maintaining the edited gene in homozygosity. In self-incompatible or dioecius perennial woody trees, biparental hemizygous Cas9/sgRNA transformation and biallelic-edited gene can be produced. Controlled crosses between male and female primary transformants with confirmed biallelic mutations should produce transgene-free, biallelic mutants in 25% of the progeny (Zhou et al., 2015). Taken together, genome engineering for targeted mutagenesis by editing genes serves as a potential strategy for generating elite cultivars of crop plants with durable climate resilience.

Concluding remarks

Global climate change is predicted to impose a severe threat to agricultural productivity worldwide, and thereby challenge food security and nutritional security. Technological advances, particularly transgene-based and molecular breeding technologies have facilitated the development of elite genotypes with durable adaptation to climate change. Genomics-assisted breeding in particular is predicted to playing a significant role in the development of climate change resilient crops. Excellent model organisms for climate change such as foxtail millet and green foxtail (for C_4 photosynthesis), *Brachypodium* (grass model) have been identified for deciphering traits that need to be decoded and introgressed in the crop plants. Advances in DNA sequencing technologies and the sequencing of CWR, along with advanced genomics tools will expedite the identification of novel genes and key regulatory regions of stress tolerance toward the development of new cultivars with durable resistance. Although, the impact of climate change on crop's resistance is difficult to predict and is likely to be variable depending on the crop and environment, genomics-assisted breeding could contribute significantly to reduce the impact of climate change on future cropping scenarios.

Author contributions

All the authors contributed equally to the manuscript. CK prepared the outline of the contents, identified the authors from the members and advisors of the International Climate Resilient Crop Genomics Consortium (www.icrcgc.org), and coordinated the drafting of the manuscript. RH, DE and RS assisted in editing. CK, MP and MM revised the manuscript and prepared the final version.

Conflict of interest statement

The authors declare that the research was conducted in the absence of any commercial or financial relationships that could be construed as a potential conflict of interest.

ACKNOWLEDGMENTS

This manuscript was researched and composed by the members of International Climate-Resilient Crop Genomics Consortium (ICRCGC). The white paper authored by the members is available athttp://www.climatechangegenomics.org/climate_change_version_1.1.pdf andhttp://www.icrcgc.org/ncc_manuscript.docx.

REFERENCES

1. Ahmad R., Parfitt D. E., Fass J., Ogundiwin E., Dhingra A., Gradziel T. M., et al. . (2011). Whole genome sequencing of peach (Prunus persica L.) for SNP identification and selection. BMC Genomics 12:569.
2. Ainsworth E. A., Ort D. R. (2010). How do we improve crop production in a warming world? Plant Physiol. 154, 526–530
3. Alm V., Busso C. S., Ergon A., Rudi H., Larsen A., Humphreys M. W., et al. . (2011). QTL analyses and comparative genetic mapping of frost tolerance, winter survival and drought tolerance in meadow fescue (Festuca pratensis huds.). Theor. Appl. Genet. 123, 369–382.
4. Alves-Freitas D., Kilian A., Grattapaglia D. (2011). Development of DArT (Diversity Arrays Technology) for high-throughput genotyping of Pinus taeda and closely related species. BMC Proc.5:22
5. Anbessa Y., Taran B., Warkentin T. D., Tullu A., Vandenberg A. (2009). Genetic analyses and conservation of QTL for ascochyta blight resistance in chickpea (Cicer arietinum L.). Theor. Appl. Genet 119, 757–765.
6. Angaji S., Septiningsih E. M., Mackill D. J., Ismail A. M. (2010). QTLs associated with tolerance of flooding during germination in rice (Oryza sativa L.). Euphytica 172, 159–168.
7. Anuradha C., Gaur P. M., Pande S., Gali K. K., Ganesh M., Kumar J., et al. (2011). Mapping QTL for resistance to botrytis grey mould in chickpea. Euphytica 182, 1–9.
8. Araus J. L., Cairns J. E. (2014). Field high-throughput phenotyping: the new crop breeding frontier.Trends Plant Sci. 19, 52–61.
9. Argout X., Salse J., Aury J. M., Guiltinan M. J., Droc G., Gouzy J., et al. . (2011). The genome ofTheobroma cacao. Nat. Genet. 43, 101–108.

10. Asfaw A., Blair M. W., Struik P. C. (2012). Multienvironment quantitative trait loci analysis for photosynthate acquisition, accumulation, and remobilization traits in common bean under drought stress. G3 (Bethesda) 2, 579–595.

11. Babu R. C., Sarkarung S., Wade L. J., Nguyen H. T., Nguyen B. D., Chamarerk V., et al. (2003).Genetic analysis of drought resistance in rice by molecular markers: association between secondary traits and field performance. Crop Sci. 43, 1457–1469.

12. Bailey-Serres J., Fukao T., Ronald P., Ismail A., Heuer S., Mackill D. (2010). Submergence tolerant rice: Sub1's journey from landrace to modern cultivar. Rice 3, 138–147.

13. Bajaj D., Saxena M. S., Kujur A., Das S., Badoni S., Tripathi S., et al. . (2014). Genome-wide conserved non-coding microsatellite (CNMS) marker-based integrative genetical genomics for quantitative dissection of seed weight in chickpea. J. Exp. Bot. 66, 1271–1290.

14. Bajaj D., Upadhyaya H. D., Khan Y., Das S., Badoni S., Shree T., et al. . (2015). A combinatorial approach of comprehensive QTL-based comparative genome mapping and transcript profiling identified a seed weight-regulating candidate gene in chickpea. Sci. Rep. 5:9264.

15. Beebe S., Rao I. M., Cajiao C., Grajales M. (2008). Selection for drought resistance in common bean also improves yield in phosphorus limited and favorable environments. Crop Sci. 48, 582–592.

16. Belhaj K., Chaparro-Garcia A., Kamoun S., Nekrasov V. (2013). Plant genome editing made easy: targeted mutagenesis in model and crop plants using the CRISPR/Cas system. Plant Methods 9:39.

17. Bennetzen J. L., Schmutz J., Wang H., Percifield R., Hawkins J., Pontaroli A. C., et al. . (2012).Reference genome sequence of the model plant Setaria. Nat. Biotechnol. 30, 555–561.

18. Bentley A. R., Horsnell R., Werner C. P., Turner A. S., Rose G. A., Bedard C., et al. . (2013). Short, natural, and extended photoperiod response in BC2F4 lines of bread wheat with different Photoperiod-1 (Ppd-1) alleles. J. Exp. Bot. 64, 1783–1793.

19. Bentley A. R., Turner A. S., Gosman N., Leigh F. J., Maccaferri M., Dreisigacker S., et al. . (2011).Frequency of photoperiod-insensitive Ppd-A1a alleles in tetraploid, hexaploid and synthetic hexaploid wheat germplasm. Plant. Breed. 130, 10–15.

20. Bidinger F. R., Napolean T., Hash C. T. (2007). Quantitative trait loci for grain yield in pearl millet under variable post flowering moisture conditions. Crop Sci. 47, 969–980.

21. Bohra A., Dubey A., Saxena R. K., Penmetsa R. V., Poornima K. N., Kumar N., et al. . (2011).Analysis of BAC-end sequences (BESs) and development of BES-SSR markers for genetic mapping and hybrid purity assessment in pigeonpea (Cajanus spp.). BMC Plant Biol. 11:56.

22. Boonekamp P. M. (2012). Are plant diseases too much ignored in the climate change debate? Eur. J. Plant Pathol. 133, 291–294.

23. Boopathi N. M., Chezhain P., Jeyaprakash P., Satheesh Kumar S., Gomez M., Suresh R., et al. (2005). QTL mapping of drought-resistance traits using indica rice (Oryza sativa L.) lines adapted to target population of environment, in Proceedings of the 2nd International Conference on Integrated Approaches to Sustain and Improve Plant Production under Drought Stress (Rome:).

24. Borrell A. K., Mullet J. E., George-Jaeggli B., van Oosterom E. J., Hammer G. L., Klein P. E., et al. . (2014). Drought adaptation of stay-green sorghum is associated with canopy development, leaf anatomy, root growth, and water uptake. J. Exp. Bot. 65, 6251–6263.

25. Brooks C., Nekrasov V., Lippman Z. B., Van Eck J. (2014). Efficient Gene editing in tomato in the first generation using the clustered regularly interspaced short palindromic repeats/CRISPR-Associated9 System. Plant Physiol. 166, 1292–1297.

26. Brown J. K. M. (2003). A cost of disease resistance: paradigm or peculiarity? Trends Genet. 19, 667–671.

27. Byrt C. S., Platten J. D., Spielmeyer W., James R. A., Lagudah E. S., Dennis E. S., et al. . (2007).HKT1;5-like cation transporters linked to Na+ exclusion loci in wheat, Nax2 and Kna1. Plant Physiol.143, 1918–1928.

28. Caldwell D. G., McCallum N., Shaw P., Muehlbauer G. J., Marshall D. F., Waugh R. (2004). A structured mutant population for forward and reverse genetics in Barley (Hordeum vulgare L.). Plant J.40, 143–150.

29. Cermak T., Doyle E. L., Christian M., Wang L., Zhang Y., Schmidt C., et al. (2011). Efficient design and assembly of custom TALEN and other TAL effector-based constructs for DNA targeting. Nucliec Acids Res. 39:e82

30. Chen K., Gao C. (2014). Targeted genome modification technologies and their applications in crop improvements Plant Cell Rep. 33, 575–583.

31. Cobos M. J., Winter P., Kharrat M., Cubero J. I., Gil J., Milian T., et al. (2009). Genetic analysis of agronomic traits in a wide cross of chickpea. Field Crops Res. 111, 130–136.

32. Cong L., Ran F. A., Cox D., Lin S., Barretto R., Habib N., et al. . (2013). Multiplex genome engineering using CRISPR/Cas systems. Science 339, 819–823.

33. Cooper M., Gho C., Leafgren R., Tang T., Messina C. (2014). Breeding drought-tolerant maize hybrids for the us corn-belt: discovery to product. J. Exp. Bot. 65, 6191–6204.

34. Courtois B., Shen L., Petalcorin W., Carandang S., Mauleon R., Li Z. (2003). Locating QTLs controlling constitutive root traits in the rice population IAC 165 × Co39. Euphytica 134, 335–345.

35. Das S., Upadhyaya H. D., Bajaj D., Kujur A., Badoni S., Laxmi, et al. . (2015). Deploying QTL-seq for rapid delineation of a potential candidate gene underlying major trait-associated QTL in chickpea.DNA Res. 22, 193–203.

36. Devreux M., Scarascia Mugnozza G. T. (1964). Effects of gamma radiation of the gametes, zygote and proembryo in Nicotiana Tabacum L. Rad. Botany 4, 373–386.

37. D'Hont A., Denoeud F., Aury J. M., Baurens F. C., Carreel F., Garsmeur O., et al. (2012). The banana (Musa acuminata) genome and the evolution of monocotyledonous plants. Nature 21, 213–217.

38. Dubey A., Farmer A., Schlueter J., Cannon S. B., Abernathy B., Tuteja R., et al. . (2011). Defining the transcriptome assembly and its use for genome dynamics and transcriptome profiling studies in pigeonpea (Cajanus cajan L.). DNA Res. 18, 153–164.

39. Dutta S., Kumawat G., Singh B., Gupta D., Singh S., Dogra V., et al. . (2011). Development of genic-SSR markers by deep transcriptome sequencing in pigeonpea [Cajanus cajan (L.) Millspaugh]. BMC Plant Biol. 11:17.

40. Dwivedi S. L., Sahrawat K. L., Upadhyaya H. D., Ortiz R. (2013). Food, nutrition and agrobiodiversity under global climate change. Adv. Agron. 120, 1–128.

41. Edwards D., Wang X. (2012). Genome sequencing initiatives in Genetics, Genomics and Breeding of Oilseed Brassicas, eds Edwards D., Batley J., Parkin I. A. P., Kole C., editors. (Enfield, NH: Science Publishers;), 152–157.

42. Edwards D., Batley J., Snowdon R. J. (2013). Accessing complex crop genomes with next-generation sequencing. Theor. Appl. Genet. 126, 1–11.

43. FAO (2014). The State of Food Insecurity in the World 2014. Strengthening the Enabling Environment for Food Security and Nutrition. Rome: FAO.

44. Feng Z. Y., Mao Y. F., Xu N. F., Zhang B. T., Wei P. L., Wang Z., et al. . (2014). Multi-generation analysis reveals the inheritance, specificity and patterns of CRISPR/Cas induced gene modifications in Arabidopsis. Proc. Natl. Acad. Sci. U.S.A. 111, 4632–4637.

45. Feng Z., Zhang B., Ding W., Liu X., Yang D.-L., Wei P., et al. . (2013). Efficient genome editing in plants using a CRISPR/Cas system. Cell Res. 23, 1229–1232.

46. Fischer R. A., Byerlee D., Edmeades G. O. (2009). Can Technology Deliver on the Yield Challenge to 2050? Rome: Food and Agriculture Organization of the United Nations.

47. Francia E., Pecchioni N., Barabaschi D., Tondelli A., Laidò G., Rizza F., et al. . (2007). Fine mapping of a HvCBF gene cluster at the frost resistance locus Fr-H2 in barley. Theor. Appl. Genet. 115, 1083–1091

48. Gaj T., Gersbach C. A., Barbas C. F. (2013). ZFN, TALEN, and CRISPR/Cas-based methods for genome engineering. Trends Biotechnol. 31, 397–405.

49. Galiba G., Quarrie S. A., Sutka J., Morgounov A., Snape J. W. (1995). RFLP mapping of the vernalization (VRN1) and Frost-resistance (FR1) genes on chromosome 5A of wheat. Theor. Appl. Genet. 90, 1174–1179.

50. Galiba G., Tuberosa R., Kocsy G., Sutka J. (1993). Involvement of chromosomes 5A and 5D in coldinduced abscisic acid accumulation and in frost tolerance of wheat calli. Plant Breed. 110, 237–242.

51. Gao C. (2015). Developing genome editing technologies for crop improvement, in Meeting on Plant Genome Engineering, Plant & Animal Genome XXIII (San Diego, CA). Available online at:https://pag.confex.com/pag/xxiii/webprogram/Paper14087.html (Accessed February 15, 2015).

52. Gasiunas G., Barrangou R., Horvath P., Siksnys V. (2012). Cas9-crRNA ribonucleoprotein complex mediates specific DNA cleavage for adaptive immunity in bacteria. Proc. Natl. Acad. Sci. U.S.A. 109, E2579–E2586.

53. Gaur R., Sethy N. K., Choudhary S., Shokeen B., Gupta V., Bhatia S. (2011). Advancing the stms genomic resources for defining new locations on the intraspecific genetic linkage map of chickpea (Cicer arietinum L.). BMC Genomics 12:117.

54. Gautami B., Foncéka D., Pandey M. K., Moretzsohn M. C., Sujay V., Qin H., et al. . (2012). An international reference consensus genetic map with 897 marker loci based on 11 mapping populations for tetraploid groundnut (Arachis hypogaea L.). PLoS ONE 7:e41213.

55. Gepts P., Aragão F. (2008). Genomics of Phaseolus beans, a major source of dietary protein and micronutrients in the tropics, in Genomics of Tropical Crop Plants, eds Moore P., Ming R., editors. (Berlin: Springer;), 113–143.

56. Gomez M., Babu R. C., Shanmugasundaram P., Kumar S. R., Suresh R., Biji K. R., et al. (2005).QTL mapping and marker assisted selection for drought tolerance in rice (Oryza sativa L.), inProceedings of the 2nd International Conference on Integrated Approaches to Sustain and Improve Plant Production under Drought Stress (Rome:).

57. Gujaria N., Kumar A., Dauthal P., Dubey A., Hiremath P., Prakash A. B., et al. . (2011).Development and use of genic molecular markers (GMMS) for construction of a transcript map of chickpea (Cicer arietinum L.). Theor. Appl. Genet. 122, 1577–1589.

58. Gupta B., Sengupta A., Saha J., Gupta K. (2013). Plant abiotic stress: 'Omics' approach. J. Plant Biochem. Physiol. 1:e108

59. Hall A. E., Ziska L. H. (2000). Crop breeding strategies for the 21st century, in Climate Change and Global Crop Productivity, eds Reddy K. R., Hodges H. F., editors. (New York, NY: CAB International;), 407–423.

60. Hayes P. M., Blake T., Chen T. H., Tragoonrung S., Chen F., Pan A., et al. . (1998). Quantitative trait loci on barley (Hordeum vulgare L.) chromosome 7 associated with components of winter hardiness.Genome 36, 66–71.

61. Hemamalini G. S., Shashidhar H. E., Hittalmani S. (2000). Molecular marker assisted tagging of morphological and physiological traits under two contrasting moisture regimens at peak vegetative stage in rice (Oryza sativa L). Euphytica 112, 69–78.

62. Hiremath P. J., Kumar A., Penmetsa R. V., Farmer A., Schlueter J. A., Chamarthi S. K., et al. . (2012). Large-scale development of cost-effective snp marker assays for diversity assessment and genetic mapping in chickpea and comparative mapping in legumes. Plant Biotechnol. J. 10, 716–732.

63. Hisas S. (2011). The Food Gap. The Impacts of Climate Change in Food Production: A 2020 Perspective. Alexandria,VA: Universal Ecological Fund.

64. Hsu P. D., Lander E. S., Zhang F. (2014). Development and applications of CRISPR-Cas9 for Genome Engineering. Cell 157, 1262–1278.

65. Hsu P. D., Scott D. A., Weinstein J. A., Ran F. A., Konermann S., Agarwala V., et al. . (2013). DNA targeting specificity of RNA-guided Cas9 nucleases. Nat. Biotechnol. 31, 827–832.

66. Huang Y. J., Qi A., King G. J., Fitt B. D. (2009). Quantitative resistance to symptomless growth ofLeptosphaeria maculans (phoma stem canker) in Brassica napus (oilseed rape). Plant Pathol. 58, 314–323.

67. Hund A., Reimer R., Messmer R. (2011). A consensus map of QTLs controlling the root length of maize. Plant Soil 344, 143–158.

68. Imelfort M., Batley J., Grimmond S., Edwards D. (2009). Genome sequencing approaches and successes. Methods Mol. Biol. 513, 345–358.

69. Ismail A. M., Ella E. S., Vergara G. V., Mackill D.J. (2009). Mechanisms associated with tolerance to flooding during germination and early seedling growth in rice (Oryza sativa). Ann. Bot. 103, 197–209.

70. Ismail A. M., Heuer S., Thomson M. J., Wissuwa M. (2007). Genetic and genomic approaches to develop rice germplasm for problem soils. Plant Mol. Biol. 65, 547–570.

71. Ismail A. M., Singh U. S., Singh S., Dar M. H., Mackill D. J. (2013). The contribution of submergence-tolerant (Sub1) rice varieties to food security in flood-prone rainfed lowland areas in asia.Field Crops Res. 152, 83–93.

72. Jain M., Misra G., Patel R. K., Priya P., Jhanwar S., Khan A. W., et al. . (2013). A draft genome sequence of the pulse crop chickpea (Cicer arietinum L.). Plant J. 74, 715–729.

73. James R. A., Davenport R. J., Munns R. (2006). Physiological characterization of two genes for Na+exclusion in durum wheat, nax1 and nax2. Plant Physiol. 142, 1537–1547.

74. Jearakongman S. (2005). Validation and Discovery of Quantitative Trait loci for Drought Tolerance in Backcross Introgression lines in Rice (Oryza sativa L.) Cultivar IR64. Ph.D. Thesis, Kasetsart University.

75. Jia H., Wang N. (2014). Targeted genome editing of sweet orange using Cas9/sgRNA. PLoS ONE9:e93806.

76. Jiang W., Zhou H., Bi H., Fromm M., Yang B., Weeks D. P. (2013). Demonstration of CRISPR/Cas9/sgRNA-mediated targeted gene modification in Arabidopsis, tobacco, sorghum and rice. Nucleic Acids Res. 41:e188.

77. Jinek M., Chylinski K., Fonfara I., Hauer M., Doudna J. A., Charpentier E. (2012). A programmable dual-RNA-guided DNA endonuclease in adaptive bacterial immunity. Science 337, 816–821.

78. Kamoshita A., Chandra Babu R., Boopathi N. M., Fukai S. (2008). Phenotypic and genotypic analysis of drought-resistance traits for development of rice cultivars adapted to rainfed environments. Field Crops Res. 109, 1–23.

79. Kamoshita A., Wade L., Ali M., Pathan M., Zhang J., Sarkarung S., et al. . (2002). Mapping QTLs for root morphology of a rice population adapted to rainfed lowland conditions. Theor. Appl. Genet. 104, 880–893.

80. Kerr R. A. (1986). Greenhouse warming still coming. Science 232, 573–575.

81. Khan Y., Suresh B. V., Muthamilarasan M., Yadav C. B., Prasad M. (2014). Comprehensive genome-wide identification and expression profiling of foxtail millet [Setaria italica (L.)] miRNAs in response to abiotic stress and development of miRNA database. Plant Cell Tiss. Organ Cult. 118, 279–292.

82. Khedikar Y. P., Gowda M. V., Sarvamangala C., Patgar K. V., Upadhyaya H. D., Varshney R. K. (2010). A QTL study on late leaf spot and rust revealed one major QTL for molecular breeding for rust resistance in groundnut (Arachis hypogaea L.). Theor. Appl. Genet. 121, 971–984.

83. Knox A. K., Dhillon T., Cheng H., Tondelli A., Pecchioni N., Stockinger E. J. (2010). Cbf gene copy number variation at frost resistance-2 is associated with levels of freezing tolerance in temperate-climate cereals. Theor. Appl. Genet. 121, 21–35.

84. Kottapalli P., Gaur P. M., Katiyar S. K., Crouch J. H., Buhariwalla H. K., Pande S., et al. (2009).Mapping and validation of QTLs for resistance to an indian isolate of ascochyta blight pathogen in chickpea. Euphytica 165, 79–88.

85. Kudapa H., Bharti A. K., Cannon S. B., Farmer A. D., Mulaosmanovic B., Kramer R., et al. . (2012).A comprehensive transcriptome assembly of pigeonpea (Cajanus cajan L.) using sanger and second-generation sequencing platforms. Mol. Plant 5, 1020–1028.

86. Kujur A., Bajaj D., Upadhyaya H. D., Das S., Ranjan R., Shree T., et al. . (2015b). A genome-wide SNP scan accelerates trait-regulatory genomic loci identification in chickpea. Sci. Rep. 5:11166.

87. Kujur A., Upadhyaya H. D., Shree T., Bajaj D., Das S., Saxena M., et al. . (2015a). Ultra-high density intra-specific genetic linkage maps accelerate identification of functionally relevant molecular tags governing important agronomic traits in chickpea. Sci. Rep. 5:9468.

88. Kulwal P. L., Thudi M., Varshney R. K. (2011). Genomics interventions in crop breeding for sustainable agriculture, in Encyclopedia of Sustainability Science and Technology, ed. Meyers R. A., editor. (New York, NY: Springer;), 2527–2540.

89. Kumari K., Muthamilarasan M., Misra G., Gupta S., Subramanian A., Parida S. K., et al. . (2013).Development of eSSR-markers in Setaria italica and their applicability in studying genetic diversity, cross-

transferability and comparative mapping in millet and non-millet species. PLoS ONE 8:e67742.

90. Lafitte H. R., Price A. H., Courtois B. (2004). Yield response to water deficit in an upland rice mapping population: associations among traits and genetic markers. Theor. Appl. Genet. 109, 1237–1246.

91. Lanceras J. C., Pantuwan G., Jongdee B., Toojinda T. (2004). Quantitative trait loci associated with drought tolerance at reproductive stage in rice. Plant Physiol. 135, 384–399.

92. Landi P., Sanguineti M. C., Salvi S. (2005). Validation and characterization of a major QTL affecting leaf ABA concentration in maize. Mol. Breed. 15, 291–303.

93. Langridge P., Fleury D. (2011). Making the most of 'omics' for crop breeding. Trends Biotechnol. 29, 33–40.

94. Lata C., Prasad M. (2012). Validation of an allele-specific marker associated with dehydration stress tolerance in a core set of foxtail millet accessions. Plant Breed. 132, 496–499.

95. Lata C., Prasad M. (2013). Association of an allele-specific marker with dehydration stress tolerance in foxtail millet suggests SiDREB2 to be an important QTL. J. Plant Biochem. Biotechnol. 23, 119–122.

96. Lata C., Bhutty S., Bahadur R. P., Majee M., Prasad M. (2011). Association of an SNP in a novel DREB2-like gene SiDREB2 with stress tolerance in foxtail millet [Setaria italica (L.)]. J. Exp. Bot. 62, 3387–3401.

97. Lata C., Gupta S., Prasad M. (2013). Foxtail millet: a model crop for genetic and genomic studies in bioenergy grasses. Crit. Rev. Biotechnol. 33, 328–343.

98. Laurie D. A., Pratchett N., Snape J. W., Bezant J. H. (1995). RFLP mapping of five major genes and eight quantitative trait loci controlling flowering time in a winter x spring barley (Hordeum vulgare L.) cross. Genome 38, 575–585.

99. Lei Y., Lu L., Liu H.-Y., Li S., Xing F., Chen L.-L. (2014). CRISPR-P: a web tool for synthetic single-guide RNA design of CRISPR-system in plants. Mol. Plant 7, 1494–1496.

100. Li J., Aach J., Norville J. E., McCormack M., Zhang D., Bush J., et al. . (2013). Multiplex and homologous recombination-mediated genome

editing in Arabidopsis and Nicotiana benthamianausing guide RNA and Cas9. Nat. Biotechnol. 31, 688–691.

101. Li Z., Mu P., Li C., Zhang H., Li Z., Gao Y., et al. . (2005). QTL mapping of root traits in a doubled haploid population from a cross between upland and lowland japonica rice in three environments.Theor. Appl. Genet. 110, 1244–1252.

102. Lilley J. M., Ludlow M. M., McCouch S. R., O'Toole J. C. (1996). Locating QTL for osmotic adjustment and dehydration tolerance in rice. J. Exp. Bot. 47, 1427–1436.

103. Liu S., Wang X., Wang H., Xin H., Yang X., Yan J., et al. . (2013). Genome-wide analysis ofZmDREB genes and their association with natural variation in drought tolerance at seedling stage ofZea mays L. PLoS Genet. 9:e1003790.

104. Lopes M. S., Rebetzke G. J., Reynolds M. (2014). Integration of phenotyping and genetic platforms for a better understanding of wheat performance under drought. J. Exp. Bot. 65, 6167–6177.

105. Lynch J. P., Chimungu J. G., Brown K. M. (2014). Root anatomical phenes associated with water acquisition from drying 1 soil: targets for crop improvement. J. Exp. Bot. 65, 6155–6166.

106. Ma Y., Qin F., Tran L. S. (2012). Contribution of genomics to gene discovery in plant abiotic stress responses. Mol. Plant 5, 1176–1178.

107. Mackay I. J., Bansept-Basler P., Barber T., Bentley A. R., Cockram J., Gosman N., et al. . (2014). An eight-parent multiparent advanced generation inter-cross population for winter-sown wheat: creation, properties and validation. G3 (Bethesda) 4, 1603–1610.

108. Mackill D. J., Ismail A. M., Singh U. S., Labios R. V., Paris T. R. (2012). Development and rapid adoption of submergence-tolerant (Sub1) rice varieties. Adv. Agron. 115, 299–352.

109. Mali P., Yang L., Esvelt K. M., Aach J., Guell M., DiCarlo J. E., et al. . (2013). RNA-guided human genome engineering via Cas9. Science 339, 823–826.

110. Mano Y., Muraki M., Fujimori M., Takamizo T., Kindiger B. (2005a). Identification of QTL controlling adventitious root formation during flooding conditions in teosinte (Zea mays ssp.huehuetenangensis) seedlings. Euphytica 142, 33–42.

111. Mano Y., Omori F., Muraki M., Takamizo T. (2005b). QTL mapping of adventitious root formation under flooding conditions in tropical maize (Zea mays L.) seedlings. Breeding Sci. 55, 343–347.

112. Mao Y., Zhang H., Xu N., Zhang B., Gao F., Zhu J. K. (2013). Application of the CRISPRCas system for efficient genome engineering in plants. Mol. Plant 6, 2008–2011.

113. McCallum C. M., Comai L., Greene E. A., Henikoff S. (2000). Targeted screening for induced mutations. Nat. Biotechnol. 18, 455–457.

114. Miao J., Guo D., Zhang J., Huang Q., Qin G., Zhang X., et al. . (2013). Targeted mutagenesis in rice using CRISPR-Cas system. Cell Res. 23, 1233–1236.

115. Miller A. K., Galiba G., Dubcovsky J. (2006). A cluster of 11 CBF transcription factors is located at the frost tolerance locus Fr-Am2 in Triticum monococcum. Mol. Genet. Genom. 275, 193–203.

116. Miro B., Ismail A. M. (2013). Tolerance of anaerobic conditions caused by flooding during germination and early growth in rice (Oryza sativa L). Front. Plant Sci. 4:269.

117. Muller H. J. (1927). Artificial transmutation of the gene. Science 66, 84–87.

118. Munns R., James R. A., Xu B., Athman A., Conn S. J., Jordans C., et al. . (2012). Wheat grain yield on saline soils is improved by an ancestral Na+ transporter gene. Nature Biotechnol. 30, 360–364.

119. Munns R., Tester M. (2008). Mechanisms of salinity tolerance. Ann. Rev. Plant Biol. 59, 651–681.

120. Muthamilarasan M., Prasad M. (2015). Advances in Setaria genomics for genetic improvement of cereals and bioenergy grasses. Theor. Appl. Genet. 128, 1–14.

121. Muthamilarasan M., Theriappan P., Prasad M. (2013). Recent advances in crop genomics for ensuring food security. Curr Sci. 105, 155–158.

122. Muthamilarasan M., Venkata Suresh B., Pandey G., Kumari K., Parida S. K., Prasad M. (2014).Development of 5123 intron-length polymorphic markers for large-scale genotyping applications in foxtail millet. DNA Res. 21, 41–52.

123. Nekrasov V., Staskawicz B., Weigel D., Jones J. D. G., Kamoun S. (2013). Targeted mutagenesis in the model plant Nicotiana benthamiana using Cas9 RNA-guided endonuclease. Nat. Biotechnol. 31, 691–693.

124. Neves L., Davis J., Barbazuk B., Kirst M. (2011). Targeted sequencing in the loblolly pine (Pinus taeda) megagenome by exome capture. BMC Proceedings 7:O48

125. Nguyen T. T. T., Klueva N., Chamareck V., Aarti A., Magpantay G., Millena A. C. M., et al. . (2004). Saturation mapping of QTL regions and identification of putative candidate genes for drought tolerance in rice. Mol. Genet. Genom. 272, 35–46.

126. Osakabe Y., Osakabe K. (2015). Genome editing with engineered nucleases in plants. Plant Cell Physiol. 56, 389–400.

127. Paliwal R., Röder M. S., Kumar U., Srivastava J. P., Joshi A. K. (2012). QTL mapping of terminal heat tolerance in hexaploid wheat (T. aestivum L.). Theor. Appl. Genet. 125, 561–575.

128. Pandey G., Misra G., Kumari K., Gupta S., Parida S. K., Chattopadhyay D., et al. . (2013). Genome-wide development and use of microsatellite markers for large-scale genotyping applications in foxtail millet [Setaria italica (L.)]. DNA Res. 20, 197–207.

129. Pingali P. L. (2012). Green revolution: impact, limits and the path ahead. Proc. Natl. Acad. Sci. U.S.A. 109, 12302–11230.

130. Pinto R. S., Reynolds M. P., Mathews K. L., McIntyre C. L., Olivares-Villegas J. J., Chapman S. C. (2010). Heat and drought adaptive QTL in a wheat population designed to minimize confounding agronomic effects. Theor. Appl. Genet. 121, 1001–1021.

131. Price A. H., Cairns J. E., Horton P., Jones H. G., Griffiths H. (2002a). Linking drought-resistance mechanisms to drought avoidance in upland rice using a QTL approach: progress and new opportunities to integrate stomatal and mesophyll responses. J. Exp. Bot. 53, 989–1004.

132. Price A. H., Townend J., Jones M. P., Audebert A., Courtois B. (2002b). Mapping QTLs associated with drought avoidance in upland rice grown in the philippines and west africa. Plant Mol. Biol. 48, 683–695.

133. Quarrie S. A., Laurie D. A., Zhu J., Lebreton C., Semikhodskii A., Steed A., et al. . (1997). QTL analysis to study the association between leaf

size and abscisic acid accumulation in droughted rice leaves and comparisons across cereals. Plant Mol. Biol. 35, 155–165.

134. Raju N., Gnanesh B., Lekha P., Jayashree B., Pande S., Hiremath P., et al. . (2010). The first set of est resource for gene discovery and marker development in pigeonpea (Cajanus cajan L.). BMC Plant Biol. 10:45.

135. Ramegowda V., Senthil-Kumar M. (2015). The interactive effects of simultaneous biotic and abiotic stresses on plants: mechanistic understanding from drought and pathogen combination. J. Plant Physiol. 176, 47–54.

136. Rebetzke G. J., Herwaarden A. F., Jenkins C., Weiss M., Lewis D., Ruuska S., et al. (2008).Quantitative trait loci for water-soluble carbohydrates and associations with agronomic traits in wheat.Aust. J. Agric. Res. 59, 891–905.

137. Rehman A. U., Malhotra R. S., Bett K., Tar'an B., Bueckert R., Warkentin T. D. (2011). Mapping QTL associated with traits affecting grain yield in chickpea (Cicer arietinum L.) under terminal drought stress. Crop Sci. 51, 450–463.

138. Ren Z., Gao J., Li L. G., Cai X., Huang W., Chao D., et al. . (2005). A rice quantitative trait locus for salt tolerance encodes a sodium transporter. Nat. Genet. 37, 1141–1146.

139. Reynolds M. P., Hays D., Chapman S. (2010). Breeding for adaptation to heat and drought stress, inClimate Change and Crop Production, ed Reynolds M. P., editor. (Oxfordshire: CABI;), 71–91.

140. Reynolds M., Tuberosa R. (2008). Translational research impacting on crop productivity in drought-prone environments. Curr. Opin. Plant Biol. 11, 171–179.

141. Robin S., Pathan M. S., Courtois B., Lafitte R., Carandang S., Lanceras S., et al. . (2003). Mapping osmotic adjustment in an advanced back-cross inbred population of rice. Theor. Appl. Genet. 107, 1288–1296.

142. Ruta N., Liedgens M., Fracheboud Y., Stamp P., Hund A. (2010). QTLs for the elongation of axile and lateral roots of maize in response to low water potential. Theor. Appl. Genet. 120, 621–631.

143. Salem K. F. M., Roder M. S., Borner A. (2007). Identification and mapping quantitative trait loci for stem reserve mobilisation in wheat (Triticum aestivum L.). Cereal Res. Commun. 35, 1367–1374.

144. Salvi S., Sponza G., Morgante M., Tomes D., Niu X., Fengler K. A., et al. . (2007). Conserved noncoding genomic sequences associated with a flowering-time quantitative trait locus in maize. Proc. Natl. Acad. Sci. U.S.A. 104, 11376–11381.

145. Sarvamangala C., Gowda M. V. C., Varshney R. K. (2011). Identification of quantitative trait loci for protein content oil content and oil quality for groundnut (Arachis hypogaea L.). Field Crops Res. 122, 49–59.

146. Saxena K. (2008). Genetic improvement of pigeonpea-a review. Trop. Plant Biol. 1, 159–178.

147. Saxena R., Prathima C., Saxena K., Hoisington D., Singh N., Varshney R. (2010). Novel SSR markers for polymorphism detection in pigeonpea (Cajanus spp.). Plant Breed. 129, 142–148.

148. Schmutz J., McClean P. E., Mamidi S., Wu G. A., Cannon S. B., Grimwood J., et al. . (2014). A reference genome for common bean and genome-wide analysis of dual domestications. Nat. Genet. 46, 707–713.

149. Schneider K. A., Brothers M. E., Kelly J. D. (1997). Marker-assisted selection to improve drought tolerance in common bean. Crop Sci. 37, 51–60.

150. Seki M., Narusaka M., Ishida J., Nanjo T., Fujita M., Oono Y., et al. . (2002). Monitoring the expression profiles of 7000 Arabidopsis genes under drought, cold and high-salinity stresses using a full-length cDNA microarray. Plant J. 31, 279–292.

151. Septiningsih E. M., Ignacio J. C., Sendon P. M., Sanchez D. L., Ismail A. M., Mackill D. J. (2013).QTL mapping and confirmation for tolerance of anaerobic conditions during germination derived from the rice landrace Ma-Zhan Red. Theor. Appl. Genet. 126, 1357–1366.

152. Serraj R., Kumar A., McNally K. L., Slamet-Loedin I., Bruskiewich R., Mauleon R., et al. (2009).Improvement of drought resistance in rice. Adv. Agron. 103, 41–99.

153. Setter T. M., Waters I. (2003). Review of prospects for germplasm improvement for waterlogging tolerance in wheat, barley and oats. Plant Soil 253, 1–34.

154. Shan Q., Wang Y., Li J., Zhang Y., Chen K., Liang Z., et al. . (2013). Targeted genome modification of crop plants using a CRISPR-Cas system. Nat. Biotechnol. 31, 686–688.

155. Shavrukov Y., Gupta N. K., Miyazaki J., Baho M. N., Chalmers K. J., Tester M., et al. . (2010).Hvnax3-a locus controlling shoot sodium exclusion derived from wild barley (Hordeum vulgare ssp spontaneum). Funct. Integr. Genomics 10, 277–291.

156. Silva G., Poirot L., Galetto R., Smith J., Montoya G., Duchateau P., et al. . (2011). Meganucleases and other tools for targeted genome engineering: perspectives and challenges for gene therapy. Curr. Gene Ther. 11, 11–27.

157. Singh S. P. (2007). Drought resistance in the race durango dry bean landraces and cultivars.Agronomy J. 99, 1219–1225.

158. Singh S., Mackill D. J., Ismail A. M. (2009). Responses of Sub1 rice introgression lines to submergence in the field: yield and grain quality. Field Crops Res. 113, 12–23.

159. Skinner J. S., Szucs P., Zitzewitz J.v., Marquez-Cedillo L., Filichkin T., Stockinger E. J., et al. . (2006). Mapping of barley homologs to genes that regulate low temperature tolerance in arabidopsis.Theor. Appl. Genet. 112, 832–842.

160. Snape J. W., Foulkes J., Simmonds J., Leverington M., Fish L. J., Wang Y., et al. (2007). Dissecting gene × environmental effects on wheat yields via QTL and physiological analysis. Euphytica 154, 401–408.

161. Stockinger E. J., Gilmour S. J., Thomashow M. F. (1997). Arabidopsis thaliana CBF1 encodes an AP2 domain-containing transcriptional activator that binds to the C-repeat/DRE, a cis-acting DNA regulatory element that stimulates transcription in response to low temperature and water deficit. Proc. Natl. Acad. Sci. U.S.A. 94, 1035–1040.

162. Sujay V., Gowda M. V., Pandey M. K., Bhat R. S., Khedikar Y. P., Nadaf H. L., et al. . (2012).Quantitative trait locus analysis and construction of consensus genetic map for foliar disease resistance based on two recombinant inbred line populations in cultivated groundnut (Arachis hypogaea L.).Mol. Breed. 30, 773–788.

163. Suresh B. V., Muthamilarasan M., Misra G., Prasad M. (2013). FmMDb: a versatile database of foxtail millet markers for millets and bioenergy grasses research. PLoS ONE 8:e71418.

164. Thomas H., James A. R. (1993). Freezing tolerance and solute changes in contrasting genotypes ofLolium perenne L. acclimated to cold and drought. Ann. Bot. 72, 249–254.

165. Thomson M. J., Ocampo M.d., Egdane J., Rahman M. A., Sajise A. G., Adorada D. L., et al. (2010).Characterizing the saltol quantitative trait locus for salinity tolerance in rice. Rice 3, 148–160.

166. Thudi M., Bohra A., Nayak S. N., Varghese N., Shah T. M., Penmetsa R. V., et al. . (2011). Novel SSR markers from bac-end sequences, dart arrays and a comprehensive genetic map with 1, 291 marker loci for chickpea (Cicer arietinum L.). PLoS ONE 6:e27275.

167. Townsend J. A., Wright D. A., Winfrey R. J., Fu F., Maeder M. L., Joung J. K., et al. . (2009). High-frequency modification of plant genes using engineered zinc-finger nucleases. Nature 459, 442–445.

168. Tripathy J. N., Zhang J., Robin S., Nguyen T. T., Nguyen H. T. (2000). QTLs for cell-membrane stability mapped in rice (Oryza sativa L.) under drought stress. Theor. Appl. Genet. 100, 1197–1202.

169. Tuberosa R. (2012). Phenotyping for drought tolerance of crops in the genomics era. Front. Physiol.3:347.

170. Tuberosa R., Salvi S., Sanguineti M. C., Frascaroli E., Conti S., Landi P. (2011). Genomics of root architecture and functions in maize, in Root genomics, eds Costa de Oliveira A., Varshney R., editors. (Berlin: Springer;), 179–204.

171. Uga Y., Sugimoto K., Ogawa S., Rane J., Ishitani M., Hara N., et al. . (2013). Control of root system architecture by deeper rooting 1 increases rice yield under drought conditions. Nat. Genet. 45, 1097–1102.

172. Urrea C. A., Yonts C. D., Lyon D. J., Koehler A. E. (2009). Selection for drought tolerance in dry bean derived from the Mesoamerican gene pool in western Nebraska. Crop Sci. 49, 2005–2010.

173. Vadez V., Krishnamurthy L., Thudi M., Anuradha C., Colmer T. D., Turner N. C., et al. (2012).Assessment of ICCV 2 × JG 62 chickpea progenies shows sensitivity of reproduction to salt stress and reveals QTL for seed yield and yield components. Mol. Breed. 30, 9–21.

174. Vagujfalvi A., Galiba G., Cattivelli L., Dubcovsky J. (2003). The cold-regulated transcriptional activator Cbf3 is linked to the frost-tolerance

locus Fr-A2 on wheat chromosome 5A. Mol. Genet. Genom. 269, 60–67.

175. Varshney R. K., Chen W., Li Y., Bharti A. K., Saxena R. K., Schlueter J. A., et al. . (2012). Draft genome sequence of pigeonpea (Cajanus cajan), an orphan legume crop of resource-poor farmers.Nat. Biotechnol. 30, 83–89.

176. Varshney R. K., Graner A., Sorrells M. E. (2005). Genomics-assisted breeding for crop improvement.Trends Plant Sci. 10, 621–630.

177. Varshney R. K., Mohan S. M., Gaur P. M., Gangarao N. V., Pandey M. K., Bohra A., et al. . (2013a).Achievements and prospects of genomics-assisted breeding in three legume crops of the semi-arid tropics. Biotechnol. Adv. 31, 1120–1134.

178. Varshney R. K., Penmetsa R. V., Dutta S., Kulwal P. L., Saxena R. K., Datta S., et al. . (2009).Pigeonpea genomics initiative (PGI): an international effort to improve crop productivity of pigeonpea (Cajanus cajan L.). Mol. Breed. 26, 393–408.

179. Varshney R. K., Song C., Saxena R. K., Azam S., Yu S., Sharpe A. G., et al. . (2013b). Draft genome sequence of kabuli chickpea (Cicer arietinum): genetic structure and breeding constraints for crop improvement. Nat. Biotechnol. 31, 240–246.

180. Varshney R. K., Terauchi R., McCouch S. R. (2014). Harvesting the promising fruits of genomics: applying genome sequencing technologies to crop breeding. PLoS Biol. 12:e1001883.

181. Velasco R., Zharkikh A., Affourtit J., Dhingra A., Cestaro A., Kalyanaraman A., et al. . (2010). The genome of the domesticated apple (Malus x domestica Borkh.). Nat. Genet. 42, 833–839.

182. Velasco R., Zharkikh A., Troggio M., Cartwright D. A., Cestaro A., Pruss D., et al. . (2007). A high quality draft consensus sequence of the genome of a heterozygous grapevine variety. PLoS ONE2:e1326.

183. Wang H., Penmetsa R. V., Yuan M., Gong L., Zhao Y., Guo B., et al. . (2012). Development and characterization of BAC-end sequence derived SSRs, and their incorporation into a new higher density genetic map for cultivated peanut (Arachis hypogaea L.). BMC Plant Biol. 12:10.

184. Wang Y., Cheng X., Shan Q., Zhang Y., Liu J., Gao C., et al. . (2014). Simultaneous editing of three homoeoalleles in hexaploid bread wheat confers heritable resistance to powdery mildew. Nat. Biotechnol. 32, 947–951.

185. Wassmann R., Jagadish S. (2009). Regional vulnerability of climate change impacts on Asian rice production and scope for adaptation. Adv. Agron. 102, 91–133.

186. Wasson A. P., Richards R. A., Chatrath R., Misra S. C., Prasad S. V., Rebetzke G. J., et al. . (2012).Traits and selection strategies to improve root systems and water uptake in water-limited wheat crops.J. Exp. Bot. 63, 3485–3498.

187. Welch J. R., Vincent J. R., Auffhammer M., Moya P. F., Dobermann A., Dawe D. (2010). Rice yields in tropical/subtropical Asia exhibit large but opposing sensitivities to minimum and maximum temperatures. Proc. Natl. Acad. Sci. U.S.A. 107, 14562–14567.

188. Xie K., Yang Y. (2013). RNA-Guided Genome Editing in Plants Using a CRISPR-Cas System. Mol. Plant 6, 1975–1980.

189. Xu J. L., Lafitte H. R., Gao Y. M., Fu B. Y., Torres R., Li Z. K. (2005). QTLs for drought escape and tolerance identified in a set of random introgression lines of rice. Theor. Appl. Genet. 111, 1642–1650.

190. Xu K., Xu X., Fukao T., Canlas P., Maghirang-Rodriguez R., Heuer S., et al. . (2006). Sub1A is an ethylene-response-factor-like gene that confers submergence tolerance to rice. Nature 442, 705–708.

191. Xu Q., Chen L. L., Ruan X., Chen D., Zhu A., Chen C., et al. . (2013). The draft genome of sweet orange (Citrus sinensis). Nat. Genet. 45, U59–U92.

192. Yadav C. B., Bonthala V. S., Muthamilarasan M., Pandey G., Khan Y., Prasad M. (2015). Genome-wide development of transposable elements-based markers in foxtail millet and construction of an integrated database. DNA Res. 22, 79–90.

193. Yadav C. B., Muthamilarasan M., Pandey G., Khan Y., Prasad M. (2014). Development of novel microRNA-based genetic markers in foxtail millet for genotyping applications in related grass species.Mol. Breed. 34, 2219–2224.

194. Yadav R. S., Sehgal D., Vadez V. (2011). Using genetic mapping and genomics approaches in understanding and improving drought tolerance in pearl millet. J. Exp. Bot. 62, 397–408.

195. Yang J., Zhang J., Liu K., Wang Z., Liu L. (2007). Abscisic acid and ethylene interact in rice spikelets in response to water stress during meiosis. J. Plant Growth Regul. 26, 318–328.

196. Yue B., Xue W., Xiong L., Yu X., Luo L., Cui K., et al. . (2006). Genetic basis of drought resistance at reproductive stage in rice: separation of drought tolerance from drought avoidance. Genetics 172, 1213–1228.

197. Zhang G., Liu X., Quan Z., Cheng S., Xu X., Pan S., et al. . (2012). Genome sequence of foxtail millet (Setaria italica) provides insights into grass evolution and biofuel potential. Nat. Biotechnol. 30, 549–554.

198. Zhang H., Zhang J., Wei P., Zhang B., Gou F., Feng Z., et al. . (2014a). The CRISPR/Cas9 system produces specific and homozygous targeted gene editing in rice in one generation. Plant Biotechnol. J.12, 797–807.

199. Zhang J., Nguyen B. D., Sarkarung S., Blum A., Nguyen H. T., Zheng H. G., et al. (2001a). Locating genomic regions associated with components of drought resistance in rice: Comparative mapping within and across species. Theor. Appl. Genet. 103, 19–29.

200. Zhang S., Tang C., Zhao Q., Li J., Yang L., Qie L., et al. (2014b). Development of highly polymorphic simple sequence repeat markers using genome-wide microsatellite variant analysis in Foxtail millet [Setaria italica (L.) P. Beauv]. BMC Genomics 2:15

201. Zhang W. P., Shen X. Y., Wu P., Hu B., Liao C. Y. (2001b). QTLs and epistasis for seminal root length under a different water supply in rice (oryza sativa l.). Theor. Appl. Genet. 103, 118–123.

202. Zheng B. S., Yang L., Zhang W. P., Mao C. Z., Wu Y. R., Yi K. K., et al. . (2003). Mapping QTLs and candidate genes for rice root traits under different water-supply conditions and comparative analysis across three populations. Theor. Appl. Genet. 107, 1505–1515.

203. Zheng H. G., Babu R. C., Pathan M. S., Ali L., Huang N., Courtois B., et al. . (2000). Quantitative trait loci for root-penetration ability and root thickness in rice: comparison of genetic backgrounds.Genome 43, 53–61.

204. Zhou X., Jacobs T. B., Xue L.-J., Harding S. A., Tsai C.-J. (2015). Exploiting SNPs for biallelic CRISPR mutations in the outcrossing woody perennial Populus reveals 4-coumarate:CoA ligase specificity and redundancy. New Phytol.

Chapter 11

Influence of Mutation Rate on Estimators of Genetic Differentiation - Lessons From *arabidopsis Thaliana*

Ilkka Kronholm,[1] Olivier Loudet,[2] and Juliette de Meaux[1]

[1]*Department of Plant Breeding and Genetics, Max-Planck Institute for Plant Breeding Research, Carl-von-Linné-Weg 10, 50829 Cologne, Germany*

[2]*Institut Jean-Pierre Bourgin, UMR1318 INRA-AgroParisTech, F-78000 Versailles, France*

BACKGROUND

Characterisation of population structure is a pivotal task in population genetics. It is important for inferring the evolutionary history of a species, assisting in conservation studies [1] and measuring dispersal [2-4]. In association mapping and forensic applications, population structure must be corrected for [see [5]]. The characterisation of population structure is also crucial for studies tracking the signature of local adaptation i. e. when natural selection maintains differences in allele frequencies between populations. Indeed, limited migration or population bottlenecks can also cause allele frequencies to diverge. A proper characterisation of population structure is absolutely required to infer the action of local selection on outlier loci with reasonable confidence [6].

One way to quantify population structure is to use the summary statistic F_{ST}, which measures population differentiation. If allele frequencies are similar in a set of populations F_{ST} will be low, and conversely, if allele frequencies are different F_{ST} will be high. Basically, F_{ST} and its hierarchical extensions quantify how genetic diversity is partitioned within and between populations or groups of populations (see [5,7-9] for a review of the large body of theory on F_{ST}). F_{ST} can also be defined in terms of the ratio of coalescence times, for pairs of alleles from the same population over pairs of alleles from different populations [7]. If we consider only identity by descent, the topology of the genealogy is independent of the number of mutations.

F_{ST} can be sometimes related to quantities describing migration [2], although this approach has its caveats [4]. In the context of selection, by using many presumably neutral markers one can build a distribution of expected F_{ST} values and then compare these to F_{ST} values of genes that are hypothesised to be subject to selection [9]. This can also be done for phenotypes using Q_{ST}, a measure of genetic differentiation in quantitative traits. It has been shown that under neutrality Q_{ST} is equivalent to a single locus F_{ST}, and this relationship is robust to different demographic scenarios [10-12]. This property can be used to compare the Q_{ST} of a quantitative trait to the distribution of F_{ST} from neutral markers and infer whether the trait is influenced by selection [reviewed in [13]].

Some recent studies have raised concerns about the reliability of F_{ST} for characterisation of population structure using markers with high mutation rates, such as microsatellites [14-18]. High levels of within population diversity bias F_{ST} estimates downwards, because F_{ST} is estimated using heterozygosities or genetic variances. If a locus has multiple alleles, classical F_{ST} can be low even if populations share no alleles [15,16,19]. In addition to the classical F_{ST}, there are other estimators that have been proposed over the years. An analogous estimator to F_{ST}, Φ_{ST}, takes into account the distances between alleles thereby correcting for mutation rate [7,20]. Classical F_{ST}, estimated in the framework of Weir & Cockerham [21] considers only allele identity while Φ_{ST} considers distances between the alleles, be it differences in repeat number (e.g. in the case of microsatellite) or number of pairwise differences between DNA haplotypes. Another measure, F'_{ST}, standardises the observed F_{ST} value with the maximum possible value that F_{ST} could attain given the amount of observed diversity [15]. Finally, Jost derived recently a new measure of genetic differentiation, D[16] to replace G_{ST} (or its equivalent for empirical studies F_{ST}). D measures allelic differentiation by partitioning heterozygosity into within and between population components [16]. D is defined as $D = \left(\frac{H_T - H_S}{1 - H_S}\right)\left(\frac{n}{n-1}\right)$, where H_T is total heterozygosity, H_S subpopulation heterozygosity and n is the number of subpopulations [16]. These estimators provide improved measures of allelic differentiation between populations, yet their usefulness for estimating genetic differentiation in the context of migration or local adaptation is not clearly established. Most previous studies addressed these concerns using G_{ST} instead of F_{ST}. When G_{ST} is used, the mathematics are less complicated. In practise, however, F_{ST} has better statistical properties when empirical studies are concerned [see [5]]. For this reason, F_{ST} is the statistic of interest in this study. Moreover, in practise the difference between these two statistics is small.

Here, we characterize the relationship between different estimators of genetic differentiation and mutation rate using simulations as well as experimental data collected from natural populations. Our study system is *Arabidopsis thaliana* (L.) Heyhn. (Brassicaceae). We use simulations to characterize the relationship between mutation rate and population differentiation for both molecular markers and quantitative traits and ask the following questions: i) are all estimators of genetic differentiation dependent on mutation rate, ii) is the estimator of differentiation for quantitative traits dependent on mutation rate, iii) does the mating system influence the relationship between mutation rate and estimators of population differentiation. We provide empirical data from *Arabidopsis thaliana*, a long-standing model for plant molecular biology that has recently become a model in plant population genetics [22]. We show that in *A. thaliana*, F_{ST} is biased downwards for high mutation rate loci. We discuss these results and their implications for various applications of F_{ST} analysis, including the reliable detection of variable selection.

METHODS

Data analysis - Computer simulations

In order to investigate the behaviour of F_{ST}, F'_{ST}, Φ_{ST} and D under high mutation rates, computer simulations using EasyPop 1.8 [23] were performed. The simulation scheme was set to 10 populations with 500 individuals each, 20 freely recombining loci and random mating hermaphrodites. All loci were set initially as polymorphic by allowing EasyPop to generate the genotypes in the first generation. Populations followed an island model of migration. Migration rates (probability that a given individual will migrate in each generation), m ranged from 0.1 to 0.00001 and mutation rates (probability that a given allele will mutate in each generation), μ from 0.00001 to 0.01. In order to simulate microsatellite loci we first examined a pure single step mutation model. Then we relaxed this assumption by using a mixed mutation model in which the loci followed a single step mutation model but with the probability of 0.2 to mutate to any state. The number of possible allelic states was set to 30. The effect of self-fertilisation was examined by doing simulations with proportion of self-fertilisation set to 0.9. Simulations were run for 2000 generations. In our simulations this was enough for F_{ST} to reach equilibrium. To simulate a realistic sampling situation, 30 individuals were finally sampled from each population for parameter estimation. Each simulation was repeated 5 times for a given set of parameter values. For each simulated dataset we calculated the estimators of genetic differentiation as described below.

Next we examined how mutation rate at underlying QTL affects Q_{ST}. We used quantiNEMO [24], with the same settings as described above for neutral markers with the following exceptions: the number of QTL underlying the variation in the quantitative trait was 10 and there were 21 possible allelic states for each QTL. We used the random mutation model in quantiNEMO for the QTL alleles, in this model, allelic effects are drawn from a normal distribution, all effects are additive in our simulations. We also ran the simulations using the incremental mutation model, where the allelic effect of a new mutation resembles its ancestor. Variance of allelic effects was set to 0.1. The simulated quantitative trait was neutral and did not have any effect on fitness. The simulation was started at a state where all loci were monomorphic, the number of generations was 4000. The time to reach equilibrium was longer for low migration and mutation rates and in these cases number of generations was 6000. This is longer than for neutral markers because simulations had to be started from a monomorphic state; otherwise distribution of allelic effects becomes unrealistic. Variance components for Q_{ST} were estimated from genotypic values, which are returned by quantiNEMO as output, using R-scripts written by IK. The statistical model was a mixed-effect model with populations as random factors; REML-estimates of variance components were used. This was done in order to calculate Q_{ST} also in the presence of self-fertilisation, which quantiNEMO does not calculate as a standard output. Q_{ST} was estimated from the equation $Q_{ST} = \frac{(1+F_{IS})\sigma^2_{BP}}{(1+F_{IS})\sigma^2_{BP} + 2\sigma^2_{WP}}$ [25], where σ^2_{BP} is the between populations genetic variance component, σ^2_{WP} is the within populations genetic variance component and F_{IS} is reduction of heterozygosity within individuals due to inbreeding. Our method of estimating Q_{ST} gives the same results as the standard output of quantiNEMO when mating is random (results not shown). For each parameter set, 50 replicates of simulations with a single quantitative trait were run.

We also performed coalescent simulations to investigate the effect of different marker types on F_{ST} calculations. We investigated DNA haplotypes (these would be derived by re-sequencing short fragments, one locus is one fragment), independent single SNP markers and microsatellite markers following a single step mutation model. All coalescent simulations were performed using the program ms [26]. We simulated an island model of population structure with 10 populations, 20 individuals were sampled from each population. For sequence haplotypes and microsatellites 30 independent loci were simulated, for SNP markers we simulated 100 independent SNPs. For single SNPs and haplotypes, multiple hits were not permitted. The microsatellite mutation model was implemented via R-script. In the program ms migration and mutation rate are expressed in terms of effective population size, *4Nm* and *4Nμ* respectively. We set up the simulations so that the effective population size was 1000 for each population and then parameters m and μ ranged from 0.0001 to 0.1 for m and 0.00001 to 0.001 for μ. Each simulation was repeated 5 times for each parameter combination.

Data analysis - Genetic diversity and population structure

All statistical analyses were done using the statistical environment R [27] unless otherwise stated. Methods not implemented by R-packages were implemented via R-scripts written by IK and are available upon request.

Measure of genetic diversity, Nei's gene diversity (*Hs*) was calculated using FSTAT 2.9.3 [28]. The microsatellite population mutation rate, ϑ, is the product of effective population size and mutation rate at a locus was calculated following equation 15 of Kimmel *et al.* [29]. The performance of this summary statistic based method has been shown to be comparable to likelihood-based methods [30]. ϑ was calculated for each locus within each region. For SNP data the minor allele frequency was calculated for each locus in each region.

F_{ST} was estimated according to Weir & Cockerham [21] for microsatellites and SNP markers, using the R-package "hierfstat" [31]. All other genetic differentiation methods were implemented via R-scripts written by IK. For microsatellites the standardised genetic differentiation measure, F'_{ST} [15], was estimated using the maximised variance component method of Meirmans [32]. In order to take the distance between the microsatellite alleles or sequence haplotypes into account [20] we estimated Φ_{ST} using the method of Michalakis and Excoffier [33]. Differentiation indices between regions were calculated in a hierarchical setting, taking into account the partition of variation between populations within regions [7]. Confidence intervals for different measures of genetic differentiation were generated by bootstrapping over loci. An estimator for *D* was calculated following Jost [16].

The expected F_{ST} was calculated for the forward and coalescent simulations using the relationship between F_{ST} and coalescence times in the island model. The expected F_{ST} was calculated as $F_{ST} = (\bar{t}_1 - \bar{t}_0) / \bar{t}_1$ [7], where $\bar{t}_1$ = average coalescence time of alleles from different populations and $\bar{t}_0$ = average coalescence time of two alleles from the same population. For an island model of population structure, $\bar{t}_0 = 2Nd$ and $\bar{t}_1 = 2Nd + \frac{d-1}{2m}$, where N = population size, d = number of populations and m = migration rate [34]. This provides us with an analytical estimate of the expected value of F_{ST}. We used this as a baseline when comparing the different estimators.

To check whether $\Phi_{ST} > F_{ST}$ for the microsatellite loci, we used a permutation test [35]; if $\Phi_{ST} > F_{ST}$ stepwise mutations may contribute to genetic differentiation, thereby providing one explanation for the difference. The test was implemented in the program SPAGeDi 1.2 [36]. This is done by permuting microsatellite allele sizes among allelic states to test if stepwise mutations contribute to genetic differentiation.

Population samples

In total 289 individuals from 41 populations were genotyped. Detailed information about the populations can be found in the supplementary material (Additional file 1). We analysed 7, 15, 13 and 6 populations from Spain, France, Norway and Central Asia, respectively. Number of sampled individuals from each population ranges from 3 to 11 with a mean of 7. Three regions in Western Europe: Spain, France, Norway create a South - North cline. The Central Asian region is composed of populations from Kyrgyzstan and Tajikistan. The Spanish populations are described in Pico *et al.* [37]. French populations were collected by Valerie Le Corre and some of them are described in Le Corre [38]. The Norwegian populations were kindly provided by Odd-Arne Rognli through NARC (Norway). Populations from Central Asia were collected by OL and are described at http://www.inra.fr/vast/collections.htm. Field collected plants were subjected to one or two generations of self-fertilisation in the greenhouse before DNA extraction.

Genotyping

DNA was extracted from young leaves using BioSprint 96 robot and BioSprint 96 DNA Plant Kit (Qiagen) according to manufacturer's instructions.

Plants were genotyped at 20 microsatellite loci. Details of the microsatellite loci used and genotyping procedures can be found in the supplementary material (Additional file 2). Microsatellites were amplified using standard PCR methods and allele sizes were determined using capillary electrophoresis. To determine the actual number of repeats in each allele, the accession Col-0 was genotyped for each locus as a reference. Using the genome sequence and the Col-0 PCR product size the actual number of repeats was deduced for each allele. The Spanish accessions had already been genotyped previously for some of the loci used here, as described in Pico *et al.* [37]. We verified that our allele sizes corresponded to the allele sizes reported previously by re-genotyping a subsample at selected alleles.

The plants were also genotyped for a set of 149 single nucleotide polymorphism (SNP) markers [developed by [39]] by Sequenom, inc. (San Diego, CA). Detailed description of the SNP markers is found in supplementary material (Additional file 3). Out of the 149 SNP markers, 12 loci had to be excluded because the genotyping failed or because there was no variation, leaving 137 loci used in the analysis. Microsatellite and SNP genotypes can be found in supplementary material (Additional file 4)

Data analysis - SNP ascertainment bias

The SNP markers we used were discovered in the study of Nordborg et al. [40] and subsequently developed for mapping purposes [39]. The criterion for ascertainment was that one allele of a SNP had to be carried by the standard laboratory accession Col-0 and the other allele had to be in high frequency in the rest of the sample. Hereafter we refer to the publicly available dataset of Nordborg et al. as Nordborg data. To investigate the ascertainment bias in our sample we calculated minor allele frequency distribution for all of the SNPs in the Nordborg data and the 137 SNPs used in this study. Because the sample in which the SNPs were discovered is different from the sample in this study ascertainment bias cannot be corrected in a straightforward manner. Therefore, we examined the effect of ascertainment bias on F_{ST} in the dataset from which they were selected. We used the 137 SNPs used in our study to calculate F_{ST} between genetic clusters defined by Nordborg *et al.* [40] in the Nordborg dataset. Then we sampled 137 SNPs at random from the Nordborg dataset 1000 times and calculated F_{ST} between the genetic clusters for each sampled dataset.

RESULTS

Computer simulations

We used forward population genetic simulations to investigate the behaviour of different estimators with varying migration and mutation rates. The best estimator, in the context of local adaptation, should be robust to mutation rate to allow comparisons between different marker types.

For a low mutation rate, F_{ST} and Φ_{ST} were reasonably good, but F'_{ST} and D differed from the expected value (Table (Table1),1), although for extremely small migration rates the situation was reversed. Results of forward population genetic simulations showed that F_{ST} tended to zero when mutation rate increased (Figure (Figure1).1). Replicate simulations clustered very well showing that there was little variance among replicates. These results follow the analytical expectation presented in Hedrick [15] and the results of Balloux & Goudet [18]. If mutations followed a pure single step model, Φ_{ST} was essentially independent from mutation rate (Figure (Figure1)1) as expected; this is true for all migration rates. F'_{ST} and D are not independent from mutation rate. In our simulations, we observed that when migration rate was very low, increasing mutation rate up to 0.01 also caused F'_{ST} and D to go downward (Figure (Figure1,1, panels C and D). If the assumptions of single step mutation model were relaxed, Φ_{ST} had the same trend as F_{ST} although the effect was somewhat weaker (Figure (Figure2).2). A self-fertilisation rate of 0.9 did not qualitatively alter the results (Additional file 5). We note that, when mutation rate was increased up to 0.01, all estimators decreased (Figure (Figure1).1). This is not a cause for concern, since this rate is biologically unrealistic. This effect is explained by the limited number of alleles in the simulations, which at high mutation rate can cause homoplasy.

We then examined the effect of mutation rate at QTL on Q_{ST}. We observed the same behaviour as for F_{ST}. When migration rates are low, increasing the mutation rate at underlying QTLs caused Q_{ST} to take lower values (Figure (Figure3).3). We also performed the simulations using an increment mutation model for the QTL, where each new mutation had an allelic value that is close to the allelic value of its ancestor. In this case Q_{ST} behaved like Φ_{ST} and was not affected by mutation rate (Figure (Figure3,3, panel B). Here again, a self-fertilisation rate of 0.9 did not qualitatively alter the results (Additional file 6).

Table 1. Comparison of expected and observed values for F_{ST}

Forward simulations					
m	E(F_{ST})	O(F_{ST})	O(Φ_{ST})	O(F'_{ST})	O(D)
0.1	0.0045	0.0040	0.0049	0.0190	0.0148
0.01	0.0431	0.0440	0.0414	0.1920	0.1549
0.001	0.3104	0.3104	0.3067	0.7581	0.6492
0.0001	0.8182	0.7570	0.7667	0.9831	0.9307
0.00001	0.9783	0.8339	0.8500	0.9931	0.9580
Coalescent simulations					
m	E(F_{ST})	O(SNP F_{ST})	O(MSAT Φ_{ST})		O(DNA Φ_{ST})
0.1	0.0023	0.0277	0.0287		0.0299
0.01	0.0220	0.0400	0.0420		0.0488
0.001	0.1837	0.2057	0.2003		0.2127
0.0001	0.6923	0.6950	0.7124		0.6933

Expected F_{ST} was calculated as described in the methods, observed values are the simulation means for the different estimators or marker types. For the coalescent simulations MSAT means microsatellite markers and DNA means DNA sequence haplotypes. Simulation values are shown for a single mutation rate, μ = 0.00001.

Next we examined the effect of mutation rate on different marker types. We simulated DNA haplotypes (as would be derived by re-sequencing short fragments from multiple individuals), microsatellite markers and single SNP markers. Results from the simulations are presented in figure figure4.4. We calculated Φ_{ST} that takes into account distance between different haplotypes or microsatellite alleles. Applying this method to both haplotypes and microsatellites gave essentially the same results (Figure (Figure44 and Table Table1)1) and Φ_{ST} was independent from mutation rate for both marker types.

Single SNP markers also gave F_{ST} values that were nearly identical to the ones obtained with other types of markers (Table (Table1).1). This is in accordance with DNA haplotype and microsatellite markers (Figure (Figure4).4). Therefore, Φ_{ST} for DNA haplotypes, microsatellites (following single step mutation model) and F_{ST} for single SNPs (free of ascertainment bias) gave comparable estimates of differentiation.

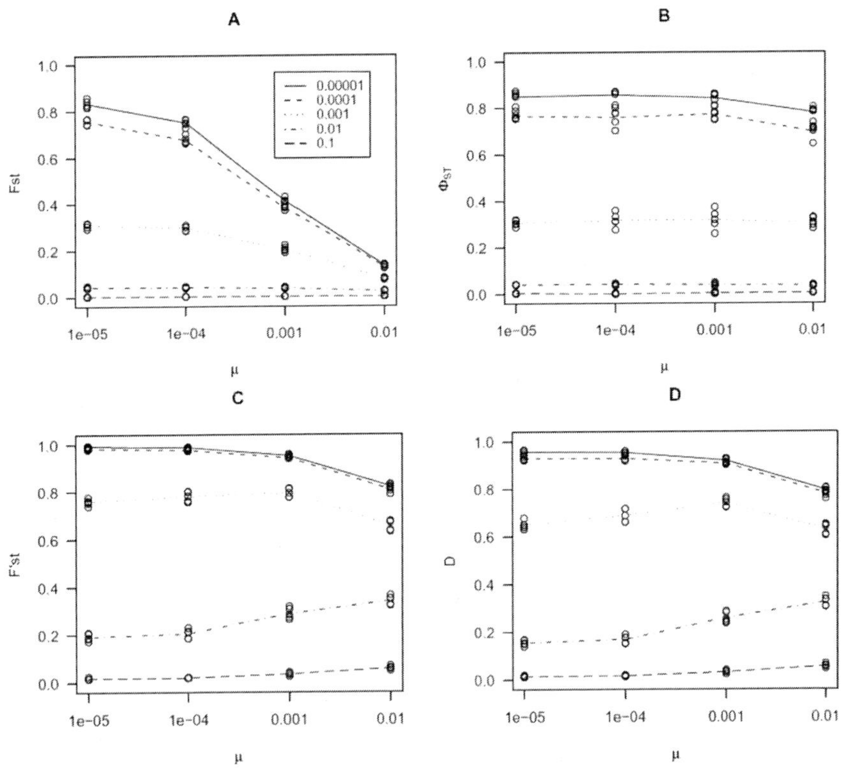

Figure 1. Results of computer simulations for single step mutation model. Different estimators of genetic differentiation were plotted against mutation rate. Different lines represent different migration rates. Migration rates 0.1, 0.01, 0.001, 0.0001 and 0.00001 ...

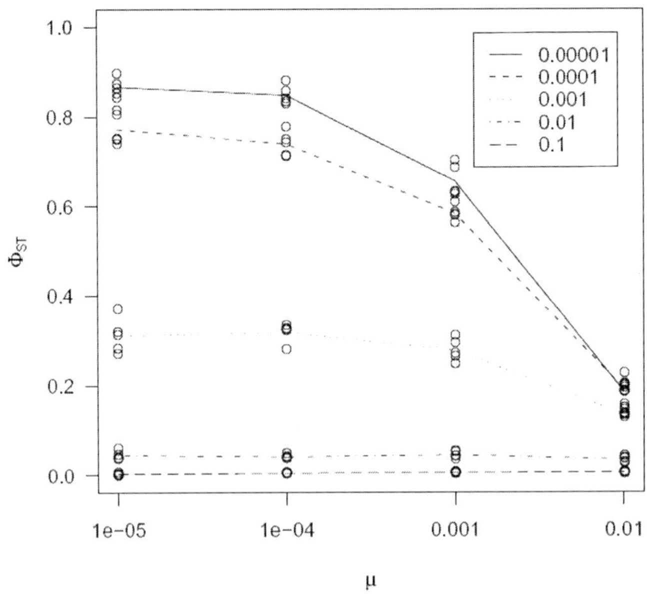

Figure 2. Results of computer simulations for mixed mutation model. The effect of mutation rate on genetic differentiation calculated from Φ_{ST}using mixed mutation model. In this model, there was a probability of 0.2 that when a mutation occurs the allele ...

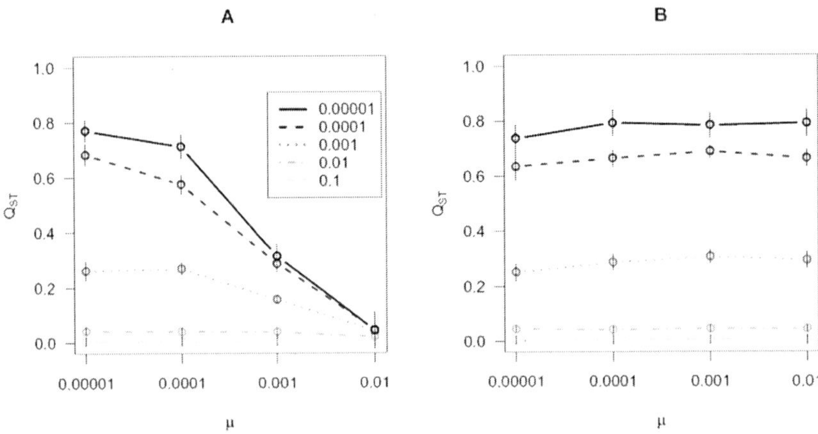

Figure 3. Results for computer simulations for Q_{ST}. The effect of mutation rate and migration rate on Q_{ST}. Different lines represent different migration rates. Migration rates 0.1, 0.01, 0.001, 0.0001 and 0.00001 correspond to different lines as indicated by the ...

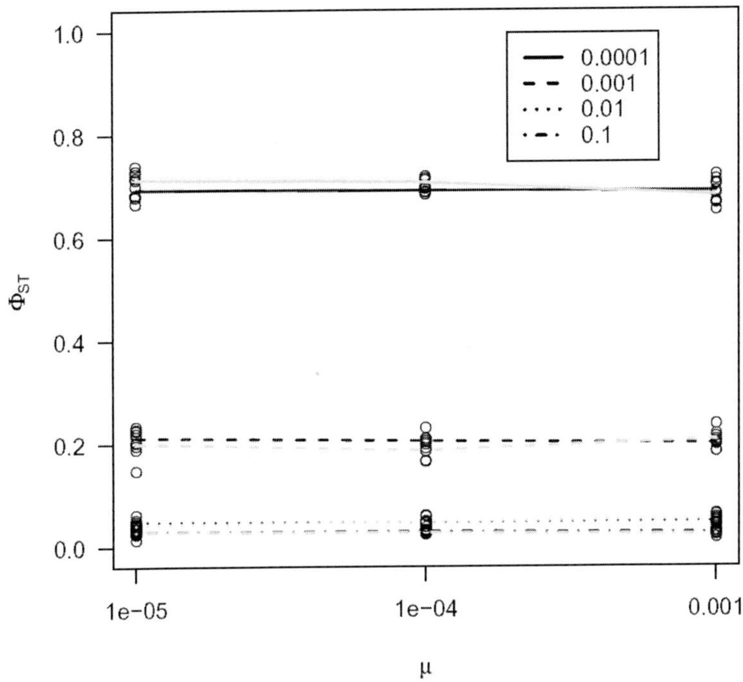

Figure 4. Results of coalescent simulations for different marker types. The effect of mutation rate on genetic differentiation, calculated from Φ_{ST}. Black lines represent estimates from DNA haplotypes, grey lines are estimates from microsatellite alleles. ...

Genetic differentiation in *Arabidopsis thaliana*

We characterised diversity within 41 populations distributed in four broad geographic regions. We observed that genetic differentiation (F_{ST}) for microsatellite loci correlates with gene diversity (*Hs*) (Figure (Figure5).5). For instance, in the Spanish populations the correlation between *Hs* and F_{ST} was r = -0.862 (95% CI = -0.944 - -0.678) with p < 0.001 (Table (Table2).2). We further examined the correlation between diversity and various alternative estimators of differentiation. There was positive albeit non-linear relationship between *Hs* and F'_{ST}, (r = 0.479, [95% CI = 0.076 - 0.760], p = 0.033). Φ_{ST} was not correlated with *Hs*, (r = -0.294, [95% CI = -0.652 - 0.170], p = 0.208). A similar pattern was observed when the population mutation rate (ϑ) was used instead of *Hs*. For ϑ and F_{ST} r = -0.682, p < 0.001, for ϑ and F'_{ST} r = 0.500, p = 0.025 and for ϑ and Φ_{ST} r = -0.301, p = 0.197, respectively. Φ_{ST} is independent from genetic diversity and mutation rate in our data, except in Central Asian populations (Table (Table2,2, Figure Figure55).

Table 2. Correlations between genetic diversity and genetic differentiation

	Hs		*θ*	
Spanish populations	r (95% CI)	p	r (95% CI)	p
F_{ST}	-0.862 (-0.944 - -0.678)	<0.001	-0.682 (-0.864 - -0.342)	<0.001
F'_{ST}	0.479 (0.046 - 0.760)	0.033	0.500 (0.074 - 0.771)	0.025
Φ_{ST}	-0.294 (-0.652 - 0.170)	0.208	-0.301 (-0.656 - 0.163)	0.197
D	0.765 (0.488 - 0.902)	<0.001	0.655 (0.300 - 0.851)	0.002
French populations				
F_{ST}	-0.867 (-0.948 - -0.681)	<0.001	-0.625 (-0.840 - -0.238)	0.004
F'_{ST}	0.645 (0.270 - 0.850)	0.003	0.705 (0.370 - 0.878)	<0.001
Φ_{ST}	-0.260 (-0.639 - 0.220)	0.282	-0.144 (-0.561 - 0.332)	0.557
D	0.876 (0.700 - 0.952)	<0.001	0.756 (0.460 - 0.901)	<0.001
Norwegian populations				
F_{ST}	-0.916 (-0.968 - -0.791)	<0.001	-0.599 (-0.828 - -0.198)	0.007
F'_{ST}	0.199 (-0.280 - -0.599)	0.413	0.446 (-0.011 - 0.748)	0.056
Φ_{ST}	-0.109 (-0.536 - 0.364)	0.658	-0.021 (-0.471 - 0.437)	0.931
D	0.631 (0.248 - 0.843)	0.004	0.717 (0.390 - 0.0.884)	<0.001
Central Asian populations				
F_{ST}	-0.801 (-0.928 - -0.506)	<0.001	-0.494 (-0.795 - 0.002)	0.052
F'_{ST}	-0.116 (-0.578 - 0.403)	0.669	0.125 (-0.395 - 0.584)	0.645
Φ_{ST}	-0.628 (-0.857 - -0.192)	0.009	-0.132 (-0.589 - 0.389)	0.625
D	0.565 (-0.013 - 0.860)	0.055	0.484 (-0.124 - 0.828)	0.111

Correlations between genetic diversity in markers and genetic differentiation between populations in different regions. Correlation coefficients are given with 95% confidence intervals are in parenthesis. *Hs* is subpopulation heterozygosity and θ is an estimate of microsatellite mutation rate.

The 137 polymorphic SNP markers used in this study were biased towards high frequency as shown in Additional file 7, where minor allele frequency is plotted for each SNP in different regions. This was apparent also from comparing the minor allele frequency spectrum of the full Nordborg data and the 137 SNP set (Additional file 7). Using the 137 SNP set we obtained F_{ST} = -0.0018 between clusters defined previously in the Nordborg data [40]. Sampling 137 SNPs at random from the Nordborg data gave us a 95% confidence interval of F_{ST} - 0.0051 - 0.0271. This suggests that ascertainment bias perhaps has only a minor effect in our dataset.

Next we calculated measures of genetic differentiation for microsatellites and SNP markers between populations within regions and between regions (Figure (Figure6).6). For microsatellites, genetic differentiation between populations was the lowest in Spain (F_{ST} = 0.2900, Φ_{ST} = 0.3556), intermediate for France (F_{ST} = 0.4937, Φ_{ST} = 0.6818) and for Asia (F_{ST} = 0.6026, Φ_{ST} = 0.3101) and the highest in Norway (F_{ST} = 0.8004, Φ_{ST} = 0.8128). A similar trend was observed for both microsatellites and SNP markers (Figure (Figure6).6). However, it should be noted that the confidence intervals were sometimes broad (Figure (Figure6),6), especially in Central Asia. Genetic differentiation between geographic regions was smaller than between populations within regions (Figure (Figure6).6). Differentiation measured by F'_{ST} from microsatellites was F'_{ST} = 0.7208 for Spain, 0.8115 for France, 0.9436 for Norway and 0.8413 for Asia. Values for D were 0.6393, 0.6509, 0.7241 and 0.6334 for the Spanish, French, Norwegian and Cental Asian populations respectively.

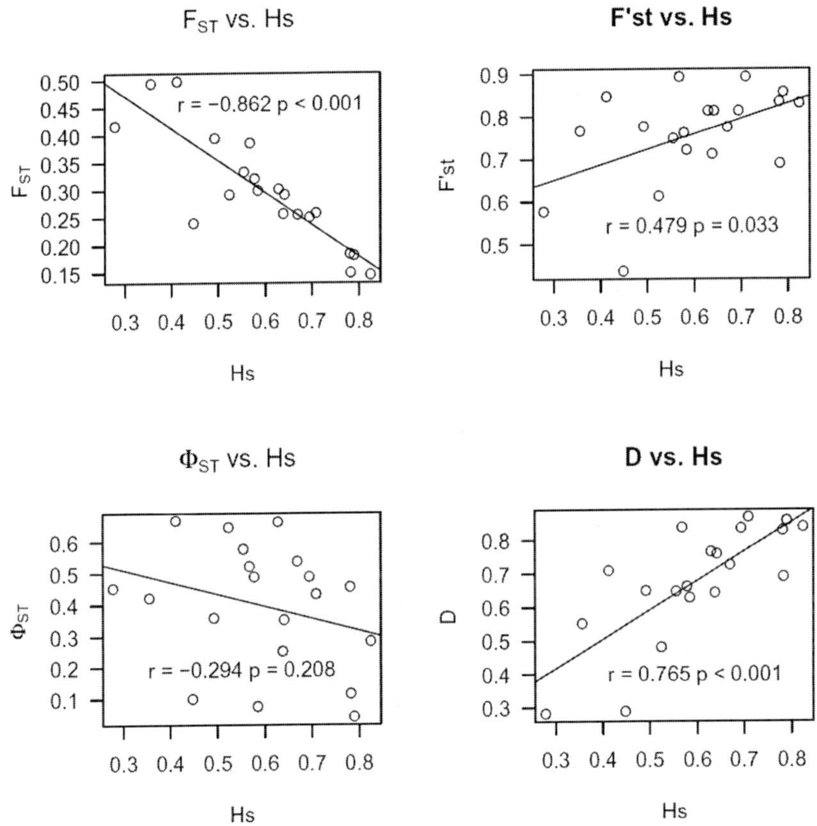

Figure 5. Correlations between genetic diversity and genetic differentiation in Spanish populations. Gene diversity, Hs, was calculated for each locus and was plotted against different estimators of genetic differentiation. A) F_{ST} B) F'_{ST} C) Φ_{ST} D) D

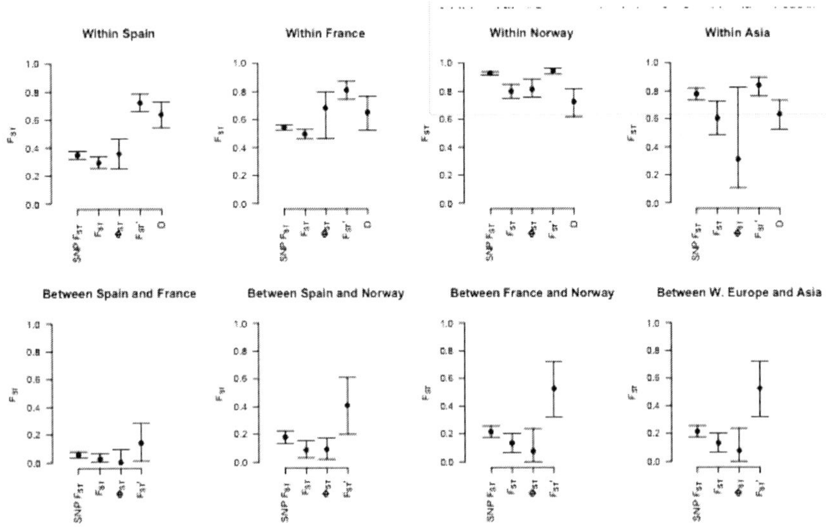

Figure 6 .Genetic differentiation within and between regions. Genetic differentiation between populations within regions and between regions. Differentiation was estimated from microsatellite markers using four different estimators (for explanation, see methods). ...

We tested whether Φ_{ST} was higher than F_{ST} by using a permutation test that permutes allele sizes between different alleles [35]. If Φ_{ST} is higher than F_{ST}, one possibility is that stepwise mutations contribute to differentiation [35]. Within Spanish populations, the difference is suggestive albeit not significant, 2-sided test p = 0.0629. Within French populations, the difference was significant p = 0.0210. In Norwegian and Asian populations, differences were not significant (p = 0.1009 and p = 0.8561 respectively). This suggests that stepwise mutations may contribute to genetic differentiation in Spain and France. In the Norwegian and Asian populations instead, microsatellite loci possibly exhibit some departure from a stepwise mutation model. However, this does not prove that the loci follow SSM model.

DISCUSSION

Implications for detecting local adaptation

We used computer simulations to analyze the behaviour of different estimators of genetic differentiation as a function of various migration and mutation rates. Our simulations show that *D* or F'_{ST} both depend on mutation rate (Figure (Figure1).1). For *D*, this is lucidly shown by Jost [16] but the fact that F'_{ST} is dependent on mutation rate is not made clear by Hedrick [15]. New mutations increase differentiation between populations, especially if migration rate is low. *D* and F'_{ST} measure actual allelic differentiation, regardless of the process that generates these differences [41]. This means that F'_{ST} or *D* are useful for studies where the amount of genetic differentiation is of interest *per se*, such as in conservation studies [15,16]. Instead, these estimators become problematic for studies interested in comparing levels of differentiation across loci, as in studies of local adaptation. Also, they cannot be directly related to coalescence times (Table (Table1).1). This is not necessarily a criticism of F'_{ST} and *D*, since *D* was derived to measure true allelic differentiation. These measures address the question of the partition of genetic diversity among populations, irrespective of the mechanisms by which this pattern arises [16]. Further work is needed to characterise the properties of these estimators.

Our results show that, because it takes distances between different alleles into account, Φ_{ST} is the only estimator that is completely independent from mutation rate (Figure (Figure1)1) [18,20]. We also showed that if the assumptions of Φ_{ST} are met, both DNA haplotypes and microsatellites give estimates comparable to F_{ST} calculated for bi-allelic SNPs (Figure (Figure4).4). In a large empirical dataset in humans [42], where microsatellites analysed using a stepwise mutation model and SNPs gave comparable results, a similar conclusion was reached. In *A. thaliana* as in many other species, microsatellite loci were often shown to deviate from pure single step mutation model [43-46]. Our simulations further show that this estimator became dependent on mutation rate, if there was some deviation from the single step mutation model (Figure (Figure2).2). This is also seen in our dataset in some geographic regions, as we find that Φ_{ST} is not different from F_{ST}. Yet, Φ_{ST} might still be preferred over F_{ST}. Even if there are deviations from the SMM model, Φ_{ST} still performs better than F_{ST}(Figure (Figure1,1, Figure Figure22).

SNP markers are unlikely to suffer from these issues in F_{ST} estimation because mutation rate for SNP markers is considerably lower than for microsatellites. However, they may have some ascertainment bias. This effect, however, was minor in the 96 genotypes in which our SNP-marker set was ascertained. Although it is not certain that this behaviour would be the same in our set of 41 populations, estimates of differentiation based on SNPs are roughly concordant with Φ_{ST} estimates based on microsatellites (Figure (Figure6).6). In Norway and central Asia however, microsatellite Φ_{ST} seems to be lower than F_{ST} for SNP markers (Figure (Figure6).6). In these regions, we do not have any evidence that microsatellites follow a stepwise mutation model, so this seems a likely cause for the discrepancy. However, ascertainment bias cannot be completely excluded. In many cases, confidence intervals for Φ_{ST} are broad in our data, a likely consequence of the high sampling variance displayed by Φ_{ST} type estimators [47]. This consideration suggests that, whenever possible, DNA haplotypes generated by re-sequencing should be used in studies of local adaptation.

It is known that there is considerable variation in mutation rates between different genes due to evolutionary constraints [48]. In order to directly compare differentiation across genes and correctly assign loci subject to local adaptation, mutation rate has to be taken into account. This problem has been considered earlier in the context of detecting loci that have outlier F_{ST} values [49]. The method of Beaumont & Nichols jointly considers heterozygosity and F_{ST}. This method was shown to be robust to mutation rate variation among loci [49]. Yet if the types of markers combined in one analysis are different and mutation rate has a strong influence for one type of marker because of low migration rates, results have to be interpreted with caution. Indeed, this method gave non sensical results with our *A. thaliana* dataset combining microsatellite and SNP markers (data not shown).

Our results have further implications for studies based on Q_{ST} vs. F_{ST} comparisons, aiming at detecting local adaptation in quantitative traits. A recent meta-analysis of F_{ST} vs. Q_{ST} studies [50] noted that using F'_{ST} would generally change the conclusions of F_{ST} vs. Q_{ST} studies. However, our study shows that using F'_{ST} or D in Q_{ST} studies is not appropriate, because these measures of genetic differentiation are not independent from the high mutation rate of microsatellites. Yet, this is a concern only for those cases where migration rate is low.

Another concern is the fact that Q_{ST}, the estimate for quantitative traits, is affected by the mutation rate at underlying QTLs (Figure (Figure3).3). In retrospect, Q_{ST} being an equivalent of F_{ST} in the absence of selection [10], this result is not completely surprising. More interesting however, is the parameter range at which this bias is observed. In our simulations we varied the parameter μ, the genic mutation rate. It should be noted that the overall polygenic mutation rate, or variance contributed by mutation in each generation (assuming additive effects), of a quantitative trait is $V_m = 2\mu N_L V_\alpha$ [51], where N_L is the number of QTL, 2 accounts for diploidy and V_α is the variance of allelic effects. In our simulations $N_L = 10$ and $V_\alpha = 0.1$ were held constant in all simulations. This translates into V_m of twice as large as genic mutation rate, so that for $\mu = 0.0001$ $V_m = 0.0002$. In our simulations environmental variance, V_E, was set to 1 so mutational variance scaled by environmental variance, $h^2_m = V_m / V_E$, [51,52] was equal to V_m. In our simulations when genic mutation rate of QTL was $\mu = 0.001$, effects on Q_{ST} were rather large when migration rates were low (Figure (Figure3).3). This corresponded to $h^2_m = 0.002$. It is interesting to note that empirical estimates of mutational heritabilities frequently fall around this value [52]. For instance, h^2_m for the well studied trait of bristle number in*Drosophila* seems to be around 0.0035 or 0.0043. Schultz et al. [53] estimated h^2_m for few life-history traits in *A. thaliana* and found that they fall around 0.003. Estimating mutational heritability empirically is not easy, but empirical data from several sources suggests that for many traits it is around 0.002 - 0.003 [52].

Interestingly, if new alleles display incremental changes of function, Q_{ST} becomes independent of the mutation rate. However, alleles of large effect contributing to quantitative trait variation have been observed frequently in natural populations [reviewed in [54]]. A prominent example is provided by the FRIGIDA locus, a gene contributing to quantitative variation in flowering time in *A. thaliana*. Several loss-of-function mutations were reported to segregate in natural populations [38]. Thus QTL are likely to frequently deviate from a purely incremental mutation model.

A comforting fact is that high mutation rates will bias Q_{ST} downwards, so if the goal is to study local adaptation, that is, the case where $Q_{ST} > F_{ST}$ is of interest, then the test will remain conservative. Traditionally it was thought that $Q_{ST} < F_{ST}$ would indicate the action of homogenising selection [13], that is, selection for the same phenotypic optima in different populations. Recently this view has been challenged. Firstly, non-additive gene action seems to bias Q_{ST} downwards [11,55,56]. Secondly, the large variance in estimatingQ_{ST} and in evolutionary outcomes of single locus (trait) may be a even more serious problem [56-58]. Intuitively, systems where F_{ST} is large, because migration is low, offer potentially greatest statistical power to detect $Q_{ST} < F_{ST}$. Our results instead suggest that the bias in Q_{ST} may be the greatest in such cases and that the utility of $Q_{ST} - F_{ST}$ comparisons may be limited to detecting diversifying selection.

Correlating estimates of population differentiation to levels of diversity has highlighted the effect of mutation rate on estimates of differentiation in *Arabidopsis thaliana*. To date, several studies have investigated patterns of population structure of *A. thaliana*, [37,59-63] but none has examined the effect of mutation rates on estimates of differentiation. We observe this effect in the four distinct geographical regions. Therefore, mutation rates seem to significantly impact estimates of population differentiation in this species. This relationship has also been found in *Arabidopsis lyrata*, a relative of *A. thaliana* exhibiting a markedly different life-history and more genetic diversity than *A. thaliana* [64,65]. Some fish species also exhibit this relationship. For example, in two subspecies of the triplefin fish, F_{ST} was low and negatively correlated with expected heterozygosity (r = -0.9) [66]. Similarly O'Reilly *et al.* [67] found a relationship between heterozygosity and F_{ST} in the fish walleye pollock, which they attributed to homoplasy. As shown by our simulations, this problem is more severe in systems where diversity is very high and migration between populations is low (Figure (Figure3).3). These examples show that a wide variety of organisms are in the parameter space where variation in F_{ST} reflects variation in both migration and mutation rates. Therefore, studies of population differentiation should systematically investigate this effect.

Effect of the mating system

Simulations were performed under random-mating as well as with a self-fertilizing rate of 0.9 and yielded essentially the same results. From a population genetics perspective, self-fertilisation should reduce effective population size and thus coalescence times of alleles within populations [68]. This is precisely what we observe in our simulations: absolute F_{ST} values are higher, but the relationship of the statistics to mutation rate remains qualitatively the same (Additional file 5). This is also true for Q_{ST} (Additional file 6). Thus the effect of self-fertilisation is mainly to increase F_{ST} values, but it does not alter the effect of mutation rates on its estimates. Although a rate of 0.9 is slightly lower than selfing rates reported for *A. thaliana* so far, we believe that our simulations reflect faithfully the situation in this species [59,69,70]. Indeed, there is one study that found the rate to be closer to 0.9 in one particular population [61] and recently Pico et al. [37] found rates to be variable, from 0.97 down to 0.93 in Iberian populations. Considering that some of these studies used a very limited set of markers, it is conceivable that true rates of self-fertilisation are closer to 0.9 than 0.99. This is also evident when more markers and more individuals from a single population are sampled, because much more heterozygosity is uncovered (IK, unpublished data).

CONCLUSIONS

In this study, we lift the ambiguity lying over the effect of variable mutation rate on estimates of population differentiation and show that virtually all estimates show a dependence on higher mutation rates if migration is low. Only Φ_{ST}, which takes mutation rate into account, controls this bias. However, in practise, Φ_{ST} assumes a stepwise mutation model for microsatellites, which may not be correct in all cases. Several microsatellite mutation models have been proposed [46]. Future studies could examine the possibility of incorporating different models of mutation in Φ_{ST} estimation. For this, the rules by which distance matrices between alleles or haplotypes are calculated could be modified. For sequence data, changes would be straightforward to incorporate different models of sequence evolution. More importantly, we show that Q_{ST} estimates are not independent of the mutation rate of QTLs underlying quantitative traits for low migration rates. This result is obtained with a model simulating traits with realistic mutational heritabilities.

Therefore $Q_{ST} < F_{ST}$ cannot be interpreted as strong evidence for homogenising selection in systems where genetic differentiation is high. Our study in *A. thaliana* shows that caution is needed when tracking the signature of local adaptation on loci controlling adaptive phenotypes in this and other model species for genetics. Further studies will eventually be needed to address the properties of these various estimates in non-equilibrium populations or to investigate the impact of the genetic architecture of a trait on Q_{ST}.

AUTHORS' CONTRIBUTIONS

The study was conceived by IK and JdM, IK performed the experiments and analysed the data. OL contributed samples. The manuscript was written by IK with input from all other authors. All authors read and approved the final manuscript.

ACKNOWLEDGEMENTS

We thank Valerie Le Corre for providing the French populations, Odd-Arne Rognli and Anna Monika Lewandowska for providing the Norwegian populations and Carlos Alonso-Blanco and Xavier Pico for providing the Spanish populations. We thank Sinead Collins, Maarten Koornneef, Maria Clauss and Marie-Hélène Muller for comments on an earlier version of the manuscript, Samuel Neuenschwander gave advice on using the program quantiNEMO. We also thank an anonymous reviewer for constructive comments. Funding was provided by Max Planck Gesellschaft and by a grant from DFG to JdM within the collaborative research network SFB680.

REFERENCES

1. Avise JC. Molecular markers, natural history and evolution. 2. Sunderland: Sinauer Associates Inc; 2004.

2. Rousset F. Genetic Differentiation and Estimation of Gene Flow from F-Statistics Under Isolation by Distance. Genetics. 1997;145(4):1219–1228.

3. Ouborg NJ, Piquot Y, van Groenendael JM. Population genetics, molecular markers and the study of dispersal in plants. J Ecol. 1999;87:551–568.

4. Whitlock MC, McCauley DE. Indirect measures of gene flow and migration: FST ≠ 1/(4Nm + 1)Heredity. 1999;82(Pt 2):117–125.

5. Holsinger KE, Weir BS. Genetics in geographically structured populations: defining, estimating and interpreting FST. Nat Rev Genet. 2009;10:639–650.

6. Storz JF. Using genome scans of DNA polymorphism to infer adaptive population divergence. Mol Ecol. 2005;14(3):671–688.

7. Excoffier L. In: Handbook of statistical genetics. Balding DJ, Bishop M, Cannings C, editor. Chichester, Wiley; 2007. Analysis of population subdivision; pp. 271–307.

8. Rousset F. In: Handbook of statistical genetics. 3. Balding DJ, Bishop M, Cannings C, editor. Chichester, Wiley; 2007. Inferences from spatial population genetics; pp. 945–979. full_text.

9. Beaumont MA. Adaptation and speciation: what can FST tell us. Trends Ecol Evol. 2005;20(8):435–440.

10. Lande R. Neutral theory of quantitative genetic variance in an island model with local extinction and colonization. Evolution. 1992;46(2):381–389. doi: 10.2307/2409859.

11. Whitlock MC. Neutral additive genetic variance in a metapopulation. Genet Res. 1999;74(3):215–221.

12. Spitze K. Population structure in Daphnia obtusa: quantitative genetic and allozymic variation.Genetics. 1993;135(2):367–374.

13. Merilä J, Crnokrak P. Comparison of genetic differentiation at marker loci and quantitative traits. J Evol Biol. 2001;14(6):892–903. doi: 10.1046/j.1420-9101.2001.00348.x.

14. Balloux F, Brunner H, Lugon-Moulin N, Hausser J, Goudet J. Microsatellites can be misleading: an empirical and simulation study. Evolution. 2000;54(4):1414–1422.

15. Hedrick PW. A standardized genetic differentiation measure. Evolution. 2005;59(8):1633–1638.

16. Jost L. GST and its relatives do not measure differentiation. Mol Ecol. 2008;17(18):4015–4026.

17. Hedrick PW. Highly variable loci and their interpretation in evolution and conservation. Evolution.1999;53(2):313–318.

18. Balloux F, Goudet J. Statistical properties of population differentiation estimators under stepwise mutation in a finite island model. Mol Ecol. 2002;11:771–783. doi: 10.1046/j.1365-294X.2002.01474.x.

19. Kalinowski ST. Evolutionary and statistical properties of three genetic distances. Mol Ecol.2002;11(8):1263–1273.

20. Slatkin M. A Measure of Population Subdivision Based on Microsatellite Allele Frequencies.Genetics. 1995;139(1):457–462.

21. Weir BS, Cockerham CC. Estimating F-statistics for the analysis of population structure. Evolution.1984;38(6):1358–1370.

22. Mitchell-Olds T, Schmitt J. Genetic mechanisms and evolutionary significance of natural variation in Arabidopsis. Nature. 2006;441(7096):947–952. doi: 10.1038/nature04878.

23. Balloux F. A computer program for the simulation of population genetics. J Heredity. 2001;92:301–302.

24. Neuenschwander S, Hospital F, Guillaume F, Goudet J. quantiNemo: an individual-based program to simulate quantitative traits with explicit genetic architecture in a dynamic metapopulation.Bioinformatics. 2008;24(13):1552–1553.

25. Bonnin I, Prosperi JM, Olivieri I. Genetic markers and quantitative genetic variation in Medicago truncatula (Leguminosae): a comparative analysis of population structure. Genetics.1996;143(4):1795–1805.

26. Hudson RR. Generating samples under a Wright-Fisher neutral model of genetic variation.Bioinformatics. 2002;18(2):337–338. doi: 10.1093/bioinformatics/18.2.337.

27. R Development Core Team. R: A language and environment for statistical computing. R Foundation for Statistical Computing, Vienna, Austria; 2006. http://www.r-project.org

28. Goudet J. FSTAT, a program to estimate and test gene diversities and fixation indices (version 2.9.3) 2001. http://www2.unil.ch/popgen/softwares/fstat.htm

29. Kimmel M, Chakraborty R, King JP, Bamshad M, Watkins WS, Jorde LB. Signatures of Population Expansion in Microsatellite Repeat Data. Genetics. 1998;148(4):1921–1930.

30. Roychoudhury A, Stephens M. Fast and Accurate Estimation of the Population-Scaled Mutation Rate, {theta}, From Microsatellite Genotype Data. Genetics. 2007;176(2):1363–1366.

31. Goudet J. hierfstat, a package for R to compute and test hierarchical F-statistics. Mol Ecol Notes.2005;5(1):184–186.

32. Meirmans PG. Using the AMOVA framework to estimate a standardized genetic differentiation measure. Evolution. 2006;60(11):2399–2402.

33. Michalakis Y, Excoffier L. A generic estimation of population subdivision using distances between alleles with special reference for microsatellite loci. Genetics. 1996;142(3):1061–1064.

34. Slatkin M. Inbreeding coefficients and coalescence times. Genet Res. 1991;58:167–175.

35. Hardy OJ, Charbonnel N, Freville H, Heuertz M. Microsatellite allele sizes: a simple test to assess their significance on genetic differentiation. Genetics. 2003;163(4):1467–1482.

36. Hardy OJ, Vekemans X. SPAGeDi: a versatile computer program to analyse spatial genetic structure at the individual or population levels. Mol Ecol Notes. 2002;2:618–620.

37. Pico FX, Mendez-Vigo B, Martinez-Zapater JM, Alonso-Blanco C. Natural Genetic Variation ofArabidopsis thaliana is Geographically Structured in the Iberian Peninsula. Genetics. 2008;180:1009–1021.

38. Le Corre V. Variation at two flowering time genes within and among populations of Arabidopsis thaliana: comparison with markers and traits. Mol Ecol. 2005;14(13):4181–4192.

39. Warthmann N, Fitz J, Detlef W. MSQT for choosing SNP assays from multiple DNA alignments.Bioinformatics. 2007;23(20):2784–2787.

40. Nordborg M, Hu TT, Ishino Y, Jhaveri J, Toomajian C, Zheng H, Bakker E, Calabrese P, Gladstone J, Goyal R. The pattern of polymorphism in Arabidopsis thaliana. PLoS Biol. 2005;3(7):e196.

41. Jost L. D vs. GST: Response to Heller and Siegismund (2009) and Ryman and Leimar (2009) Mol Ecol. 2009;18(10):2088–2091.

42. Sun JX, Mullikin JC, Patterson N, Reich DE. Microsatellites Are Molecular Clocks That Support Accurate Inferences about History. Mol Biol Evol. 2009;26(5):1017–1027. doi: 10.1093/molbev/msp025.

43. Symonds VV, Lloyd AM. An analysis of microsatellite loci in Arabidopsis thaliana: mutational dynamics and application. Genetics. 2003;165(3):1475–1488.

44. Ellegren H. Microsatellites: simple sequences with complex evolution. Nat Rev Genet. 2004;5(6):435–445. doi: 10.1038/nrg1348.

45. Calabrese P, Sainudiin R. In: Statistical methods in Molecular Evolution. Nielsen R, editor. New York: Springer; 2005. Models of microsatellite evolution; pp. 289–305.

46. Bhargava A, Fuentes FF. Mutational dynamics of microsatellites. Molecular Biotechnology.2010;44:250–266. doi: 10.1007/s12033-009-9230-4.

47. Balloux F, Lugon-Moulin N. The estimation of population differentiation with microsatellite markers.Mol Ecol. 2002;11(2):155–165.

48. Clark RM, Schweikert G, Toomajian C, Ossowski S, Zeller G, Shinn P, Warthmann N, Hu TT, Fu G, Hinds DA. Common Sequence Polymorphisms Shaping Genetic Diversity in Arabidopsis thaliana.Science. 2007;317(5836):338–342.

49. Beaumont MA, Nichols RA. Evaluating loci for use in the genetic analysis of population structure.Proc R Soc Lond B Biol Sci. 1996;263:1619–1626.

50. Leinonen T, O'Hara RB, Cano JM, Merilä J. Comparative studies of quantitative trait and neutral marker divergence: a meta-analysis. J Evol Biol. 2008;21(1):1–17.

51. Lynch M. The rate of polygenic mutation. Genet Res. 1988;51:137–148.

52. Lynch M, Walsh B. Genetics and the analysis of quantitative traits. Sunderland: Sinauer Associates, Inc.; 1998.

53. Schultz ST, Lynch M, Willis JH. Spontaneous deleterious mutation in Arabidopsis thaliana. Proc Nat Acad Sci USA. 1999;96(20):11393–11398.

54. Orr HA. The genetic theory of adaptation: a brief history. Nat Rev Genet. 2005;6(2):119–127. doi: 10.1038/nrg1523.

55. Goudet J, Büchi L. The effects of dominance, regular inbreeding and sampling design on QST, an estimator of population differentiation for quantitative traits. Genetics. 2006;172(2):1337–1347.

56. Goudet J, Martin G. Under neutrality, QST ≤ FST when there is dominance in an island model.Genetics. 2007;176:1371–1374.

57. Miller JR, Wood BP, Hamilton MB. FST and QST Under Neutrality. Genetics. 2008;180(2):1023–1037.

58. Whitlock MC. Evolutionary inference from QST. Mol Ecol. 2008;17(8):1885–1896..

59. Bakker EG, Stahl EA, Toomajian C, Nordborg M, Kreitman M, Bergelson J. Distribution of genetic variation within and among local populations of Arabidopsis thaliana over its species range. Mol Ecol.2006;15(5):1405–1418.

60. He F, Kang D, Ren Y, Qu LJ, Zhen Y, Gu H. Genetic diversity of the natural populations ofArabidopsis thaliana in China. Heredity. 2007;99(4):423–431.

61. Bergelson J, Stahl E, Dudek S, Kreitman M. Genetic Variation Within and Among Populations ofArabidopsis thaliana. Genetics. 1998;148(3):1311–1323.

62. Beck JB, Schmuths H, Schaal BA. Native range genetic variation in Arabidopsis thaliana is strongly geographically structured and reflects Pleistocene glacial dynamics. Mol Ecol. 2008;17(3):902–915.

63. Stenoien HK, Fenster CB, Tonteri A, Savolainen O. Genetic variability in natural populations ofArabidopsis thaliana in northern Europe. Mol Ecol. 2005;14(1):137–148.

64. Muller MH, Leppälä J, Savolainen O. Genome-wide effects of postglacial colonization in Arabidopsis lyrata. Heredity. 2007;100(1):47–58.

65. Clauss MJ, Mitchell-Olds T. Population genetic structure of Arabidopsis lyrata in Europe. Mol Ecol.2006;15(10):2753–2766.

66. Carreras-Carbonell J, Macpherson E, Pascual M. Population structure within and between subspecies of the Mediterranean triplefin fish

Tripterygion delaisi revealed by highly polymorphic microsatellite loci. Mol Ecol. 2006;15:3527–3539.

67. O'Reilly P, Canino M, Bailey K, Bentzen P. Inverse relationship between FST and microsatellite polymorphism in the marine fish walleye pollock (Theragra chalcogramma): implications for resolving weak population structure. Mol Ecol. 2004;13:1799–1814.

68. Nordborg M, Donnelly P. The coalescent process with selfing. Genetics. 1997;146(3):1185–1195.

69. Abbott RJ, Gomes MF. Population genetic structure and outcrossing rate of Arabidopsis thaliana (L.) Heynh. Heredity. 1989;62:411–418.

70. Hoffmann MH, Bremer M, Schneider K, Burger F, Stolle E, Moritz G. Flower Visitors in a Natural Population of Arabidopsis thaliana. Plant Biol. 2003;5:491–494.

Chapter 12

Sensitivity of Methods for Estimating Breeding Values Using Genetic Markers to the Number of QTL and Distribution of QTL Variance

Albart Coster,[1] John WM Bastiaansen,[1] Mario PL Calus,[2] Johan AM van Arendonk,[1] and Henk Bovenhuis[1]

[1]Animal Breeding and Genomics Centre, Wageningen University, PO Box 338, 6700 AH, Wageningen, The Netherlands

[2]Animal Breeding and Genomics Centre, Animal Science Group, Lelystad, The Netherlands

ABSTRACT

The objective of this simulation study was to compare the effect of the number of QTL and distribution of QTL variance on the accuracy of breeding values estimated with genomewide markers (MEBV). Three distinct methods were used to calculate MEBV: a Bayesian Method (BM), Least Angle Regression (LARS) and Partial Least Square Regression (PLSR). The accuracy of MEBV calculated with BM and LARS decreased when the number of simulated QTL increased. The accuracy decreased more when QTL had different variance values than when all QTL had an equal variance. The accuracy of MEBV calculated with PLSR was affected neither by the number of QTL nor by the distribution of QTL variance. Additional simulations and analyses showed that these conclusions were not affected by the number of individuals in the training population, by the number of markers and by the heritability of the trait. Results of this study show that the effect of the number of QTL and distribution of QTL variance on the accuracy of MEBV depends on the method that is used to calculate MEBV.

BACKGROUND

In current breeding programs, estimation of breeding values is based on phenotypes of selection candidates and their relatives, often measured after animals reach to a certain age. This leads to a moderate to long generation interval, substantial costs and complex logistics for phenotypic recording [1]. Comparatively, breeding values estimated with genomewide distributed markers (MEBV) will increase annual genetic gain due to a reduced generation interval and improved accuracy, at lower costs [2,1].

Calculation of MEBV requires a population with information on genetic markers and phenotypes, called the*training* population. Phenotypic performance of the training population is used to estimate effects for the genetic markers which can be used to calculate MEBV of individuals with only marker information, called the *evaluation* population. Accuracy of MEBV depends on the heritability of the trait, the size of the training population, the method used to estimate marker effects and linkage disequilibrium (LD) between markers and quantitative trait loci (QTL) [2-6].

Linkage disequilibrium between markers and QTL is a function of the distance between markers and QTL and of the effective population size [7]. A large number of markers, distributed over the whole genome, is required to achieve high LD between markers and QTL when number and location of QTL on the genome are unknown. Simulation studies have shown that accuracy of MEBV increases when LD increases [2,8,9,4].

The accuracy of MEBV also depends on the variance of individual QTL since the ability to detect a QTL is related to its size. The size of a QTL, measured as the proportion of the genetic variance explained by that QTL, depends on its variance and on the genetic variance. Genetic variance, in turn, is a function of the number of QTL and of the variance of the individual QTL. Hayes and Goddard [10] have estimated parameters of a Gamma distribution describing the QTL effects found in published QTL detection experiments. This gamma distribution has been used in simulation studies to model the distribution of QTL effects [2,8,4,9,6]. Even though the distribution of QTL effects can vary considerably between different traits, the effect of the number of QTL on the accuracy of MEBV has been addressed only by Daetwyler [11] and the effect of distribution of QTL variance on the accuracy of MEBV has not been studied.

An important problem when estimating marker effects is the large number of markers relative to the number of phenotypes in the training data [2]. Meuwissen et al. [2] have solved this by using a Bayesian method (BM) that uses a sampling algorithm to obtain a posterior distribution of the marker effects. This Bayesian method is used in many simulation studies and in practical breeding programs, e.g. [12]. The Bayesian setup enables to incorporate a prior for the number of QTL and for the distribution of QTL effects [2]. Goddard [5] has found higher accuracies when a prior distribution for QTL effects reflecting the gamma (or exponential) distribution of QTL effects was used, compared to using a normal prior distribution for QTL effects. For many quantitative traits, however, the true distribution of the QTL effects is unknown.

Two other methods that might be suitable for estimating MEBV are Least Angle Regression (LARS) and Partial Least Square Regression (PLSR). LARS is a penalized regression method which identifies predictor variables that are highly correlated to the response variable and includes these in a regression model [13]. Park and Casella have shown similarities between LASSO, a variant of LARS, and Bayesian regression models [14]. They have shown that the posterior mode of a Bayesian model, similar to that proposed by Meuwissen et al. [2], and the regression coefficients estimated using LASSO are equal. Thus, LARS is a nonbayesian alternative to BM.

Regardless of the number of genetic markers, the rank of the matrix of marker data will be less or equal than the number of individuals in the training data. This implies the existence of correlations between marker genotypes. These correlations can be used to calculate MEBV by regressing the phenotypes on linear combinations of the markers. Partial Least Square Regression (PLSR) is a method that builds orthogonal linear combinations of the markers that have a maximum correlation with the phenotypes and regresses the phenotypes on these linear combinations, which are also called components [15]. Since components are orthogonal, regression coefficients of the components are independent. Datta et al. [16] have used PLSR in gene expression studies, Moser et al. [17] and Solberg et al. [6] have used PLSR to calculate MEBV.

Although BM and PLSR have been used independently to calculate MEBV, the accuracy of these methods when the number of QTL and the distribution of QTL variance varies is unknown. Therefore, the objective of this study is to investigate the effect of number of QTL and distribution of QTL variance on the accuracy of MEBV estimated with methods BM, LARS and PLSR.

METHODS

Simulation of data

Each simulated genome consisted of four chromosomes of 1 Morgan each. Ten thousand loci were equally distributed over each chromosome, there were thus 40,000 loci distributed over the whole genome. In the base population, 4,000 of these loci, equally distributed over the genome, were made biallelic with allele frequency equal to 0.50. The remaining 36,000 loci were monomorphic in the base population. Two hundred gametes for the base population were simulated assuming linkage equilibrium and were randomly combined to create 100 individuals. Five thousand generations were simulated to generate LD between loci and to reach a mutation-drift equilibrium. Each individual in each generation contributed two gametes to the next generation with the objective of maintaining a population size of 100 individuals with Ne equal to 199 (the simulated population structure was thus different from a Wright-Fisher scenario). Each gamete transmitted to the offspring was simulated as an independent meiotic event. The number of recombinations for each chromosome was drawn from a Poisson(1) distribution, reflecting the size of the chromosomes in Morgan. The positions of the recombinations were sampled assuming no interference between recombinations.

Mutation rate for the 40,000 loci was set at 10^{-5}. A mutation switched the allelic status; mutation of a 0 allele produced a 1 allele and mutation of a 1 allele produced a 0 allele.

Each individual in generation 5,000 contributed 10 gametes to generation 5,001, resulting in 50 fullsib families of 10 individuals each. Each individual in generation 5,001 contributed two offspring to generation 5,002, resulting in 250 fullsib families of 2 individuals each. Generation 5,001 was used as the training population and generation 5,002 was used as the evaluation population. Mutation rate was set to 0 in generations 5,001 and 5,002 to avoid the introduction of a large number of new alleles with a low Minor Allele Frequency (MAF). We simulated sixty replicates.

Table 1. Scenarios with different number of QTL and distribution of QTL variance.

Scenario	Number of QTL	Distribution of QTL variance
1	low	unequal
2	intermediate	unequal
3	high	unequal
4	low	equal
5	intermediate	equal
6	high	equal

Scenarios were numbered from 1 to 6, according to the number of QTL contributing 90% of the genetic variance.

To simulate a range of QTL distributions, six scenarios were generated which were combinations of three levels for number of QTL and two distributions of QTL variance (Table (Table1).1). Depending on the scenario, up to fifty percent of the loci with a MAF greater than 0.10 were selected to become QTL in generation 5,001. QTL scenarios were numbered from 1 to 6, with increasing number of QTL accounting for 90% of the total genetic variance. Biallelic loci that were not selected as QTL in any scenario were used as biallelic markers. Within a replicate, this resulted in the same marker set across all QTL scenarios. Each QTL scenario was applied to all 60 replicates.

The number of QTL contributing to the trait was changed by letting 5% (*low number of QTL*), 25% (*intermediate number of QTL*) or 50% (*high number of QTL*) of all loci with a MAF greater than 0.10 contribute to the trait. QTL for the scenarios with low and intermediate numbers of QTL were uniformly selected from the 50% of loci selected as QTL in the scenario with high number of QTL.

The variances of all QTL contributing to the trait were equal (*equal QTL variance*), or unequal (*unequal QTL variance*). The additive effects of QTL were calculated based on the specified QTL variance and the allele frequency of each QTL. For the scenarios of equal QTL variance, variance of each QTL was set to 1. For the scenarios of unequal QTL variance, variance of every tenth QTL was set to 81 and variances of the other 9 QTL were set to 1. In this way 10% of the QTL were responsible for 90% of the total additive genetic variance. The QTL effects were assigned to each QTL after the QTL were selected and therefore the same QTL were present in scenarios of equal and unequal QTL variance.

The true breeding value (TBV) of each individual was calculated as the sum of the allelic effects. Additive genetic variance, σ_a^2, was calculated as the variance of the TBV in generation 5,001. Deviates from a N(0, σ_e^2) distribution were added to TBV and σ_e^2 was equal to σ_a^2 to simulate phenotypes with a heritability of 0.50.

In addition to the QTL scenarios, we studied the effect of heritability, pre-selection of markers based on MAF, and size of the training population on the accuracy of the MEBV calculated with the three methods. In the first alternative, heritability of the trait was reduced from 0.50 to 0.25. In the second alternative, markers with a MAF lower than 0.10 in the training population were excluded from the marker data. In the third alternative, the size of the training population was increased from 500 to 1,000 individuals by adding 10 fullsibs to each family while the size of the evaluation population was maintained at 500 individuals. Each alternative was applied to all six QTL scenarios and to the 60 replicates.

The simulations were performed with HaploSim [18], a package for R [19] which is available from the R repository CRAN http://cran.r-project.org/package=HaploSim. The simulations and computations were run on a system with a dual core Intel 2.33 Ghz processor and a Fedora Core 10 operating system.

Analysis of population data

To validate and characterize the simulations, we determined the number of biallelic markers, heterozygosity of biallelic markers, linkage disequilibrium between adjacent markers and coefficient of determination of QTL. Heterozygosity of a population is the average number of heterozygous loci of an individual. Expected heterozygosity in a situation of mutation-drift equilibrium, expressed as a fraction of the total number of loci, is a function of mutation rate (u) and effective population size (Ne) [20]:

$$H = \frac{4 \cdot Ne \cdot u}{1 + 4 \cdot Ne \cdot u} \tag{1}$$

In our simulations, where effective population size was 199 (Crow and Kimura, Equation 3.13.5 [20]) and mutation rate was 10^{-5}, expected H is $7.90 \cdot 10^{-3}$. For a genome consisting of 40,000 loci, the expected number of heterozygous loci in an individual is 316.

Linkage disequilibrium between adjacent markers was calculated as the squared correlation between adjacent markers and was expressed as r^2.

The coefficient of determination of a QTL, expressed as R^2, is the proportion of variance of that QTL explained by a set of markers. $R^2$was calculated using the equation $R^2 = c'K^{-1}c$, where c is a vector of correlation coefficients between the markers and the QTL, and K is the matrix of pairwise correlations of the markers. When the absolute correlation between a pair of markers exceeded 0.95, only one of these two markers was used to avoid singularity of matrix K. R^2 was calculated as the mean of R^2 between each QTL and the 50 markers in highest LD with that QTL and provided an estimate of the upper limit of the accuracy of MEBV that could be obtained based on this number of markers.

Calculation of breeding values

We used three methods to estimate marker effects in the training population. The methods differed in how they estimated the additive effects of individual marker loci, but used an identical approach to calculate MEBV after these effects were estimated:

$$MEBV = Xa, \tag{2}$$

where **MEBV** is the vector of breeding values estimated with the marker genotypes, **X** is an incidence matrix that relates genotypes to individuals, and **a** is the vector of additive effects for the markers, which is estimated by each method.

BM

The Bayesian Model (BM) used was proposed by Meuwissen and Goddard [2]. In this model, the additive effects of the markers are considered as independent random normal variables. The additive effect of markers which are considered to be associated to a QTL are sampled from a N(0, σ_1^2) distribution. The additive effects of markers with are considered not to be associated to a QTL are sampled from a N(0, σ_1^2/100) distribution, which has a lower variance. The method requires a prior for the number of QTL and a prior for QTL variance σ_1^2. The prior for the number of QTL was set at 50 in all scenarios, regardless of the true number of QTL in that simulation scenario. The prior for QTL variance was set at 0.20, regardless of the simulation scenario.

BM uses Gibbs sampling to numerically integrate over the posterior distribution of the model. The sampler was run for 10,000 iterations and the first 1,000 iterations were discarded as burn-in. Regression coefficients of the markers were calculated as the means of their posterior distributions.

LARS

Least Angle Regression is a penalized regression method where predictor variables are included sequentially in the model [13]. Regression coefficients of all markers are zero at the start of the algorithm. LARS builds the model in sequential steps, in each step the marker that has the highest correlation with the residual is added to the model and the model proceeds in a direction of equal angle between all markers included in the model and the sequentially added marker [13]. After n steps, there are n markers in the model. We used thelars function in the lars package [21] of R and used cross validation on the training data to find the number of markers that minimized prediction error.

PLSR

Partial Least Square regression reduces the dimensions of the regression model by building orthogonal linear combinations of markers that have a maximal correlation with the response variable [15]. The trait is subsequently regressed on the linear combinations of markers, or components. Cross validation was used to find the number of components that minimized the prediction error.

To reduce the computation time required to fit the PLSR models, the algorithm to find the optimal number of components was modified as follows. In a first step, a model was fitted with ten components. Cross validation was used to find the optimal number of components. If the optimal number of components was below ten, this optimal number of components was used and the algorithm was stopped. If the optimal number of components was ten, a next iteration was performed with 20 components. If the optimal number of components, found by cross validation, was below 20, this number of components was used. Otherwise, the procedure was repeated with 30 components, and so on, until the number of components was equal to the number of observations or to the number of marker loci. The plsr function in the pls package [22] of R was used to fit and cross validate the models in each iteration. Cross validation was performed on the training data.

Comparison of methods to calculate breeding values

The performance of each method was assessed based on the accuracy and the Mean Square Error of Prediction (MSEP) of MEBV. Accuracy of MEBV is the correlation between MEBV and TBV. Mean Square Error of Prediction is the average of the squared prediction errors of MEBV. Accuracy and MSEP were calculated based on individuals in the evaluation population.

Computation time of each method was recorded in all six QTL scenarios for ten replicates. The time recorded included the time required to fit the model on the training population, the time required for cross validation when using LARS and PLSR, and the time required to calculate MEBV for the evaluation population.

Results

Characteristics of simulated populations

Average heterozygosity was equal to 0.0110 in generation 1,000 and stabilized after 4,000 generation at 0.0076, corresponding to 304 heterozygous markers. This is slightly below the expected number based on Equation 1. The average number of biallelic markers in the data was 1,431 (Table (Table2).2). Eighty percent of these markers had a MAF below 0.10, reflecting an L-shaped distribution of MAF.

Table 2. Average (standard error) of number of polymorphic markers (nSNP), LD between adjacent markers (r^2), number of QTL (nQTL), and average coefficient of determination of QTL (R^2).

Situation	nSNP	r^2	nQTL	R^2
low nQTL	1431 (5.3)	0.048 (< 0.001)	35 (0.2)	0.806 (0.003)
low nQTL MAF > 0.10	374 (2.1)	0.145 (0.002)	35 (0.2)	0.715 (0.004)
int. nQTL	1431 (5.3)	0.048 (< 0.001)	172 (1.0)	0.811 (0.002)
int. nQTL MAF > 0.10	374 (2:1)	0.145 (0.002)	172 (1.0)	0.717 (0.002)
high nQTL	1431 (5.3)	0.048 (< 0.001)	343 (2.0)	0.811 (0.001)
high nQTL MAF > 0.10	374 (2.1)	0.145 (0.002)	343 (2.0)	0.717 (0.001)

The simulated number of QTL was low, intermediate (int.) or high and markers with a MAF lower than 0.10 were either or not included in the marker data. The table summarizes 60 replicated simulations.

Average LD between all adjacent markers, measured as r^2, was 0.048 (Table (Table2).2). Expected LD, based on Equation 7 of Sved [7], is 0.31 (assuming an average distance between markers of 4/1431 Morgan). When markers with a MAF lower than 0.10 were excluded from the data, average LD between adjacent markers increased to 0.146 (Table (Table2).2). The expected LD based on Sved [7] is 0.11, however, does not account for mutations. To compare the LD obtained in our simulations with its expectation, we calculated the average LD between adjacent markers which were introduced in generation 0 and remained polymorphic in generation 5,000. On average, there were 174 of these markers and average LD between these markers was 0.036 which is close to the expected LD of 0.052 (assuming an equal distance between markers of 4/174 Morgan).

The average number of QTL was 35 in the scenarios with a low number of QTL and increased to 343 in the scenarios with a high number of QTL (Table (Table2).2). The average coefficient of determination of the QTL (R^2) was 0.80 when all markers were used and 0.71 when markers with a MAF above 0.10 were used to calculate R^2 (Table (Table22).

Based on the average number of QTL (Table (Table2),2), the estimated number of QTL accounting for 90% of the total genetic variance ranged from 3, in scenario 1 (low number of QTL, unequal QTL variance), to 309, in scenario 6 (high number of QTL, equal QTL variance). The number of QTL accounting for 90% of the genetic variance in scenario 3 (high number of QTL, unequal QTL variance, approx. 31 QTL) was similar to that in scenario 4 (low number of QTL, equal QTL variance, approx. 35 QTL).

Characteristics of MEBV

The average accuracy of MEBV calculated with BM and LARS decreased when the number of QTL increased and was stronger in the scenarios of unequal QTL distribution than in the scenarios of equal QTL distribution (Table (Table33 and Figure Figure1).1). The highest accuracies using BM and LARS were in scenario 1 (low number of QTL and unequal distribution of QTL variance) (Table (Table3).3). The highest accuracy using PLSR was in scenario 4, but with this method there was not a clear trend of accuracies between scenarios (see Table Table33and Figure Figure1).1). Overall, accuracies of BM were highest except in scenario 3 (Table (Table33).

Table 3. Average (standard error) accuracy of MEBV for individuals in the evaluation population.

Method	unequal QTL variance			equal QTL variance		
	low nQTL	int. nQTL	high nQTL	low nQTL	int. nQTL	high nQTL
	sc. 1	sc. 2	sc. 3	sc. 4	sc. 5	sc. 6
BM	0.77 (0.009)	0.67 (0.010)	0.60 (0.012)	0.71 (0.004)	0.67 (0.005)	0.67 (0.006)
LARS	0.75 (0.009)	0.67 (0.005)	0.65 (0.004)	0.65 (0.005)	0.63 (0.006)	0.63 (0.006)
PLSR	0.66 (0.009)	0.66 (0.007)	0.67 (0.007)	0.68 (0.006)	0.67 (0.006)	0.66 (0.007)

The MEBV were calculated with methods BM, LARS and PLSR. Simulated number of QTL was low (low nQTL), intermediate (int. nQTL) or high (high nQTL). The simulated variance of every tenth QTL was 81 times larger than variance of the remaining QTL (unequal QTL variance) or equal for all QTL (equal QTL variance). The averages and standard deviations were calculated using 60 replicated simulations.

Additional simulations were done with a number of QTL ranging between the intermediate and high number of QTL and using an unequal distribution of QTL variance to investigate the strong decrease of accuracies of BM from scenario 2 to scenario 3 (Table (Table3).3). Results of these additional simulations, confirm the decrease of accuracy of MEBV with BM between scenarios 2 and 3 (Figure (Figure11).

The accuracy of MEBV decreased when heritability was reduced from 0.50 to 0.25 in the three methods (Table (Table4).4). In the scenarios with a low number of QTL (scenarios 1 and 4), BM was the most accurate (combining Tables Tables33 and and4).4). In the scenarios with an intermediate and high number of QTL, PLSR was the most accurate (combining Tables Tables33 and and44).

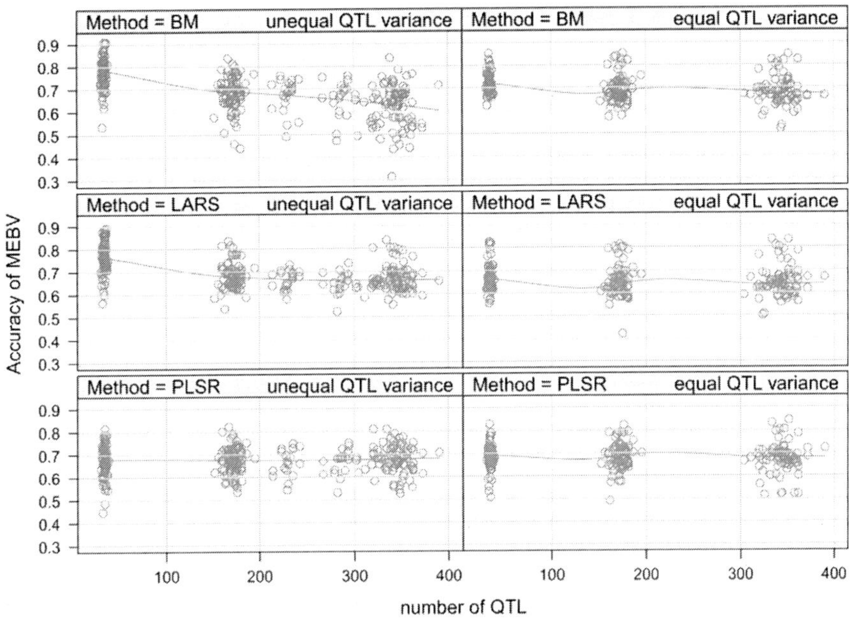

Figure 1. Plot of the accuracies of MEBV calculated with BM, LARS and PLSR as affected by the simulated number of QTL. The plots display the accuracies of 60 replicated simulations for number of QTL around 35, 172 and 343 plus the accuracies of 10 replicated simulation...

Table 4. Average (standard error) change of accuracy of MEBV for individuals in the evaluation population as affected by alternative simulation situations.

Method	unequal QTL variance			equal QTL variance		
	low QTL sc. 1	int. QTL sc. 2	high QTL sc. 3	low QTL sc. 4	int. QTL sc. 5	high QTL sc. 6
$h^2 = 0.25$						
BM	-0.14 (< 0.01)	-0.16 (0.01)	-0.08 (0.01)	-0.12 (< 0.01)	-0.16 (< 0.01)	-0.18 (< 0.01)
LARS	-0.14 (< 0.01)	-0.16 (< 0.01)	-0.15 (< 0.01)	-0.15 (< 0.01)	-0.14 (< 0.01)	-0.14 (< 0.01)
PLSR	-0.10 (< 0.01)	-0.11 (< 0.01)	-0.11 (< 0.01)	-0.11 (< 0.01)	-0.12 (< 0.01)	-0.11 (< 0.01)
nTR = 1,000						
BM	0.05 (< 0.01)	0.11 (0.01)	0.16 (0.01)	0.06 (< 0.01)	0.08 (< 0.01)	0.07 (< 0.01)
LARS	0.04 (< 0.01)	0.07 (< 0.01)	0.06 (< 0.01)	0.07 (< 0.01)	0.07 (< 0.01)	0.07 (< 0.01)
PLSR	0.07 (< 0.01)	0.06 (< 0.01)	0.06 (< 0.01)	0.06 (< 0.01)	0.06 (< 0.01)	0.07 (< 0.01)
MAF >0.1						
BM	-0.03 (< 0.01)	-0.01 (< 0.01)	0.02 (0.01)	-0.03 (< 0.01)	-0.03 (< 0.01)	-0.04 (< 0.01)
LARS	-0.02 (< 0.01)	0.00 (< 0.01)	-0.01 (< 0.01)	0.00 (< 0.01)	0.00 (< 0.01)	0.00 (< 0.01)
PLSR	-0.02 (< 0.01)	-0.01 (< 0.01)	-0.01 (< 0.01)	-0.03 (< 0.01)	-0.02 (< 0.01)	-0.01 (< 0.01)

Simulated heritability was reduced from 0.5 to 0.25 ($h^2 = 0.25$); the size of the training population was increased from 500 to 1,000 individuals (nTR = 1,000); only markers with a MAF above 0.10 were used to fit the models (MAF > 0.1). The simulated number of QTL was low (low nQTL), intermediate (int. nQTL) or high (high nQTL). The simulated variance of every tenth QTL was 81 times larger than variance of remaining QTL (unequal QTL variance) or equal for all QTL (equal QTL variance). Methods BM, LARS and PLSR were used to calculate the MEBV. The averages and standard deviations were calculated using 60 replicated simulations.

The accuracy of MEBV calculated with all methods increased when the size of the training population was increased from 500 to 1,000 individuals (Table (Table4)4) and BM was the most accurate method in all scenarios (combining Tables Tables33 and and44).

The accuracies of MEBV calculated with BM and PLSR decreased when markers with a MAF lower than 0.10 were excluded from the data, except for BM in scenario 3 (Table (Table4).4). Accuracies of MEBV calculated with LARS were not clearly affected by excluding markers with a MAF lower than 0.10. There was no clear effect of QTL scenario on the change of accuracies due to this exclusion (Table (Table4).4). The decrease of accuracies calculated with BM and PLSR when markers with a MAF lower than 0.10 were excluded was in line with the decrease of R^2 (Table (Table22).

Mean Square Error of Prediction of MEBV calculated with the three methods increased when the number of QTL increased (Table (Table5).5). The average MSEP of MEBV calculated with BM were low in all scenarios, except in scenario 3 where it was highest (Table (Table55).

Table 5. Average (standard error) of Mean Square Error of Prediction (MSEP) of MEBV for individuals in the evaluation population.

Method	unequal QTL variance			equal QTL variance		
	low nQTL	int. nQTL	high nQTL	low nQTL	int. nQTL	high nQTL
	sc. 1	sc. 2	sc. 3	sc. 4	sc. 5	sc. 6
BM	659 (26)	4049 (108)	10463 (343)	79 (2)	416 (6)	850 (12)
LARS	707 (24)	4019 (71)	8230 (124)	91 (2)	465 (6)	927 (12)
PLSR	993 (24)	4242 (73)	8405 (123)	93 (2)	458 (6)	922 (14)

The additive genetic variance increased when the number of QTL increased and was higher in the scenarios with unequal distribution of QTL variance (Table (Table6).6). This is due to the fact that the variance of 10% of the QTL was made 81 times larger than in the scenarios of equal QTL variance. The variance of MEBV calculated with the three methods was lower than the simulated additive genetic variance in all scenarios. The variance of MEBV calculated with PLSR was highest in all scenarios (Table (Table6).6). The variance of MEBV calculated with the three methods increased when number of QTL increased, except for the variance of MEBV calculated with BM in scenario 3 (Table (Table6).6). If MEBV were unbiased, then the variance of MEBV would be equal to $r^2\sigma_a^2$, where r^2 is the squared accuracy of MEBV (Table (Table3).3). The variance of MEBV calculated with BM was lower than this expected variance in all scenarios (combining Tables Tables33 and and6).6). The variances of MEBV calculated with LARS and PLSR were higher than the expected variance in all scenarios and this difference was greatest for method PLSR (combining Tables Tables33 and and66).

Table 6. Average (standard error) of the simulated additive genetic variance (σ_a^2) in the evaluation population, and variance of MEBV calculated for individuals in the evaluation population.

Method	unequal QTL variance			equal QTL variance		
	low QTL	int. QTL	high QTL	low QTL	int. QTL	high QTL
	sc. 1	sc. 2	sc. 3	sc. 4	sc. 5	sc. 6
σ_a^2	1623 (23)	7210 (88)	14193 (156)	158 (2)	767 (8)	1538 (18)
BM	890 (38)	2537 (168)	2032 (283)	81 (3)	327 (13)	575 (24)
LARS	914 (31)	3937 (164)	7017 (293)	75 (4)	344 (15)	715 (29)
PLSR	1249 (49)	5263 (198)	10747 (393)	129 (5)	618 (21)	1150 (46)

The methods BM, LARS and PLSR were used to calculate the MEBV. The simulated number of QTL was low (low nQTL), intermediate (int. nQTL) or high (high nQTL). The simulated variance of every tenth QTL was 81 times larger than variance of remaining QTL (unequal QTL variance) or equal for all QTL (equal QTL variance). The averages and standard deviations were calculated using 60 replicated simulations.

Table 7. Average (standard error) computation time required for fitting the MEBV models to the training population and calculating MEBV for the evaluation population, measured in seconds.

Method	Normal	$h^2 = 0.25$	nTr = 1,000	MAF > 0.10
BM	423.25 (3.73)	429.57 (3.88)	820.75 (9.05)	109.49 (1.90)
LARS	211.75 (3.28)	210.92 (2.62)	1058.38 (9.34)	57.37 (1.80)
PLSR	4.05 (0.10)	4.10 (0.18)	6.47 (0.15)	0.81 (0.02)

Situation normal: heritability equal to 0.5, size of the training population equal to 500 individuals, and all markers included in the data. Situation $h^2 = 0.25$: heritability was decreased from 0.50 to 0.25. Situation nTr = 1,000: size of training population was increased from 500 to 1,000 individuals. Situation MAF > 0.10: markers with a MAF below 0.10 were excluded from the data. The table summarizes ten simulations for the scenario of intermediate number of QTL and equal QTL variance.

The average computation time required by the three methods increased when the size of the training population increased and when the number of markers included in the data increased (Table (Table7).7). In a normal situation, where the size of the training population was 500 individuals, all the markers were included in the data, and the heritability was equal to 0.50, PLSR required approximately 4 seconds to fit, cross validate and evaluate the models. LARS required approximately 211 seconds and BM required approximately 430 seconds (Table (Table77).

Discussion and conclusions

The accuracies of MEBV calculated with the BM method in this study were compared to accuracies obtained by Calus et al. [4] and by Solberg et al. [6]. The approximate number of QTL was 75 in the simulations of Calus et al. [4], and 55 in the simulations of Solberg et al. [6]. Based on their descriptions, approximately seven QTL would account for 90% of the total genetic variance in both studies. Therefore, the simulations of Calus et al. [4] and Solberg et al. [6] are most comparable to scenario 1 (low number of QTL, unequal QTL variance), where an average of three QTL accounted for 90% of the total genetic variance.

The average accuracy of MEBV for individuals without performance data of their own by Calus et al. [4] was 0.75. The accuracy reported by Solberg et al. [6] in the scenario with a low number of markers was 0.69 with BM and 0.61 with PLSR. Accuracies in both studies, but especially in Solberg et al. [6], were lower than accuracies in scenario 1 of this study (Table (Table3).3). A lower LD between markers and QTL in the study of Solberg et al. [6] might be the reason for this lower accuracy.

The average LD between adjacent markers provides an indication for LD between markers and QTL because QTL are necessarily located somewhere between the markers. Average LD between adjacent markers can not be compared directly to expected LD based on Equation 7 of Sved [7] because mutations are expected to have a very strong impact on this LD. This strong impact is expected because a mutation will generally introduce a new marker between two markers which were previously considered adjacent. We calculated LD between adjacent markers that were polymorphic in generation 0 and still polymorphic in generation 5001. This LD can be compared to expected LD based on Sved [7] because newly mutated markers are not used and the effect of mutations on specific markers is negligible. Linkage disequilibrium, calculated in this way, was very similar to expected LD, providing evidence for the adequateness of our simulations.

Simulated QTL scenarios were numbered from 1 to 6, according to the number of QTL accounting for 90% of the genetic variance. The total number of biallelic QTL in the data is often used to describe simulations [3,4,9,6]; we think that the number of QTL accounting for a specific proportion of the genetic variance is a more appropriate description of the complexity of the genetic architecture underlying the trait. In this context, we expected similar results in scenarios 3 and 4 since the number of QTL accounting for 90% of the genetic variance were similar (34 in scenario 3 and 31 in scenario 4). Average accuracies of MEBV calculated with LARS and PLSR confirmed this expectation but accuracies with BM did not.

With method BM, higher accuracies were expected in QTL scenarios which more closely resembled the prior distributions for QTL number and distribution of QTL effects. The high accuracies with BM in scenario 1 were in line with this expectation but the stronger decrease of accuracies in scenarios 1 to 3 compared to the decrease of accuracies in scenarios 4 to 6 was not. The consistency of the decline in scenarios 1 to 3 was confirmed by additional simulations, with a number of QTL ranging between that in scenario 2 and in scenario 3. Accuracies of MEBV in these simulations confirmed this decrease (Figure (Figure11).

To investigate whether accuracies of MEBV calculated with BM were affected by the prior distribution for QTL effects, we reanalyzed the data using a prior that more closely resembled the QTL scenarios that were simulated. In each scenario, the prior for number of QTL was set equal to the average number of QTL in this scenario (Table (Table2)2) and the prior for the variance of individual QTL was set equal to the average simulated genetic variance divided by the average number of QTL in this scenario (Tables (Tables22 and and6).6). Comparison of these accuracies (Table (Table8)8) to the accuracies in Tables Tables33 and and44 shows that using a prior which is more correct does not improve average accuracy of MEBV. The accuracies of MEBV calculated with method BM in the different scenarios indicate that the highest accuracies are obtained with this method in situations were a small number of QTL accounts for a large proportion of the total genetic variance. The results in Table Table88 indicate that the accuracies with BM did not depend on the correctness of the prior for QTL distribution and, furthermore, that a prior which was closer to the actual QTL distribution even led to lower accuracies in scenarios with a high number of QTL. These results contrast the results of Goddard [5], who found higher accuracies when a exponential prior for QTL effects was compared to a normal prior for QTL effects when the QTL effects were exponentially distributed. In this study, however, we compared accuracies obtained with different prior parameters, while using the same kind of distribution. Combining the results of Goddard [5] and of this comparison, it can be stated that using a correct kind of distribution as prior of QTL effects can be important for accuracy of BM but that the exact parametrization of this prior is not important.

Table 8. Average (standard error) accuracy of MEBV for individuals in the evaluation population.

Method	unequal QTL variance			equal QTL variance		
	low nQTL	int. nQTL	high nQTL	low nQTL	int. nQTL	high nQTL
Standard	0.80 (0.007)	0.67 (0.006)	0.57 (0.007)	0.69 (0.005)	0.62 (0.006)	0.57 (0.006)
h^2 = 0.25	0.68 (0.011)	0.52 (0.006)	0.56 (0.008)	0.57 (0.004)	0.51 (0.005)	0.53 (0.006)
MAF>0.10	0.77 (0.008)	0.69 (0.006)	0.64 (0.007)	0.67 (0.006)	0.66 (0.005)	0.61 (0.004)

The simulated number of QTL was low (low nQTL), intermediate (int. nQTL) or high (high nQTL). The simulated variance of every tenth QTL was 81 times larger than variance of remaining QTL (unequal QTL variance) or equal for all QTL (equal QTL variance). The rows of the table correspond to the standard situation (h^2 = 0.5, size of training population = 500 individuals, all markers included), the situation with h^2 = 0.25, and the situation where markers with MAF < 0.10 were excluded from the data. Method BM was used to calculate the. The prior number of QTL was 35 QTL in the scenarios with a low number of QTL, 172 QTL in the scenarios with an intermediate number of QTL, and 343 QTL in the scenarios with a high number of QTL. The prior for QTL variance was the ratio of the total genetic variance (Table 2) and the number of QTL. The averages and standard deviations were calculated using 60 replicated simulations

The number of QTL contributing to a trait is unknown in real situations. The scenarios of unequal QTL variance were motivated by the real situation where a few QTL contribute an important proportion of the total genetic variance. Examples of these situations include the DGAT1 gene and the SCD gene on bovine chromosomes 14 and 26 which contribute a large proportion of the genetic variation of milk fat content [23,24] and the IGF2 gene on porcine chromosome 2, which contributes a large proportion of the genetic variation of muscle mass in pigs [25]. Simulations and analyses that use a distribution similar to the one estimated by Hayes and Goddard [10] implicitly assume this situation. The scenarios with an equal QTL variance were motivated by the situations where many QTL contribute a small proportion of the total genetic variation of an individual trait, e.g. height in humans [26-28]. This study shows that accuracy and MSEP of distinct methods to calculate MEBV are affected by the distribution of QTL underlying a trait. Results of this study also show that the good performance of a method in one specific QTL scenario does not guarantee a good performance in other QTL scenarios.

Characteristics of the methods used to fit the MEBV models differed. Methods BM and LARS attempt to identify markers highly correlated with QTL and estimate effects for these markers. Results confirmed that the approach used by both BM and LARS, was advantageous when few QTL accounted for a large proportion of the total genetic variance. Method PLSR builds orthogonal, linear combinations of the predictor data (marker genotypes) that are highly correlated with the response and regresses the response on these components. The advantage of this method was that accuracies were almost not affected by the QTL scenario that was simulated; this was especially clear when comparing the decline of accuracies obtained with BM in scenarios 1 to 3 to the constant level of accuracies obtained with PLSR in scenarios 1 to 3 (Table(Table3).3). In this study, PLSR was advantageous over BM and LARS in situations where a large number of QTL contributed to the genetic variation of the trait of interest but methods BM and LARS performed better than PLSR in situations where few QTL contributed to the trait. An alternative method, not evaluated in this study, is GBLUP [2,29]. In this method, markers are used to estimate the relationship matrix of the individuals in the data and this relationship matrix is subsequently used to estimate breeding values with BLUP. Daetwyler [11] have reported that accuracy of GBLUP is not affected by the number of QTL in the data. In situations where few QTL contribute to the trait, accuracies obtained with BM are higher than accuracies obtained with GBLUP but at high number of QTL these accuracies are identical [11] suggesting that BM will always perform equally or better than GBLUP. When [5] derived accuracies for GBLUP and BM he showed that higher accuracy can be obtained with BM because this method better takes into account the variable contribution of individual QTL. Based on this, BM should be preferred over GBLUP. Since the number of QTL contributing to the trait is generally unknown, using the method PLSR can be a secure alternative for method BM. A pragmatic solution to overcome the problem of ignoring the number of QTL is cross validation [17]. For cross validation, a subset of individuals with highly reliable EBV can be used to evaluate the accuracy of MEBV obtained with BM, LARS and PLSR. The method which gives the highest accuracies can subsequently be used for the genetic evaluation of individuals with unknown breeding values.

Assignment of QTL by giving additive effects to biallelic loci was deferred to generation 5001. There were two reasons for not doing this earlier in the simulations. The first reason was to control the number of QTL that contributed to the trait. With QTL assigned in generation zero, the number of QTL will vary between replicates due to drift and mutations. The second reason was to reduce computing resources required for simulation. Simulating QTL is computationally more expensive than simulating loci because QTL require handling the additive effects in addition to the biallelic genotypes.

The six QTL scenarios were created after all generations were simulated, to ensure that QTL variance was the only difference between scenarios of equal and unequal QTL variance. The QTL scenarios were designed with the objective of identifying the effect of number of QTL and distribution of QTL variance on accuracy of MEBV with the distinct methods. A deterministic approach was used to assign the number of QTL contributing to the trait and to calculate the additive effect of each QTL contributing to the trait. This approach was very different from the random approach used to simulate QTL in other simulation studies (for example [2,30,4,9,6]) where QTL effects were drawn from a distribution similar to the gamma distribution for QTL effects estimated by Hayes and Goddard [10].

An important disadvantage of drawing QTL effects from any distribution is that randomness is introduced in the simulations that does not contribute to the research question because it is difficult to control the resulting distribution of QTL effects. The research question in our study concerned the effect of QTL distribution on the estimation of MEBV; hence distinct QTL scenarios covering a range of QTL distributions were simulated.

Strength of LD between a pair of loci is constrained by the difference between MAF of both loci [31]. In addition, variance of QTL with a low MAF is likely to be low, because the variance of QTL is a function of the allele frequency [32]. Excluding markers with a MAF below a specific threshold from the data, as done by Calus et al. [4], therefore seems reasonable. Results of this study, however, show that accuracy of MEBV was consistently lower when markers with a low MAF were excluded from the data (Table (Table4).4). These lower accuracies were supported by the lower R^2 when markers with a MAF below 0.10 were excluded (Table (Table2).2). Based on results of this study, using all markers to calculate MEBV is recommended.

This study reveals that method BM should be recommended in situations were few QTL are expected to account for a large proportion of the total genetic variance. When the number of QTL accounting for the genetic variance is larger or unknown, method PLSR is recommended.

COMPETING INTERESTS

The authors declare that they have no competing interests.

AUTHORS' CONTRIBUTIONS

All authors were involved in the design of the study. AC and JB programmed the simulations and wrote the manuscript. All authors read and approved the manuscript.

ACKNOWLEDGEMENTS

The work of AC was funded by Technologiestichting STW. The work of JB and MC was funded by the EU project Robustmilk. AC acknowledges Gus Rose for reading through the manuscript. We acknowledge the anonymous reviewers for reviewing this manuscript.

REFERENCES

1. Schaeffer LR. Strategy for applying genome-wide selection in dairy cattle. J Anim Breed Genet.2006;123:218–223.

2. Meuwissen THE, Hayes BJ, Goddard ME. Prediction of total genetic value using genome-wide dense marker maps. Genetics. 2001;157:1819–1829.

3. Calus M, Veerkamp R. Accuracy of breeding values when using and ignoring the polygenic effect in genomic breeding value estimation with a marker density of one SNP per cM. J Anim Breed Genet.2007;124:362–368.

4. Calus MPL, Meuwissen THE, de Roos APW, Veerkamp RF. Accuracy of genomic selection using different methods to define haplotypes. Genetics. 2008;178:553–561.

5. Goddard M. Genomic selection: prediction of accuracy and maximisation of long term response.Genetica. 2008;136:245–257.

6. Solberg T, Sonesson A, Woolliams J, Meuwissen T. Reducing dimensionality for prediction of genome-wide breeding values. Genet Sel Evol. 2009;41:29.

7. Sved JA. Linkage disequilibrium and homozygosity of chromosome segments in finite populations.Theor Popul Biol. 1971;2:125–141.

8. Muir W. Comparison of genomic and traditional BLUP-estimated breeding value accuracy and selection response under alternative trait and genomic parameters. J Anim Breed Genet.2007;124:342–355.

9. Solberg TR, Sonesson AK, Woolliams JA, Meuwissen THE. Genomic selection using different marker types and densities. J Anim Sci. 2008;86:2447–2454.

10. Hayes B, Goddard ME. The distribution of the effects of genes affecting quantitative traits in livestock.Genet Sel Evol. 2001;33:209–229.

11. Daetwyler H. PhD thesis. Wageningen University, Wageningen, The Netherlands; 2009. Genome-Wide Evaluation of Populations.

12. De Roos A, Schrooten C, Mullaart E, Beek S van der, de Jong G, Voskamp W. Genomic selection at CRV. Interbull Bull. 2009;39:47–50.

13. Efron B, Hastie T, Tibshirani R. Least angle regression. Ann Stat. 2004. pp. 407–499.

14. Park T, Casella G. The bayesian lasso. J Am Stat Assoc. 2008;103:681–686.

15. de Jong S. SIMPLS: An alternative approach to partial least squares regression. Chemom Intell Lab Syst. 1993;18:251–263.

16. Datta S, Le-Rademacher J, Datta S. Predicting patient survival from microarray data by accelerated failure time modeling using partial least squares and LASSO. Biometrics. 2007;63:259–271.

17. Moser G, Tier B, Crump R, Khatkar M, Raadsma H. A comparison of five methods to predict genomic breeding values of dairy bulls from genome-wide SNP markers. Genet Sel Evol. 2009;41:56.

18. Coster A, Bastiaansen J. HaploSim: HaploSim. 2009. [R package version 1.8]

19. R Development Core Team. R: A Language and Environment for Statistical Computing. R Foundation for Statistical Computing, Vienna, Austria; 2009. [ISBN 3-900051-07-0]

20. Crow JF, Kimura M. An introduction to population genetics theory. Alpha Editions; 1970.

21. Hastie T, Efron B. lars: Least Angle Regression, Lasso and Forward Stagewise. 2007. [R package version 0.9-7]

22. Wehrens R, Mevik BH. pls: Partial Least Squares Regression (PLSR) and Principal Component Regression (PCR) 2007. [R package version 2.1-0]

23. Grisart B, Coppieters W, Farnir F, Karim L, Ford C, Berzi P, Cambisano N, Mni M, Reid S, Simon P, Spelman R, Georges M, Snell R. Positional candidate cloning of a QTL in dairy cattle: identification of a missense mutation in the bovine DGAT1 gene with major effect on milk yield and composition.Genome Res. 2002;12:222–231.

24. Mele M, Conte G, Castiglioni B, Chessa S, Macciotta N, Serra A, Buccioni A, Pagnacco G, Secchiari P. Stearoyl-coenzyme A desaturase gene polymorphism and milk fatty acid composition in Italian Holsteins. J Dairy Sci. 2007;90:4458. doi: 10.3168/jds.2006-617.

25. Jeon J, Carlborg Ö, Törnsten A, Giuffra E, Amarger V, Chardon P, Andersson-Eklund L, Andersson K, Hansson I, Lundstrom K, Andersson L. A paternally expressed QTL affecting skeletal and cardiac muscle mass in pigs maps to the IGF2 locus. Nature. 1999;21:157–158.

26. Weedon MN, Lango H, Lindgren CM, Wallace C, Evans DM, Mangino M, Freathy RM, Perry JRB, Stevens S, Hall AS, Samani NJ, Shields B, Prokopenko I, Farrall M, Dominiczak A, Initiative DG, Consortium TWTCC, TJ, Bergmann S, Beckmann JS, Vollenweider P, Waterworth DM, Mooser V, Palmer CNA, Morris AD, Ouwehand WH, Consortium G, Caulfield M, Munroe PB, Hattersley MI, McCarthy AT, Frayling M. Genome-wide association analysis identifies 20 loci that influence adult height. Nat Genet. 2008;40:575–583.

27. Gudbjartsson DF, Walters GB, Thorleifsson G, Stefansson H, Halldorsson BV, Zusmanovich P, Sulem P, Thorlacius S, Gylfason A, Steinberg S, Helgadottir A, Ingason A, Steinthorsdottir V, Olafsdottir EJ, Olafsdottir GH, Jonsson T, Borch-Johnsen K, Hansen T, Andersen G, Jorgensen T, Pedersen O, Aben KK, Witjes JA, Swinkels DW, Heijer Md, Franke B, Verbeek ALM, Becker DM, Yanek LR, Becker LC, Tryggvadottir L, Rafnar T, Gulcher J, Kiemeney LA, Kong A, Thorsteinsdottir U, Stefansson K. Many sequence variants affecting diversity of adult human height.Nat Genet. 2008;40:609–615.

28. Lettre G, Jackson A, Gieger C, Schumacher F, Berndt S, Sanna S, Eyheramendy S, Voight B, Butler J, Guiducci C, T I, Hackett R, Heid KB, Jacobs IM, Lyssenko V, Uda M. Initiative TDG, FUSION, KORA, Colorectal TPL, Trial OCS, Study TNH. Sardi NIA, Boehnke M, Chanock SJ, Groop LC, Hu FB, Isomaa B, Kraft P, Peltonen L, Salomaa V, Schlessinger D, Hunter DJ, Hayes RB, Abecasis GR,

Wichmann HE, Mohlke KL, Hirschhorn JN. Identification of ten loci associated with height highlights new biological pathways in human growth. Nat Genet. 2008;40:584–591.

29. Hayes BJ, Visscher PM, Goddard ME. Increased accuracy of artificial selection by using the realized relationship matrix. Gen Res. 2009;91:47–60.

30. Grapes L, Dekkers JCM, Rothschild MF, Fernando RL. Comparing linkage disequilibrium-based methods for fine mapping quantitative trait loci. Genetics. 2004;166:1561–1570.

31. Wray N. Allele frequencies and the r2 measure of linkage disequilibrium: impact on design and interpretation of association studies. Twin Res Hum Genet. 2005;8:87–94.

32. Falconer DS, Mackay TFC. Quantitative Genetics. England: Pearson Education Limited; 1996.

Index

N

P

Q

R

R